HARCOURT

Math

Intervention Strategies and Activities

Grade 5

Orlando • Boston • Dallas • Chicago • San Diego
www.harcourtschool.com

Printed in the United States of America

ISBN 0-15-320765-5

3 4 5 6 7 8 9 10 082 2003 2002

CONTENTS

Skills *(continued)*

Number Sense: Whole Number Multiplication

Number Sense: Whole Number Division

Number Sense: Fractions

Skills *(continued)*

Number Sense: Decimals

Algebra and Functions

Measurement and Geometry: Measurement

Skills *(continued)*

Measurement and Geometry: Geometry

Statistics, Data Analysis, and Probability

Using the Intervention Strategies and Activities

The *Intervention Strategies and Activities* will help you accommodate the diverse skill levels of students in your class and will help you better prepare students to work successfully on grade-level content by targeting the prerequisite skills for each chapter in the program. The following questions and answers will help you make the best use of this rich resource.

How can I determine which skills or strategies a student or students should work on?

Before beginning each chapter, have students complete the "Check What You Know" page in the Pupil Book. This page targets the prerequisite skills necessary for success in the chapter. A student's performance on this page will allow you to diagnose skill weaknesses and prescribe appropriate interventions. Intervention strategies and activities are tied directly to each of the skills assessed. A chart at the beginning of each chapter correlates the skill assessed to the appropriate intervention materials. The chart appears in the Harcourt Math Teacher's Edition.

In what format are the intervention materials?

Intervention materials are available in different formats, including the following:

A. Cards—which provide skill development on one side and skill practice on the other side. These cards can be used by students in an independent setting such as a learning center. The kit that houses the cards also contains the *Intervention Strategies and Activities Teacher's Guide with Copying Masters.* Please note that students should not write on the Cards.

B. Copying masters—which provide the skill development and skill practice on reproducible pages. These pages in the *Intervention Strategies and Activities Teacher's Guide with Copying Masters* can be used by individual students or small groups. You can also allow students to record their answers on copies of the pages. This guide also provides teaching suggestions for skill development, as well as an alternative teaching strategy for students who continue to have difficulty with the skill.

C. CD-ROM—which provides the skill development and practice in an interactive format. Teaching suggestions and alternative teaching strategies are provided as printable PDF files.

Are manipulative activities included in the intervention strategies?

The teaching strategies in the teacher's materials for the *Intervention Strategies and Activities* do require manipulatives, easily gathered classroom objects, or copying masters from the *Teacher's Resource Book.* Since these activities are designed for only those students who show deficits in their skill development, the quantity of manipulatives will be small. For many activities, you may substitute materials, such as squares of paper for counters, coins for two-color counters, and so on.

How can I organize my classroom so that I have time and space to help students who show a need for these intervention strategies and activities?

You may want to set up a Math Skill Center with a folder for each of your students. Based on a student's performance on the *Check What You Know* page, assign appropriate skills by marking the student's record folder. The student can then work through the intervention materials, record the date of completion, and place the completed work in a folder for your review. You may wish to group students in pairs or small groups, or you may wish to have a specified time during the day to meet with one or more of the individuals or small groups to assess their progress and to provide direct instruction.

How are the activities structured?

Each skill begins with a model or an explanation with a model for each skill. The first section of exercises titled *Try These* provides 2–4 exercises that allow students to move toward doing the work independently. A student who has difficulty with the *Try These* exercises might benefit from the activity for that skill described in this Teacher's Guide before they attempt the *Practice on Your Own* page. The *Practice on Your Own* page provides an additional model for the skill and scaffolded exercises, which gradually remove prompts. Scaffolding provides a framework within which the student can achieve success for the skill. At the end of the *Practice on Your Own,* there is a *Check.* The *Check* provides 3–4 problems that check the student's proficiency in the skill. Guidelines for success are provided in the teacher's materials.

Student's Name ______________________________

Individual Prerequisite Skills Checklist

Chapter	Prerequisite Skill	Prescription	Skill Mastered

Intervention Strategies and Activities Chapter Correlations

Number Sense

Place Value

Skill Number	Skill Title	Chapter Correlation
1	Place value (to hundred thousands)	1, 2
2	Read and write whole numbers (to hundred thousands)	1
3	Rounding	3, 4

Whole Number Addition

Skill Number	Skill Title	Chapter Correlation
4	Addition Facts 1–10	5
5	Addition Facts 11–19	5
6	Order and Zero Properties of Addition	5
7	Grouping Property of Addition	5

Whole Number Subtraction

Skill Number	Skill Title	Chapter Correlation
8	Subtraction across zeros	3
9	Subtraction Facts 0–9	5
10	Subtraction Facts 10–19	5

Money

Skill Number	Skill Title	Chapter Correlation
11	Multiply money	10

Whole Number Multiplication

Skill Number	Skill Title	Chapter Correlation
12	Multiplication Facts 1–5	6, 15, 26
13	Multiplication Facts 6–10	6, 15, 26
14	Multiplication properties	6
15	Distributive property	9
16	Multiply by 10 and 100	9, 14
17	Multiply by 1-digit numbers (2-, 3-digit factors)	9
18	Skip count on a number line	8
19	Multiply by 2-Digit numbers	9, 12
20	Factors	15
21	Repeated Factors	27

Whole Number Division

Skill Number	Skill Title	Chapter Correlation
22	Divide 2-digit numbers by 1-digit numbers	11
23	Check division	11
24	Divide by 10	12
25	Divide by 1-digit numbers	12, 13
26	Division Patterns (divide by multiples of 10)	13
27	Divide by 2-digit numbers	13
28	Related Facts	14
29	Division Facts 1–6	15
30	Division Facts 7–10	15

Fractions

Skill Number	Skill Title	Chapter Correlation
31	Understand fractions	16, 17, 28
32	Compare fractions	16
33	Fractions on a ruler (nearest eighth of an inch)	17

Fractions (continued)

Skill Number	Skill Title	Chapter Correlation
34	Understand mixed numbers	18, 19, 20, 30
35	Add fractions	18
36	Subtract fractions	18
37	Fractions of a whole or a group	19, 28
38	Find the GCF	20

Decimals

Skill Number	Skill Title	Chapter Correlation
39	Read and write decimals	2
40	Round decimals	4
41	Mental Math: decimals	4
42	Add and subtract decimals	4
43	Repeated addition of decimals	10
44	Equivalent decimals	13
45	Divide decimals by whole numbers	14
46	Model decimal multiplication	19
47	Understand hundredths (for percents)	29
48	Relate fractions and decimals	29

Algebra and Functions

Skill Number	Skill Title	Chapter Correlation
49	Mental math: Function tables	3, 5
50	Addition and subtraction equations	5
51	Use parentheses	6
52	Graph ordered pairs	8, 22
53	Expressions with exponents (2 and 3)	27
54	Compare whole numbers	21
55	Function tables	22

Measurement and Geometry

Measurement

Skill Number	Skill Title	Chapter Correlation
56	Customary units and tools	25
57	Metric units and tools	25
58	Read a thermometer (negative numbers)	21

Geometry

Skill Number	Skill Title	Chapter Correlation
59	Classify plane figures	23
60	Name polygons	23
61	Classify angles	24
62	Faces of solid figures	24
63	Perimeter and area	26
64	Faces, edges, vertices	27

Statistics, Data Analysis, and Probability

Skill Number	Skill Title	Chapter Correlation
65	Frequency tables	7
66	Read pictographs	7
67	Read bar graphs	7
68	Read line graphs	7
69	Certain or Impossible events	30
70	Likely, unlikely	30

Number Sense

Place Value

Place Value (to Hundred Thousands)

Using Skill 1

OBJECTIVE Identify the place value of digits to hundred thousands

MATERIALS place-value charts

Display some 4-, 5-, and 6-digit numbers. Discuss how commas are used to separate groups of three digits, or *periods*, in large numbers. Have students place commas to separate the ones and thousands periods in the numbers you have displayed. Be sure students start from the right and count three places as they move left. Review the ones period, asking how many ones, tens, and hundreds.

Draw student's attention to the place-value chart for 731,825. Review the value of each digit. Start with ones. Ask: **How many ones are there?** (5) **What is the value of the 5?** ($5 \times 1 = 5$)

How many tens? (2) **What is the value of the tens?** ($2 \times 10 = 20$)

Continue through hundred thousands. Lead students to realize that the value of each digit is 10 times the value of the digit to its right.

Have them study the expanded form of the number; focus on the addition involved. Relate it to the placement of the digits in the chart and to the standard form.

Have students write the numbers you displayed at the beginning of the lesson in a place-value chart. Ask them to give the value of each digit. Then have them write the number in expanded form.

TRY THESE Exercises 1–5 provide practice writing the value of digits in the following places:

- **Exercise 1** Ten thousands place.
- **Exercise 2** Thousands place.
- **Exercise 3** Hundreds place.
- **Exercise 4** Tens place.
- **Exercise 5** Ones place.

PRACTICE ON YOUR OWN Review the example at the top of the page. Be sure students understand what to do when there is a zero in any place of a number. In Exercises 1–8, students write the place value of digits to thousands and ten thousands. In Exercises 9–11, students identify the place value of selected digits to hundred thousands.

CHECK Determine if students can identify the place value of digits to hundred thousands. Success is indicated by 3 out of 3 correct responses.

Students who successfully complete the **Practice on Your Own** and **Check** are ready to move to the next skill.

COMMON ERRORS

- Students may confuse period names with place-value names.
- Students may not understand that each place to the left is 10 times the value of the place to the right.

Students who made more than 2 errors in the **Practice on Your Own**, or who were not successful in the **Check** section, may benefit from the **Alternative Teaching Strategy** on the next page.

Optional

Alternative Teaching Strategy

20 Minutes

Place-Value Charts to Hundred Thousands

OBJECTIVE Find the place value of digits to hundred thousands

MATERIALS place-value charts, number cards 0-9

Distribute place-value charts to students. Tell students that you will display a card and name the value that the digit represents. Students record the digit in the appropriate column in the chart. For example, hold up the digit card for 3 and state that it represents 3 tens. Have the students record the digit in the tens column.

Continue to build a 6-digit number by repeating the procedure with other digits for ones, hundreds, thousands, ten thousands, and hundred thousands until students have written a number, for example, 172,536, in their place-value charts.

Refer to the chart.

Ask: **How many tens are there?** (3)

Show students how to write 3 tens as $3 \times 10 = 30$.

Ask: **What is the value of the digit 3?** (3 tens, or 30)

Have students use their charts to write the number in expanded form:

100,000 + 70,000 + 2,000 + 500 + 30 + 6.

Thousands Period			Ones Period		
hundred thousands	ten thousands	thousands,	hundreds	tens	ones
1	7	2	5	3	6

100,000 + 70,000 + 2,000 + 500 + 30 + 6

Show the addition in the expanded form vertically:

$$\begin{array}{r} 100{,}000 \\ 70{,}000 \\ 2{,}000 \\ 500 \\ 30 \\ +\quad 6 \\ \hline 172{,}536 \end{array}$$

Repeat the activity using other 6-digit numbers. Include examples with zeros in ones, tens, hundreds, thousands, or ten thousands places.

When students have demonstrated understanding, have them work in pairs. The first student chooses a 6-digit number and writes it in the place-value chart. The second student writes the expanded form of the number.

Name ______________________ Skill ______________________

Grade 5 Skill

Place Value (to Hundred Thousands)

Write the place value of each digit in the number 731,825.

THOUSANDS PERIOD				ONES PERIOD		
hundred thousands	ten thousands	thousands	,	hundreds	tens	ones
7	3	1	,	8	2	5

Think:

$7 \times 100{,}000 = 700{,}000$

$3 \times 10{,}000 = 30{,}000$

$1 \times 1{,}000 = 1{,}000$

$8 \times 100 = 800$

$2 \times 10 = 20$

$5 \times 1 = 5$

Use a comma to separate periods.

Standard Form: 731,825

Expanded Form: 700,000 + 30,000 + 1,000 + 800 + 20 + 5

Write the value of each digit.
The value of the **7** is 700,000 or 7 hundred thousands.
The value of the **3** is 30,000 or 3 ten thousands.
The value of the **1** is 1,000 or 1 thousand.
The value of the **8** is 800 or 8 hundreds.
The value of the **2** is 20 or 2 tens.
The value of the **5** is 5 or 5 ones.

In our number system, the value of each place in a number is 10 times as great as the value of the place to its right.

Try These

Write the value of the digits.

THOUSANDS PERIOD				ONES PERIOD		
hundred thousands	ten thousands	thousands	,	hundreds	tens	ones
	2	4	,	7	0	6

Standard Form: 24,706

Expanded Form: 20,000 + 4,000 + 700 + 6

1. The value of the 2 is ________ or ________.
2. The value of the 4 is ________ or ________.
3. The value of the 7 is ________ or ________.
4. The value of the 0 is ________ or ________.
5. The value of the 6 is ________ or ________.

Go to the next side.

Name ______________________ Skill ______________________

Practice on Your Own

Think: Use a comma to separate the periods.

THOUSANDS PERIOD			ONES PERIOD		
hundred thousands	ten thousands	thousands ,	hundreds	tens	ones
		4 ,	0	9	2

The value of 4 is 4,000 or 4 thousands.

The value of 0 is 0 or 0 hundreds.

The value of 9 is 90 or 9 tens.

The value of 2 is 2 or 2 ones.

Standard Form: 4,092
Expanded Form: 4,000 + 90 + 2

Write 5,781 in the place value chart. Write the value of each digit.

1.

THOUSANDS PERIOD			ONES PERIOD		
hundred thousands	ten thousands	thousands ,	hundreds	tens	ones

2. The value of the 5 is _____ or ______________.
3. The value of the 7 is _____ or ______________.
4. The value of the 8 is _____ or ______________.

Write 12,430 in the place value chart.
Write the value of each digit.

5.

THOUSANDS PERIOD			ONES PERIOD		
hundred thousands	ten thousands	thousands ,	hundreds	tens	ones

6. The value of the 1 is _____ or ______________.
7. The value of the 2 is _____ or ______________.
8. The value of the 4 is _____ or ______________.

Write the value of the underlined digit.

9. 4<u>8</u>0 ______________
10. <u>3</u>0,876 ______________
11. <u>8</u>53,612 ______________

Check

Write the value of the underlined digit.

12. 54, 8<u>7</u>1 ______________
13. 5<u>4</u>2,087 ______________
14. <u>6</u>09,813 ______________

Read and Write Whole Numbers (to Hundred Thousands)

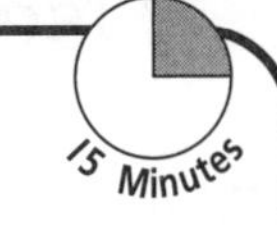

Using Skill 2

OBJECTIVE Read and write whole numbers to hundred thousands

Begin by reviewing the place-value chart and the fact that greater numbers are grouped in *periods* of three digits. Note that *ones* and *thousands* are periods. Also stress that to read a greater number, say the number in a period and then the period name. The ones period name is not said.

Direct the students' attention to the place-value chart. Stress that all periods have the same pattern of places: one, tens, and hundreds and that the periods are separated by a comma.

Direct the students' attention to the three ways the number can be written.

Say: **The expanded form shows the place value of each digit in the standard form. Use the place-value chart to help you write the expanded form.**

TRY THESE Exercises 1–3 provide the place-value chart to help students write the number in expanded form and in word form. Each exercise has a zero in a place-value position on the chart.

- **Exercise 1** 4-digit number with zero in the tens place.
- **Exercise 2** 5-digit number with zero in the ones place.
- **Exercise 3** 6-digit number with zero in the ten-thousands place.

PRACTICE ON YOUR OWN
Review the examples at the top of the page. With each example, take the students through process of writing the standard form for a number when the number is given in the expanded form or in the word form and using the place-value chart with each.

CHECK Determine if students can write the expanded form and the word form. Success is indicated by 2 out of 2 correct responses.

Students who successfully complete the **Practice on Your Own** and **Check** are ready to move to the next skill.

COMMON ERRORS

- Students may omit zeros or write too many zeros within a period when writing the number in expanded form.
- Students may omit the period name, *thousands*, when writing the word form.

Students who made more than 2 errors in the **Practice on Your Own**, or who were not successful in the Check section, may benefit from the **Alternative Teaching Strategy** on the next page.

Optional

Alternative Teaching Strategy

15 Minutes

Use a Place Value Chart for Whole Numbers to Hundred Thousands

OBJECTIVE Read whole numbers to hundred thousands and write them in standard form and expanded form using the place-value chart

MATERIALS place-value pocket chart, and multiple copies of number cards 0-9

Start the activity by showing a number in the place-value pocket chart, such as 1, 3, 7, and 2.

Say: **This number is read one thousand three hundred seventy-two. You say the name of the period, which is thousands, before you stop for the comma.**

Ask: **What are the names of the two periods?** (thousands and ones)

Remove the digit cards and show 29,042. Have a student read the number. (twenty-nine thousand forty-two). Another student writes the number on the board. Repeat the activity for a six-digit number.

Now show the first number (1,372) again in the place value pocket chart. Show the students how the chart will help them write the number in expanded form. Display the number in the expanded form.

(1000 + 300 + 70 + 2)

Repeat the activity to include five-digit and six-digit numbers.

Students work in pairs in this part of the activity. Give each pair a card with a number, such as 80,642. One partner shows the number on the place-value pocket chart and the other student reads the number. Partners take turns writing the expanded form and the word form.

Repeat the activity with numbers having zeros in different positions.

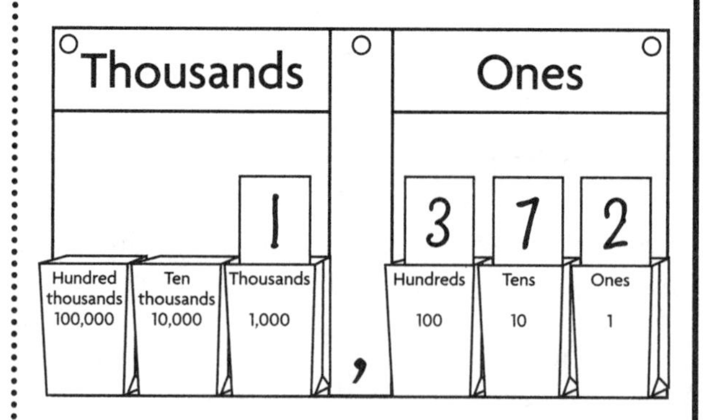

Name ______________________ Skill ______________________

Grade 5 **Skill 2**

Read and Write Whole Numbers (to Hundred Thousands)

Numbers can be written in different ways.

THOUSANDS			ONES		
Hundreds	Tens	Ones ,	Hundreds	Tens	Ones
4	7	6 ,	1	8	9
476 is in the thousands period.			189 is in the ones period.		

Standard form: 476,189 — Write only the digits.

Expanded form: 400,000 + 70,000 + 6,000 + 100 + 80 + 9 — Write the value of each digit in the number.

Word form: four hundred seventy-six thousand, one hundred eighty-nine — Write the word form.

Try These

Write the number in expanded and word form.

1. 4,908

THOUSANDS			ONES		
Hundreds	Tens	Ones	Hundreds	Tens	Ones
		4 ,	9	0	8

Expanded form:
____ + ____ + ____
Word form: ______________________
____

2. 32,760

THOUSANDS			ONES		
Hundreds	Tens	Ones	Hundreds	Tens	Ones
	3	2 ,	7	6	0

Expanded form:
____ + ____ + ____ + ____
Word form: ______________________
____

3. 709,145

THOUSANDS			ONES		
Hundreds	Tens	Ones	Hundreds	Tens	Ones
7	0	9 ,	1	4	5

Expanded form:
____ + ____ + ____ + ____ + ____
Word form: ______________________
____

Go to the next side.

Name ______________________ Skill ______________________

Practice on Your Own

Write the number in standard form.

Expanded form: 300,000 + 40,000 + 9,000 + 600 + 20 + 1

THOUSANDS			ONES		
Hundreds	Tens	Ones	Hundreds	Tens	Ones
3	4	9,	6	2	1

Standard form: 349,621

Word form: seven hundred nine thousand, two hundred fifty-eight

THOUSANDS			ONES		
Hundreds	Tens	Ones	Hundreds	Tens	Ones
7	0	9,	2	5	8

Standard form: 709,258

Write the number in standard form.

1. 40,000 + 1,000 + 500 + 90 + 3

THOUSANDS			ONES		
Hundreds	Tens	Ones	Hundreds	Tens	Ones
	4	1,	5	9	3

____

2. six hundred eight thousand, forty-nine

THOUSANDS			ONES		
Hundreds	Tens	Ones	Hundreds	Tens	Ones
6	0	8,	0	4	9

____

3. 600,000 + 50,000 + 3,000 + 900 + 20 + 6

____

4. forty-five thousand, seventy-one

____

Write the number in expanded and word form.

5. 89,652

Expanded form: ______________________

Word form: ______________________

6. 450,092

Expanded form: ______________________

Word form: ______________________

7. 54,708

Expanded form: ______________________

Word form: ______________________

8. 970,541

Expanded form: ______________________

Word form: ______________________

Check

Write the number in expanded and word form.

9. 80,944

Expanded form: ______________________

Word form: ______________________

10. 329,067

Expanded form: ______________________

Word form: ______________________

Using Skill 3

OBJECTIVE Round whole numbers to the nearest ten and hundred

Discuss rounded numbers and why they are useful. Students should realize that rounded numbers are often easier to work with than other numbers because they end in one or more zeros.

Read through Steps 1 and 2 for rounding 2,345 to the nearest ten. Ask: **To what place are you rounding?** (tens) **What digit do you underline?** (the 4 in the tens place) **Where do you draw the arrow?** (over the 5 in the ones place—the place to the right of the tens place)

Focus on rounding up in Step 3. Ask: **Is the digit in the ones place 5 or greater?** (yes) **Then what happens to the 4 in the tens place?** (It increases by 1.)

Follow Steps 1 and 2 for rounding to the nearest hundred. Ask: **To what place are you rounding?** (hundreds) **What digit do you underline?** (the 3 in the hundreds place) **Where do you draw the arrow?** (over the 4 in the tens place—the place to the right of the hundreds place)

Focus on rounding down in Step 3.

Ask: **Is the digit in the tens place 5 or greater?** (no) **Then what happens to the 3 in the hundreds place?** (It stays the same.)

Make sure students understand that numbers rounded to the nearest ten end in 1 or more zeros, and numbers rounded to the nearest hundred end in 2 or more zeros.

TRY THESE Exercises 1–4 provide practice in rounding.

- **Exercises 1 and 2** Rounding to the nearest ten.
- **Exercises 3 and 4** Rounding to the nearest hundred.

PRACTICE ON YOUR OWN Read together the examples at the top of the page. Have volunteers explain the steps for rounding.

CHECK Determine if students can round to the nearest ten and hundred. Success is indicated by 4 out of 4 correct responses.

Students who successfully complete the **Practice on Your Own** and **Check** are ready to move to the next skill.

COMMON ERRORS

- Some students may not understand that, when rounding *down*, the digit in the rounding place *stays the same*. It is not reduced by 1.
- Students may forget that, after rounding up or down, they write zeros to the right of the rounding place.

Students who made more than 3 errors in the **Practice on Your Own**, or who were not successful in the **Check** section, may benefit from the **Alternative Teaching Strategy** on the next page.

Optional

Alternative Teaching Strategy

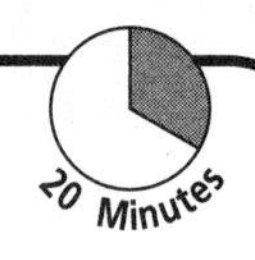

Use Number Lines to Round

OBJECTIVE Round to the nearest ten and hundred

MATERIALS number lines

Present the number line shown. Have a volunteer locate 13.

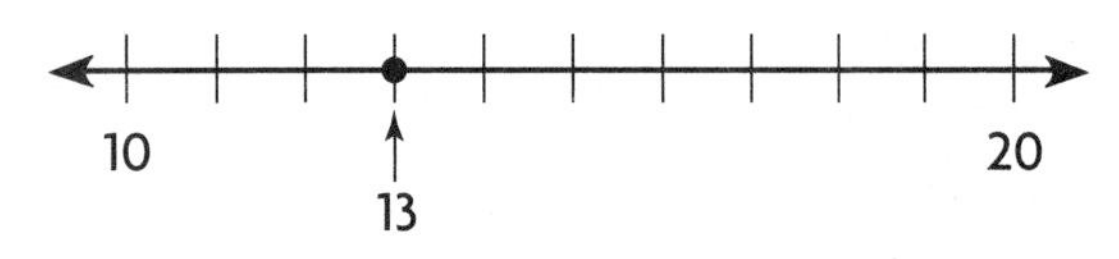

Ask: **Where is 13?** (between 10 and 20) **Which ten is it closer to?** (10)

Have students locate 18.

Ask: **Where is 18?** (between 10 and 20) **Which ten is it closer to?** (20)

Present the number line shown. Have a volunteer locate 130.

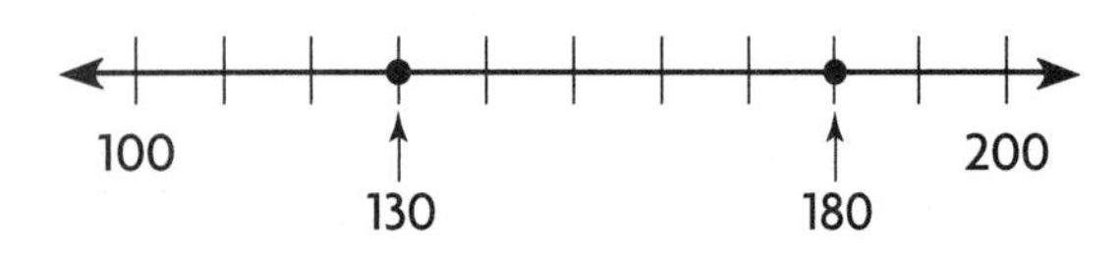

Ask: **Where is 130?** (between 100 and 200) **Which hundred is it closer to?** (100)

Have students locate 180.

Ask: **Where is 180?** (between 100 and 200) **Which hundred is it closer to?** (200)

Guide students through the process of rounding to the nearest ten and hundred. For example, to round 367 to the nearest ten, guide students:

Underline the digit in the tens place. What digit is it? (6) **Draw an arrow above the digit to the right. What digit is it?** (7) **Is the digit to the right 5 or greater, or is it less than 5?** (5 or greater) **Do you round up or down?** (up) **What does the number round to?** (370)

Be sure students understand that rounding up means the digit in the tens place increases by 1, but that rounding down means that the digit in the tens place stays the same.

Give students other numbers to round. Have them explain the steps aloud.

Name ____________________ Skill ____________________

Rounding

Grade 5 Skill 3

Round 2,345 to the nearest ten and hundred.

	Step 1	Step 2	Step 3
	Underline the digit in the place to be rounded.	Draw an arrow above the first digit to the right of the underlined digit.	• If the digit under the arrow is less than 5, the digit in the rounding place stays the same. You round down. • If the digit is 5 or greater, the digit in the rounding place increases by 1. You round up.
Round to the nearest **ten**. →	$2{,}3\underline{4}5$	$2{,}3\underline{4}\overset{\downarrow}{5}$	$2{,}3\underline{4}\overset{\downarrow}{5}$ $5 = 5$, so round up. Rounds to: 2,350
Round to the nearest **hundred**. →	$2{,}\underline{3}45$	$2{,}\underline{3}\overset{\downarrow}{4}5$	$2{,}\underline{3}\overset{\downarrow}{4}5$ $4 < 5$, so round down. Rounds to: 2,300

Try These

Round the numbers.

1. Round 91 to the nearest ten.
   $\underline{9}\overset{\downarrow}{1}$
   ____

2. Round 5,476 to the nearest ten.
   $5{,}4\underline{7}\overset{\downarrow}{6}$
   ____

3. Round 389 to the nearest hundred.
   $\underline{3}\overset{\downarrow}{8}9$
   ____

4. Round 7,285 to the nearest hundred.
   $7{,}\underline{2}\overset{\downarrow}{8}5$
   ____

Go to the next side.

Name ______________________ Skill ______________________

Practice on Your Own

Round 1,648 to the nearest ten.

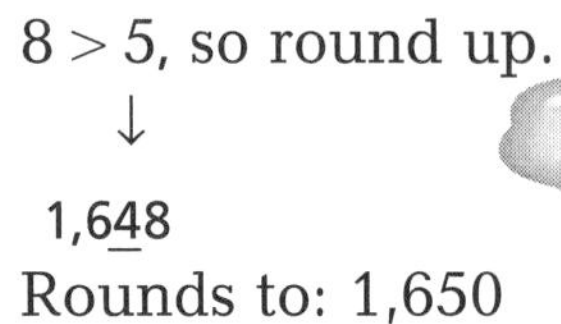

The digit in the ones place is greater than 5.

8 > 5, so round up.
↓
1,648
Rounds to: 1,650

Round 1,648 to the nearest hundred.

The digit in the tens place is less than 5.

4 < 5, so round down.
↓
1, 648
Rounds to: 1,600

Round to the nearest ten.

1. 42 ↓ 42 ______
2. 534 ↓ 534 ______
3. 6,318 ↓ 6,318 ______
4. 4,275 ↓ 4,275 ______

Round to the nearest hundred.

5. 738 ↓ 738 ______
6. 581 ↓ 581 ______
7. 7,539 ↓ 7,539 ______
8. 2,294 ↓ 2,294 ______

Round as indicated.

9. To the nearest ten
   638 ______
10. To the nearest ten
    2,701 ______
11. To the nearest hundred
    8,542 ______
12. To the nearest hundred
    3,198 ______

13. To the nearest ten
    645 ______
14. To the nearest ten
    5,783 ______
15. To the nearest hundred
    625 ______
16. To the nearest hundred
    7,498 ______

Name ______________________ Skill ______________________

Answer Card
Place Value
Grade 5

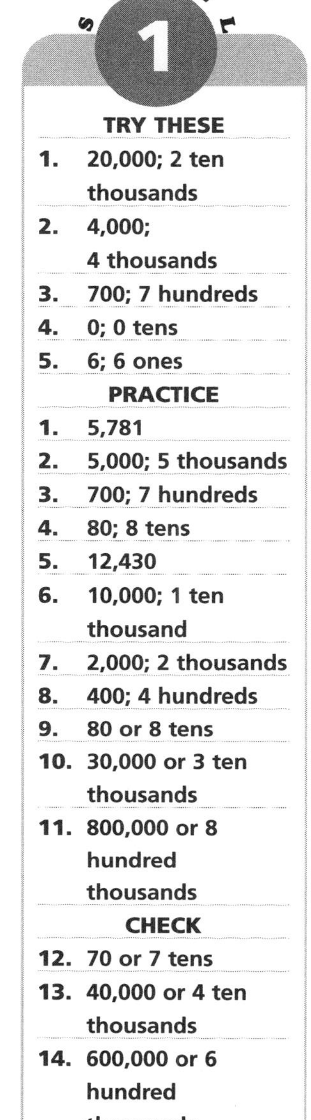

SKILL 1

TRY THESE

1. 20,000; 2 ten thousands
2. 4,000; 4 thousands
3. 700; 7 hundreds
4. 0; 0 tens
5. 6; 6 ones

PRACTICE

1. 5,781
2. 5,000; 5 thousands
3. 700; 7 hundreds
4. 80; 8 tens
5. 12,430
6. 10,000; 1 ten thousand
7. 2,000; 2 thousands
8. 400; 4 hundreds
9. 80 or 8 tens
10. 30,000 or 3 ten thousands
11. 800,000 or 8 hundred thousands

CHECK

12. 70 or 7 tens
13. 40,000 or 4 ten thousands
14. 600,000 or 6 hundred thousands

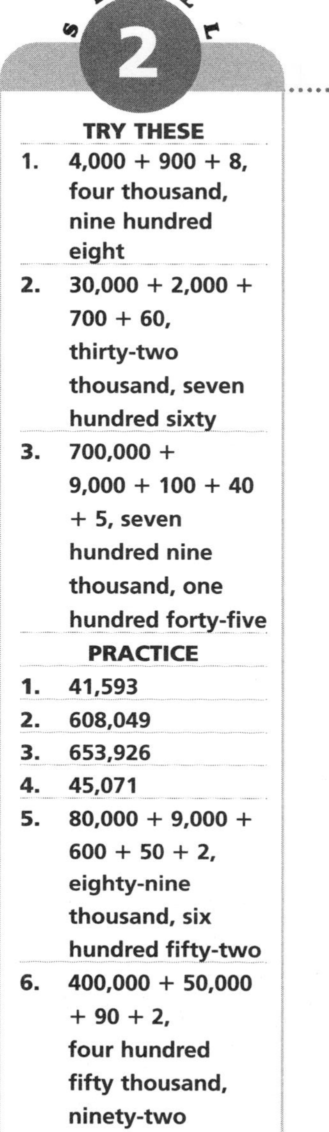

SKILL 2

TRY THESE

1. 4,000 + 900 + 8, four thousand, nine hundred eight
2. 30,000 + 2,000 + 700 + 60, thirty-two thousand, seven hundred sixty
3. 700,000 + 9,000 + 100 + 40 + 5, seven hundred nine thousand, one hundred forty-five

PRACTICE

1. 41,593
2. 608,049
3. 653,926
4. 45,071
5. 80,000 + 9,000 + 600 + 50 + 2, eighty-nine thousand, six hundred fifty-two
6. 400,000 + 50,000 + 90 + 2, four hundred fifty thousand, ninety-two

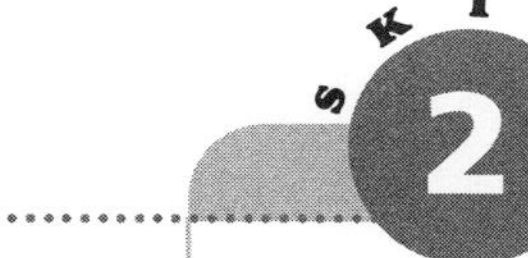

SKILL 2

7. 50,000 + 4,000 + 700 + 8, fifty-four thousand, seven hundred eight
8. 900,000 + 70,000 + 500 + 40 + 1, nine hundred seventy thousand, five hundred forty-one

CHECK

9. 80,000 + 900 + 40 + 4, eighty thousand, nine hundred forty-four
10. 300,000 + 20,000 + 9,000 + 60 + 7, three hundred twenty-nine thousand, sixty-seven

Name ____________________ Skill ____________________

Answer Card
Place Value
Grade 5

SKILL 3

TRY THESE

1. 90
2. 5,480
3. 400
4. 7,300

PRACTICE

1. 40
2. 530
3. 6,320
4. 4,280
5. 700
6. 600
7. 7,500
8. 2,300
9. 640
10. 2,700
11. 8,500
12. 3,200

CHECK

13. 650
14. 5,780
15. 600
16. 7,500

Number Sense

Whole Number Addition

Skill 4 Grade 5 Addition Facts 1–10

20 Minutes

Using Skill 4

OBJECTIVE Use strategies to remember basic addition facts for sums 1 through 10

Begin by reminding the students that they can use addition strategies to help them remember addition facts.

Call students' attention to the first example on the page, 5 + 3. Explain the first strategy "counting on." Remind students that when they count on they need to begin with the greater number.

Ask: **Why do you begin counting with the greater number?** (It is easier and faster to count on from 5 instead of 3; there are fewer numbers to count.)

Ask a student to count on from 5 as you work through the example.

In the second example, students may note that although 4 and 5 are not doubles, they are close. So, they can find 4 + 5 by thinking of doubles and then adding one or subtracting one.

Say: **You need to find the sum of 4 + 5. Suppose you use the double 4 + 4. Which strategy could you use?** (doubles plus one)

Explain. (5 is one more than 4, so I add 4 and 4 plus 1 and get the sum, 9.)

What other double could you use? (5 + 5)

Will the sum be greater or less than 4 + 5? (greater, because 5 is one more than 4)

Explain how you would use the *double minus one* strategy. (Since 4 is one less than 5, I take one away, so 5 + 5 – 1 = 9.)

TRY THESE In Exercises 1–5, students use patterns to help them understand the strategies.

- **Exercises 1–2** Doubles plus one, doubles minus one.
- **Exercises 3–5** Counting on.

PRACTICE ON YOUR OWN Review the examples at the top of the page. Be sure that the students understand when it is efficient to use the strategies.

CHECK Determine if students can quickly find sums using their memories or strategies.

Success is indicated by 4 out of 4 correct responses.

Students who successfully complete the **Practice on Your Own** and **Check** are ready to move to the next skill.

COMMON ERRORS

- Students may not know doubles facts.
- Students may count on from an arbitrary addend, instead of choosing the larger addend.

Students who made more than 4 errors in the **Practice on Your Own** or who were not successful in the **Check** section may benefit from the **Alternative Teaching Strategy** on the next page.

Optional

Alternative Teaching Strategy

Make Triangle Flash Cards

20 Minutes

OBJECTIVE Use triangle flash cards and an addition table to review addition facts to 10

MATERIALS index cards or construction paper cut into equilateral triangles, markers

Students who have difficulty remembering basic facts benefit by daily review using flash cards. In this activity students make triangle flash cards and plan a program of daily review.

Have students list the addition facts to 10. Distribute the triangle cards.

Instruct students to write the numbers for an addition fact on each card. Write an addend in each of two corners, and the sum on the third corner.

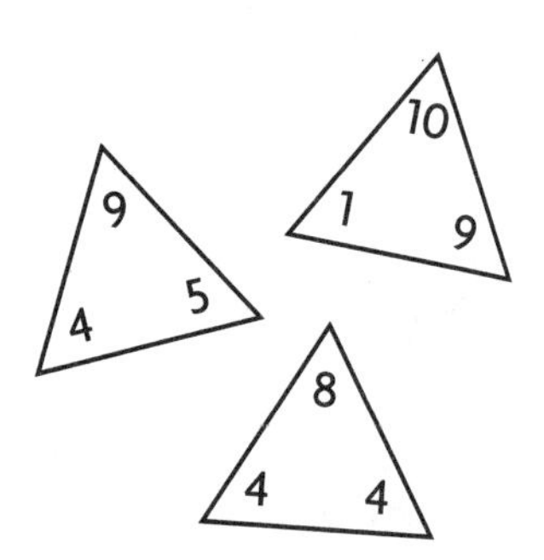

Demonstrate how a student can hold up the card, and cover a corner, so that another student can provide the missing addend or sum. (Students can also use the cards to practice subtraction facts.)

Suggest that students schedule a review time each day. Have them decide what facts they will review and record them in an addition table.

When the table is complete they will have reviewed all the facts.

Name ______________ Skill ______________

Grade 5 Skill 4

Addition Facts 1–10

You can use strategies to remember addition facts.

Counting On
Find 5 + 3.

Start with 5, then count on 3 more:

5, 6, 7, 8

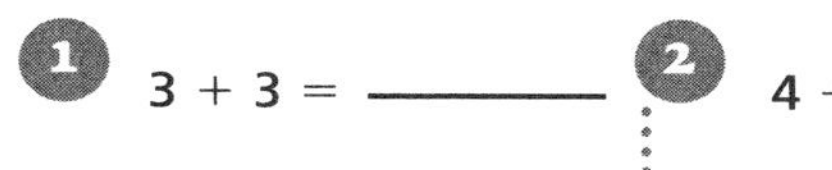

So, 5 + 3 = 8.

Doubles Plus One or Doubles Minus One
Find 4 + 5.

Think: 4 + 4 = 8

So, 4 + 5 is 1 *more*, or 9.
4 + 5 = (4 + 4) + 1 = 9

Think: 5 + 5 = 10

So, 5 + 4 is 1 *less*, or 9.
4 + 5 = (5 + 5) − 1 = 9

So, 4 + 5 = 9.

Try These

Find the sum.

1. 3 + 3 = ____
   3 + 4 = ____
   4 + 4 = ____
2. 4 + 4 = ____
   4 + 5 = ____
   5 + 5 = ____
3. 7 + 1 = ____
   7 + 2 = ____
   7 + 3 = ____
4. 6 + 2 = ____
   6 + 3 = ____
   6 + 4 = ____
5. 8 + 0 = ____
   8 + 1 = ____
   8 + 2 = ____

Go to the next side.

Name ________________________ Skill ________________________

Practice on Your Own

Find 7 + 3.
Use counting on.

Say: 7
Count: 8, 9, 10

Start with 7, then count on 3 more.
7, 8, 9, 10

So, 7 + 3 = 10.

Find 3 + 4.
Use doubles plus one or doubles minus one.
(3 + 3) + 1 = 7 or
(4 + 4) − 1 = 7

3 + 3 = 6
3 + 4 is 1 *more.*

4 + 4 = 8
3 + 4 is 1 *less.*

So, 3 + 4 = 7.

Find the sum. Use counting on.

1. 6 + 2 = ____
2. 5 + 3 = ____
3. 5 + 2 = ____
4. 6 + 3 = ____

Find the sum. Use doubles plus 1.

5. 4 + 5 = ____ double: ____
6. 3 + 4 = ____ double: ____
7. 3 + 2 = ____ double: ____
8. 5 + 4 = ____ double: ____

Find the sum. Use doubles minus 1.

9. 3 + 2 = ____ double: ____
10. 5 + 4 = ____ double: ____
11. 2 + 3 = ____ double: ____
12. 4 + 3 = ____ double: ____

Find the sum.

13. 2 + 4 = ____
14. 4 + 4 = ____
15. 6 + 2 = ____
16. 8 + 2 = ____
17. 5 + 5 = ____
18. 6 + 1 = ____
19. 3 + 4 = ____
20. 7 + 3 = ____

Check

Find the sum.

21. 4 + 5 = ____
22. 6 + 3 = ____
23. 7 + 2 = ____
24. 3 + 4 = ____

Using Skill 5

15 Minutes

OBJECTIVE Use strategies to recall addition facts

MATERIALS ten-frame, ones blocks, addition facts cards for doubles, 2 cards each labeled "+1" and "−1"

Begin by reviewing how to make a ten using the ten-frame and the ones blocks. Model the addition fact 7 + 5, by placing 7 blocks in the ten-frame.

Ask: **How many blocks are needed to make a ten?** (3) **What is another name for 5?** (4 + 1, or 3 + 2) **Which of these facts will help us make a ten?** (3 + 2)

Show the 5 blocks as a group of 3 and 2. Move the 3 blocks into the ten-frame to make a ten. Continue:

What number sentence is shown now? (10 + 2) **What is the sum for 10 + 2 and 7 + 5?** (12)

To review the *doubles plus 1* and *doubles minus 1* strategy, do an oral drill activity. First review the doubles with the facts cards. Next, show a doubles fact card and a "+ 1" card, for example, 8 + 8 and the "+ 1" card.

Ask: **What is 8 + 8?** (16) **Now add 1. What are the addends now?** (8 and 9) **What is the sum?** (17)

Repeat the activity several times with the "+ 1" card. Then do the doubles with the "−1" card in the same manner.

For the second example emphasize the idea that when one addend is 1 more than the other addend the doubles plus 1 or minus 1 is an efficient strategy to use.

TRY THESE Exercises 1–3 provide practice with each of the strategies.

- **Exercise 1** Make a ten strategy.
- **Exercise 2** Doubles plus 1 strategy.
- **Exercise 3** Doubles minus 1 strategy.

The same addition fact is used to demonstrate that facts can be recalled using different strategies. Encourage students to use the strategy that best helps them remember the facts.

PRACTICE ON YOUR OWN Review the strategy examples at the top of the page. Ask students to explain each step shown.

CHECK Determine if students can use the strategies to find an addition fact. Success is indicated by 3 out of 3 correct responses.

Students who successfully complete the **Practice on Your Own** and **Check** are ready to move to the next skill.

COMMON ERROR

- Students may be unable to use a strategy correctly. For example, given 6 + 7, students may write the answer as 10 (make a ten) or 14 (doubles plus 1)

Students who made more than 2 errors in the **Practice on Your Own**, or who were not successful in the **Check** section, may benefit from the **Alternative Teaching Strategy** on the next page.

Optional

Alternative Teaching Strategy

15 Minutes

Memorize Addition Facts 11–19

OBJECTIVE Memorize addition facts 11–19

MATERIALS number cards 0–9 plus additional number cards 6, 7, 8, and 9

There are two types of difficulties students usually have with the basic facts. First, the student may not understand the basis facts. Students may not understand the concept of the operation, or students have not memorized the facts. If students need help recalling the meaning of addition, provide concrete or pictorial representations that reflect part/whole relationships or joining action. If students need more time to memorize, provide daily practice sessions and immediate feedback for incorrect responses. These sessions should be about 10 minutes in duration.

Give each student a set of number cards. Students will use these cards to show all of the numbers that can be added for a sum for 11–19.

Say: **On your desk arrange the number cards in pairs so that each pair has a sum of 11.**

Check to see that each student shows all the combinations. Have a student list the combinations on the board.

Name ______________________ Skill ______________________

Grade 5
Skill

Addition Facts 11–19

Use strategies to recall addition facts.

One addend is one more than the other.
So, I can use doubles plus or minus one.

Make a Ten

$7 + 5 = \square$

Make a ten
$7 + 3 = 10$

$7 + 5 = 7 + 3 + 2$
$7 + 5 = (7 + 3) + 2$
$= 10 + 2$
$= 12$
So, $7 + 5 = 12$.

Doubles Plus 1

$8 + 7 = \square$
Think: Double the smaller addend. $7 + 7 = 14$
So $7 + 8$ is 1 *more*, or 15.
$8 + 7 = (7 + 7) + 1$
$= 14 + 1$
$= 15$
So, $8 + 7 = 15$.

Doubles Minus 1

$8 + 7 = \square$
Think: Double the greater addend. $8 + 8 = 16$
So $8 + 7$ is 1 *less*, or 15.
$8 + 7 = (8 + 8) - 1$
$= 16 - 1$
$= 15$
So, $8 + 7 = 15$.

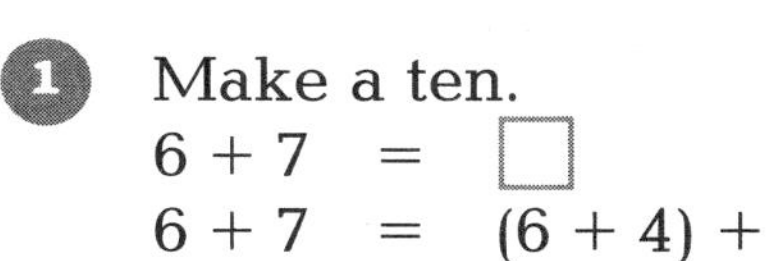

Find the sum.

1. Make a ten.
$6 + 7 = \square$
$6 + 7 = (6 + 4) + 3$
$= \_\_\_\_ + \_\_\_\_$
$= \_\_\_\_$

2. Use doubles plus 1.
$6 + 7 = \square$
$6 + 7 = (6 + 6) + 1$
$= \_\_\_\_ + \_\_\_\_$
$= \_\_\_\_$

3. Use doubles minus 1.
$6 + 7 = \square$
$6 + 7 = (7 + 7) - 1$
$= \_\_\_\_ - \_\_\_\_$
$= \_\_\_\_$

Name ______________________ Skill ______________________

Practice on Your Own

Skill 5

$8 + 6 = \square$

Make a ten to add. $8 + 2 = 10$

$8 + 6 = (8 + 2) + 4$
$= 10 + 4$
$= 14$

So, $8 + 6 = 14$.

$5 + 6 = \square$

Use doubles plus 1 or doubles minus 1.

$5 + 6 = (5 + 5) + 1 = 11$ $\quad 5 + 5 + 1 = 11$

or

$5 + 6 = (6 + 6) - 1 = 11$ $\quad 6 + 6 - 1 = 11$

So, $5 + 6 = 11$.

Make a ten to find the sum.

1. $5 + 8 = \square$
   $5 + 8 = (5 + \_\_\_\_) + \_\_\_\_$
   $= \_\_\_\_ + \_\_\_\_$
   $= \_\_\_\_$

2. $9 + 3 = \square$
   $9 + 3 = (9 + \_\_\_\_) + \_\_\_\_$
   $= \_\_\_\_ + \_\_\_\_$
   $= \_\_\_\_$

3. $6 + 8 = \square$
   $6 + 8 = (6 + \_\_\_\_) + \_\_\_\_$
   $= \_\_\_\_ + \_\_\_\_$
   $= \_\_\_\_$

Add. Use doubles plus 1.

4. $4 + 5 = \square$
   $4 + 5 = (4 + 4) + 1$
   $= \_\_\_\_ + \_\_\_\_$
   $= \_\_\_\_$

5. $7 + 8 = \square$
   $7 + 8 = (7 + 7) + 1$
   $= \_\_\_\_ + \_\_\_\_$
   $= \_\_\_\_$

6. $8 + 9 = \square$
   $8 + 9 = (8 + \_\_\_\_) + \_\_\_\_$
   $= \_\_\_\_ + \_\_\_\_$
   $= \_\_\_\_$

Add. Use doubles minus 1.

7. $3 + 4 = \square$
   $3 + 4 = (4 + 4) - 1$
   $= \_\_\_\_ - 1$
   $= \_\_\_\_$

8. $7 + 6 = \square$
   $7 + 6 = (7 + 7) - 1$
   $= \_\_\_\_ - \_\_\_\_$
   $= \_\_\_\_$

9. $8 + 7 = \square$
   $8 + 7 = (8 + 8) - 1$
   $= \_\_\_\_ - \_\_\_\_$
   $= \_\_\_\_$

Add.

10. $8 + 4 = \square$
    $8 + 4 = (\_\_\_\_ + \_\_\_\_) + \_\_\_\_$
    $= \_\_\_\_ + \_\_\_\_$
    $= \_\_\_\_$

11. $6 + 5 = \square$
    $6 + 5 = (\_\_\_\_ + \_\_\_\_) + \_\_\_\_$
    $= \_\_\_\_ + \_\_\_\_$
    $= \_\_\_\_$

12. $9 + 8 = \square$
    $9 + 8 = (\_\_\_\_ + \_\_\_\_) - \_\_\_\_$
    $= \_\_\_\_ - \_\_\_\_$
    $= \_\_\_\_$

Check

Add.

13. $8 + 3 = \square$
    $8 + 3 = (\_\_\_\_ + \_\_\_\_) + \_\_\_\_$
    $= \_\_\_\_ + \_\_\_\_$
    $= \_\_\_\_$

14. $6 + 7 = \square$
    $6 + 7 = (\_\_\_\_ + \_\_\_\_) + \_\_\_\_$
    $= \_\_\_\_ + \_\_\_\_$
    $= \_\_\_\_$

15. $7 + 5 = \square$
    $7 + 5 = (\_\_\_\_ + \_\_\_\_) + \_\_\_\_$
    $= \_\_\_\_ + \_\_\_\_$
    $= \_\_\_\_$

Skill 6 Grade 5

Order and Zero Properties of Addition

15 Minutes

Using Skill 6

OBJECTIVE Use and identify the Order Property of Addition and the Zero Property of Addition

MATERIALS dominoes

Display dominoes to represent addends in the following examples.

Have students read about the Order Property of Addition. Then ask them to count the dots on the domino that shows 6 + 5. Turn the domino to represent 5 + 6.

As you display each position, ask: **What are the addends?** (6 and 5) **What is 6 + 5?** (11) **What is 5 + 6?** (11) **Are the sums the same?** (yes) **Does the order matter?** (no) So addends can be added in any order. The sums are always the same.

Display the domino that shows 3 + 4 and 4 + 3. Have students count the dots to verify that the sum is 7.

Distribute dominoes so students can practice finding a few more sums two ways.

Then have students read about the Zero Property of Addition.

Ask: **What is the sum when you add 0 to any number?** (that number)

Have students write the addition sentences found in the addition table and continue the table to show several more examples.

TRY THESE Exercises 1–4 provide practice using the properties.

- **Exercises 1 and 3** Use and identify Order Property of Addition.
- **Exercises 2 and 4** Use and identify Zero Property of Addition.

PRACTICE ON YOUR OWN Read through the examples at the top of the page. Ask volunteers to explain the properties in their own words. Then have students write the addition examples vertically.

CHECK Determine if students can identify the Order Property and the Zero Property of Addition. Success is indicated by 4 out of 4 correct responses.

Students who successfully complete the **Practice on Your Own** and **Check** are ready to move to the next skill.

COMMON ERRORS

- When adding 0 to a number, students may write the sum as 0 instead of the number.
- Some students may understand the properties, but add incorrectly.

Students who made more than 4 errors in the **Practice on Your Own**, or who were not successful in the **Check** section, may benefit from the **Alternative Teaching Strategy** on the next page.

Optional

Alternative Teaching Strategy

Model Order and Zero Properties of Addition

20 Minutes

OBJECTIVE Model Order Property of Addition and Zero Property of Addition

MATERIALS dominoes, tiles

To review basic addition facts, have students complete an addition table such as the one below:

+	0	1	2	3	4	5	6	7	8	9
0	0	1	2	3						
1	1	2	3							
2	2	3	4							

Then model the Order Property of Addition. First, lay out a domino. Have students help you write the addition sentence.

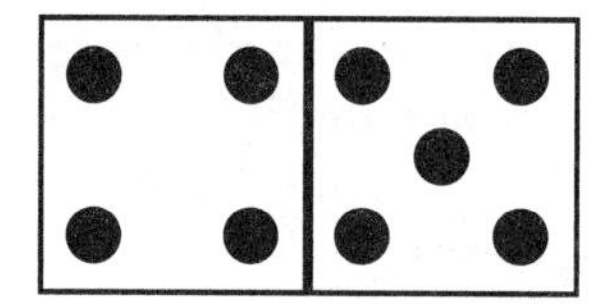

$4 + 5 = 9$

Then, reverse the domino. Have students help you write the new addition sentence.

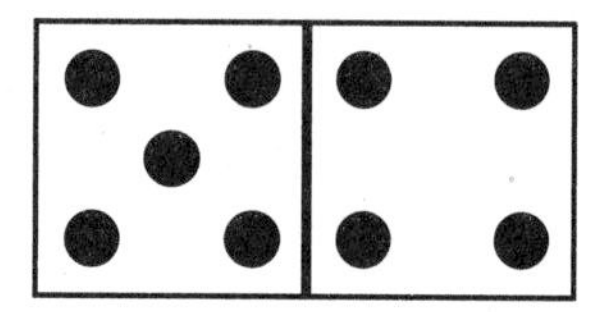

$5 + 4 = 9$

Ask: **Are the sums the same?** (yes) **Does the order of the addends matter?** (no) Have students use dominoes to model and write other pairs of addition sentences.

To illustrate the Zero Property of Addition, point to a group of 5 tiles. (or you may wish to use the domino with one blank side)

Ask: **How many tiles are there?** (5)

Tell students that you are adding 0 tiles to the group.

Ask: **Now how many tiles are there?** (5)

Have students help you write the addition sentence: $5 + 0 = 5$.

Have students start with 0 tiles and add 5. Ask them to help you write the new addition sentence: $0 + 5 = 5$.

Refer students to their addition tables, and have them read across the first row to confirm that 0 plus any number is the number.

Name ______________________ Skill ______________________

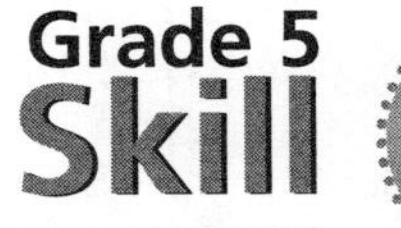

Order and Zero Properties of Addition

The Order Property of Addition is also called the Commutative Property of Addition.

Order Property of Addition
When the order of addends is changed, the sum is the same.

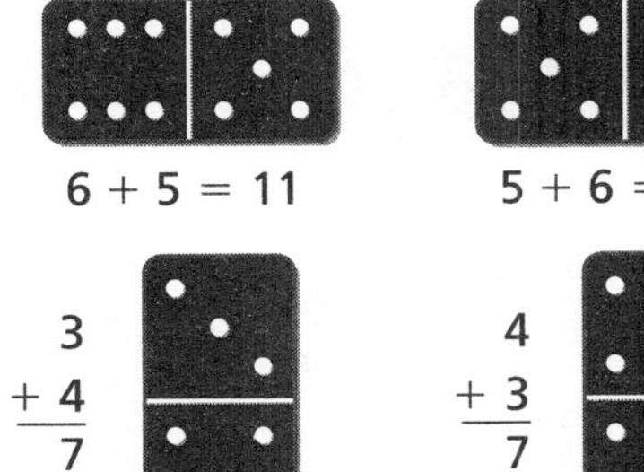
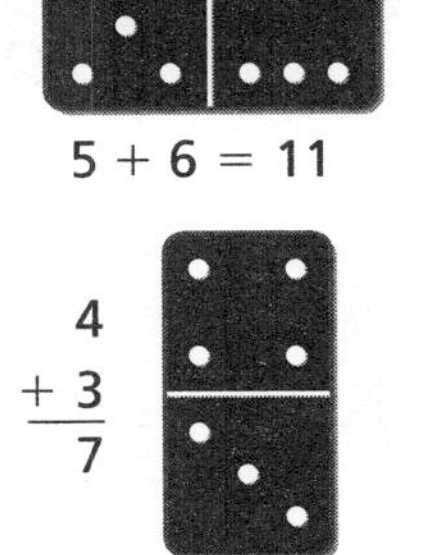

6 + 5 = 11 5 + 6 = 11

$$\begin{array}{r} 3 \\ +\,4 \\ \hline 7 \end{array} \qquad \begin{array}{r} 4 \\ +\,3 \\ \hline 7 \end{array}$$

Zero Property of Addition
When you add 0 to any number, the sum is that number.

0 + 2 = 2 2 + 0 = 2

This addition table shows the Zero Property of Addition.

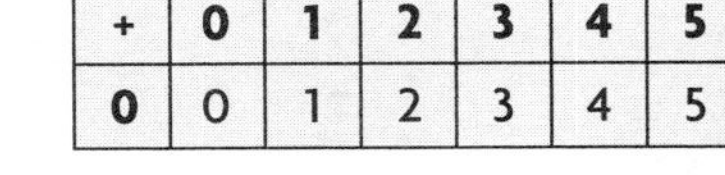

+	0	1	2	3	4	5
0	0	1	2	3	4	5

Try These

Find the sum. Write the property.

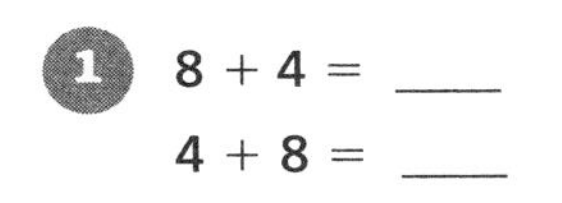

1. 8 + 4 = ____
   4 + 8 = ____
   ____

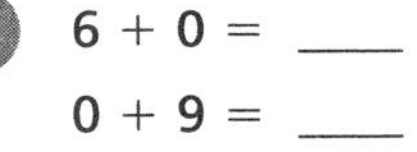
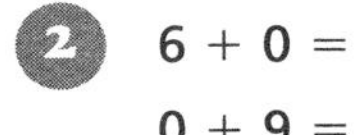

2. 6 + 0 = ____
   0 + 9 = ____
   ____

3. $\begin{array}{r} 9 \\ +2 \\ \hline \end{array} \qquad \begin{array}{r} 2 \\ +\,9 \\ \hline \end{array}$
   ____

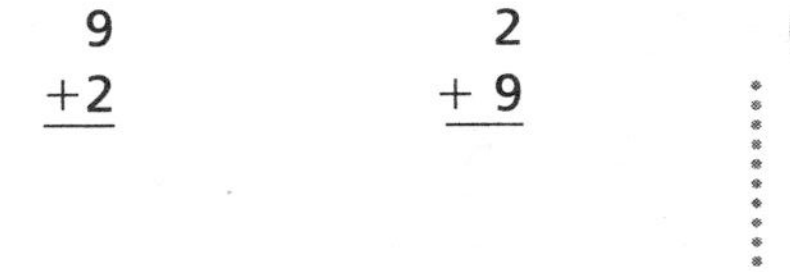

4. $\begin{array}{r} 0 \\ +\,8 \\ \hline \end{array} \qquad \begin{array}{r} 5 \\ +\,0 \\ \hline \end{array}$
   ____

Name ______________________ Skill ______________________

Practice on Your Own

Skill 6

Ask yourself:
Is the order changed?
Is one of the addends zero?

Order Property

10 + 8 = 18

8 + 10 = □

Think:
The order is changed.
The sum is the same.
So, 8 + 10 = 18.

Zero Property of Addition

32 + 0 = □

Think:
0 is one of the addends.
So, 32 + 0 = 32.

Find the sum. Write the name of the addition property you used.

1. 4 + 15 = ____
   15 + 4 = ____
   ____
2. 17 + 0 = ____
   0 + 22 = ____
   ____
3. 18 + 9 = ____
   9 + 18 = ____
   ____
4. 12 + 24 = ____
   24 + 12 = ____
   ____
5. 12 + 25 = ____ 25 + 12 = ____
   ____
6. 26 + 0 = ____ 0 + 16 = ____
   ____
7. 21 + 34 = ____ 34 + 21 = ____
   ____
8. 45 + 12 = ____ 12 + 45 = ____
   ____
9. 12 + 0 = ____
   0 + 35 = ____
   ____
10. 14 + 15 = ____
    15 + 14 = ____
    ____
11. 8 + 14 = ____
    14 + 8 = ____
    ____
12. 21 + 17 = ____
    17 + 21 = ____
    ____
13. 28 + 31 = ____ 31 + 28 = ____
    ____
14. 16 + 32 = ____ 32 + 16 = ____
    ____
15. 13 + 41 = ____ 41 + 13 = ____
    ____
16. 61 + 12 = ____ 12 + 61 = ____
    ____

Check

Find the sum. Write the name of the addition property you used.

17. 7 + 19 = ____
    19 + 7 = ____
    ____
18. 25 + 0 = ____
    0 + 24 = ____
    ____
19. 32 + 26 = ____ 26 + 32 = ____
    ____
20. 73 + 0 = ____ 0 + 76 = ____
    ____

Grouping Property of Addition

Using Skill 7

OBJECTIVE Use Grouping Property of Addition

Have students read about the Grouping Property of Addition on Skill 7. Be sure students are familiar with the words used to describe the property.

Ask: **What do you call the numbers you add?** (addends) **What do you call the result?** (sum)

Lead students through Steps 1 and 2.

Ask: **What numbers do you add first?** (the numbers in parentheses) **Why is it helpful to make a ten?** (It is easy to add 10.)

Have students compare the two sums in Step 3.

Ask: **Are the sums equal?** (yes) **Does it matter how you group the addends?** (no)

TRY THESE Exercises 1–3 model using the Grouping Property of Addition.

- **Exercises 1–3** Add inside parentheses first, find the sum, and then compare sums.

PRACTICE ON YOUR OWN Work through the example at the top of the page. Have a student explain the property. Then write $2 + (3 + 5) = (\square + 3) + 5$ and have students fill in the missing addend, 2.

CHECK Determine if students can group addends and add inside the parentheses first.

Success is indicated by 2 out of 2 correct responses.

Students who successfully complete the **Practice on Your Own** and **Check** are ready to move to the next skill.

COMMON ERRORS

- Students may understand the property but add incorrectly.
- Students may add incorrectly inside the parentheses and then use that incorrect sum to complete the addition.

Students who made more than 2 errors in the **Practice on Your Own**, or who were not successful in the **Check** section, may benefit from the **Alternative Teaching Strategy** on the next page.

Optional

Alternative Teaching Strategy

20 Minutes

Model the Grouping Property of Addition

OBJECTIVE Model Grouping Property of Addition

MATERIALS counters

Model the Grouping Property of Addition. Lay out separate piles of 4, 3, and 2 counters. Group together the piles of 4 and 3 counters. Ask how many counters. (7) Collect the remaining 2 counters and add them to the pile. Ask how many counters there are in all. (9)

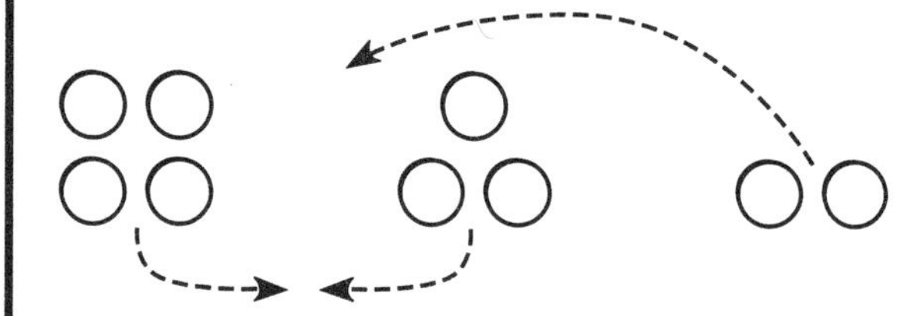

Have students help you use numbers to show the addition:

$$(4 + 3) + 2$$

$$7 + 2$$

$$9$$

Then group together the piles of 3 and 2 counters. Ask how many counters there are altogether. (5) Collect the remaining 4 counters and add them to the pile. Ask how many counters there are in all. (9)

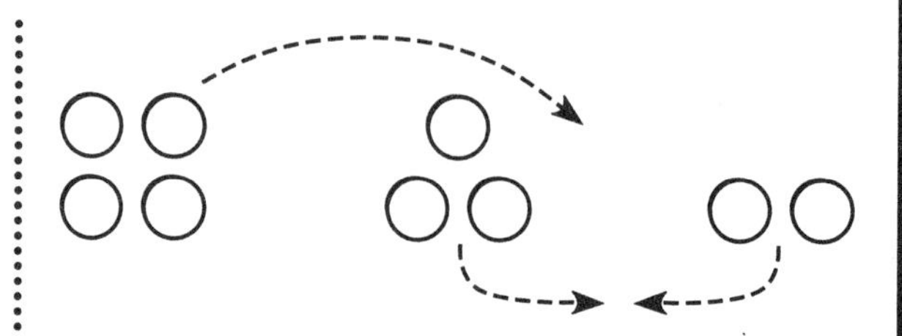

Have students help you use numbers to show the addition:

$$4 + (3 + 2)$$

$$4 + 5$$

$$9$$

Ask: **Are the sums equal?** (yes) **Does it matter how you group the addends 4, 3, and 2?** (no)

Have students help you write an addition sentence similar to those in Skill 7.

$$(4 + 3) + 2 = 4 + (3 + 2)$$

$$7 + 2 = 4 + 5$$

$$9 = 9$$

Have students use counters to model more addition examples and write addition sentences to show grouping.

Name ______________________ Skill ______________________

Grade 5 Skill 7

Grouping Property of Addition

Use the Grouping Property to help you find 7 + 3 + 5.

The Grouping Property of Addition is also called the Associative Property. It states that addends can be grouped differently, but the sum remains the same.

Step 1
Use the Grouping Property.
Group the addends differently.

$7 + (3 + 5) = (7 + 3) + 5$

Step 2
Add inside the parentheses first.

$7 + (3 + 5) = (7 + 3) + 5$

$7 + \mathbf{8} = \mathbf{10} + 5$

Step 3
Find the sums.

$7 + (3 + 5) = (7 + 3) + 5$

$7 + 8 = 10 + 5$

$15 = 15$

The sums are equal.

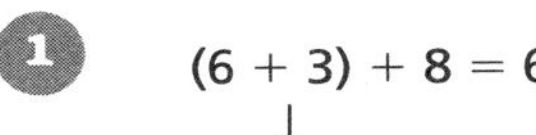

Try These

Add. Use the Grouping Property.

1. $(6 + 3) + 8 = 6 + (\_\_ + \_\_)$

   $\_\_ + 8 = 6 + \_\_$

   $\_\_ = \_\_$

   Equal sums? yes no

2. $(5 + 4) + 7 = 5 + (\_\_ + \_\_)$

   $\_\_ + 7 = 5 + \_\_$

   $\_\_ = \_\_$

   Equal sums? yes no

3. $4 + (6 + 7) = (\_\_ + \_\_) + 7$

   $4 + \_\_ = \_\_ + 7$

   $\_\_ = \_\_$

   Equal sums? yes no

Name ____________________ Skill ____________________

Practice on Your Own

Skill 7

Example: Find 2 + 3 + 5.

Think:
Grouping Property of Addition: Addends can be grouped differently, but the sum remains the same.

2 + (3 + 5) = (2 + 3) + 5 ← Group the addends differently.

2 + **8** = **5** + 5 ← Add inside the parentheses first.

✔ 10 = 10 The sums are equal.

Add. Use the Grouping Property.

1. (3 + 6) + 4 = 3 + (____ + ____)
   ____ + 4 = 3 + ____
   ____ = ____
   Equal sums? yes no

2. (8 + 7) + 3 = 8 + (____ + ____)
   ____ + 3 = 8 + ____
   ____ = ____
   Equal sums? yes no

3. (8 + 5) + 5 = ____ + (____ + ____)
   ____ = ____

4. (2 + 3) + 7 = ____ + (____ + ____)
   ____ = ____

5. 5 + 1 + 2
   (____ + ____) + ____ = ____ + (____ + ____)
   ____ = ____

6. 4 + 3 + 7
   (____ + ____) + ____ = ____ + (____ + ____)
   ____ = ____

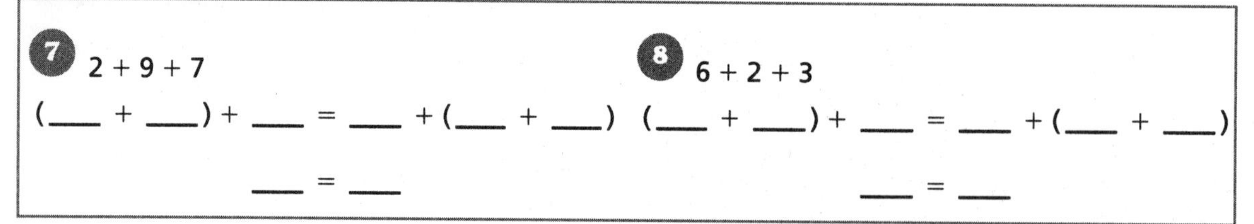

Check

7. 2 + 9 + 7
   (____ + ____) + ____ = ____ + (____ + ____)
   ____ = ____

8. 6 + 2 + 3
   (____ + ____) + ____ = ____ + (____ + ____)
   ____ = ____

Name ______________________ Skill ______________________

Answer Card

Addition

Grade 5

Skill 4

TRY THESE

1. 6, 7, 8
2. 8, 9, 10
3. 8, 9, 10
4. 8, 9, 10
5. 8, 9, 10

PRACTICE

1. 8
2. 8
3. 7
4. 9
5. 9; (4 + 4)
6. 7; (3 + 3)
7. 5; (2 + 2)
8. 9; (4 + 4)
9. 5; (3 + 3)
10. 9; (5 + 5)
11. 5; (3 + 3)
12. 7; (4 + 4)
13. 6
14. 8
15. 8
16. 10
17. 10
18. 7
19. 7
20. 10

Skill 4

CHECK

21. 9
22. 9
23. 9
24. 7

Skill 5

TRY THESE

1. 10, 3; 13
2. 12, 1; 13
3. 14, 1; 13

PRACTICE

1. 5; 3; 10, 3; 13
2. 1; 2; 10; 2; 12
3. 4, 4; 10, 4; 14
4. 8, 1; 9
5. 14, 1; 15
6. 8; 1; 16, 1; 17
7. 8, 7
8. 14, 1; 13
9. 16, 1; 15
10. 8, 2, 2; 10, 2; 12
11. 6, 4, 1, or 5, 5, 1; 10, 1; 11
12. 9, 9, 1; 18, 1; 17

CHECK

13. 8, 2, 1 or 3, 7, 1; 10, 1; 11
14. 6, 6, 1; 12, 1; 13, or 6, 4, 3; 10, 3; 13
15. 7, 3, 2, or 5, 5, 2; 10, 2; 12

Name ______________________ Skill ______________________

Answer Card

Addition

Grade 5

SKILL 6

TRY THESE

1. 12; 12; Order
2. 6; 9; Zero
3. 11; 11; Order
4. 8; 5; Zero

PRACTICE

1. 19; 19; Order
2. 17; 22; Zero
3. 27; 27; Order
4. 36; 36; Order
5. 37; 37; Order
6. 26; 16; Zero
7. 55; 55; Order
8. 57; 57; Order
9. 12, 35; Zero
10. 29; 29; Order
11. 22; 22; Order
12. 38; 38; Order
13. 59; 59; Order
14. 48; 48; Order
15. 54; 54; Order
16. 73; 73; Order

CHECK

17. 26; 26; Order
18. 25; 24; Zero
19. 58; 58; Order
20. 73; 76; Zero

SKILL 7

TRY THESE

1. 3, 8; 9, 11; 17, 17; yes
2. 4, 7; 9, 11; 16, 16; yes
3. 4, 6; 13, 10; 17, 17; yes

PRACTICE

1. 6, 4; 9, 10; 13, 13; yes
2. 7, 3; 15, 10; 18, 18; yes
3. 8, 5, 5; 18, 18
4. 2, 3, 7; 12, 12
5. 5, 1, 2; 5, 1, 2; 8, 8
6. 4, 3, 7; 4, 3, 7; 14, 14

CHECK

7. 2, 9, 7; 2, 9, 7; 18, 18
8. 6, 2, 3; 6, 2, 3; 11, 11

Number Sense

Whole Number Subtraction

Skill 8 Grade 5

Subtraction Across Zeros

Using Skill 8

OBJECTIVE Subtract across zeros

MATERIALS base-ten blocks

You may wish to work through Steps 1–4 with the students using base-ten blocks. Explain to the students that they are asked to subtract from 3-digit numbers that have one or two zeros.

Begin by pointing to Step 1 and noting the zeros in the ones and tens places. Ask: **Do you have enough ones to subtract?** (no) **Can you regroup tens as ones?** (No, there are no tens)

Guide them to recognize that first they must regroup 1 hundred as 10 tens, and then they can regroup 1 ten as 10 ones. Ask: **After you regroup 1 hundred as 10 tens, how many hundreds are left?** (4) **After you regroup 1 ten as 10 ones, how many tens are left?** (9)

Lead students through the subtraction steps, subtracting the ones in Step 2, the tens in Step 3 and the hundreds in Step 4. Emphasize to students that, in this example, they must go to the hundreds place and regroup hundreds as tens before they can regroup tens as ones.

TRY THESE In Exercises 1–4 students regroup first in the tens place only, then in both the tens and hundreds places.

- **Exercises 1–2** Regroup tens.
- **Exercises 3–4** Regroup tens and hundreds.

PRACTICE ON YOUR OWN Review the example at the top of the page. Ask students to explain the subtraction and regrouping process.

CHECK Determine if students can regroup once or more than once as they subtract from 3-digit numbers with zeros.

Success is indicated by 3 out of 4 correct responses.

Students who successfully complete the **Practice on Your Own** and **Check** are ready to move to the next skill.

COMMON ERRORS

- Students may forget to write the regrouped digits, and may thus subtract from the original numbers.
- Students may not regroup, but simply subtract zero from the bottom number.

Students who made more than 4 errors in the **Practice on Your Own**, or who were not successful in the **Check** section, may benefit from the **Alternative Teaching Strategy** on the next page.

Optional

Alternative Teaching Strategy

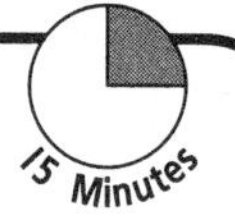

Use Models to Subtract Across Zeros

OBJECTIVE Use base-ten blocks to subtract from 2- and 3-digit numbers that contain zeros

MATERIALS base-ten blocks

Students may benefit from working with a partner. One student models the subtraction with the base-ten blocks while the other student records each step with paper and pencil.

Distribute the base-ten blocks and present this example:

$$\begin{array}{r} 50 \\ -\underline{23} \end{array}$$

Have students use the base-ten blocks to model the number 50.

Guide students to recognize that they cannot subtract ones until they regroup 1 ten as 10 ones. Then remind them to cross out the 5 as they record the regrouping, to show that there are 4 tens left. Have them count the blocks and record the difference as 27.

Then present this example:

$$\begin{array}{r} 400 \\ -\underline{135} \end{array}$$

Help students recognize that they cannot subtract either ones or tens until they first regroup 1 hundred as 10 tens, and then regroup 1 ten as 10 ones.

Remind the students to cross out the zero in the tens place, and write 10 above it to show the regrouped tens. Then cross out the 10 and write a 9 above it to show regrouping 1 ten as 10 ones.

$$\begin{array}{r} \;\;9\;\;\; \\ 3\;\cancel{10}\;10 \\ \cancel{4}\;\;\cancel{0}\;\;0 \\ -1\;\;3\;\;5 \\ \hline 2\;\;6\;\;5 \end{array}$$

Have students exchange roles as they repeat the activity with similar examples. When the students show understanding of the regrouping process, remove the base-ten blocks and have them try an exercise using only paper and pencil. Ask students to explain each step as they complete the subtraction across zeros.

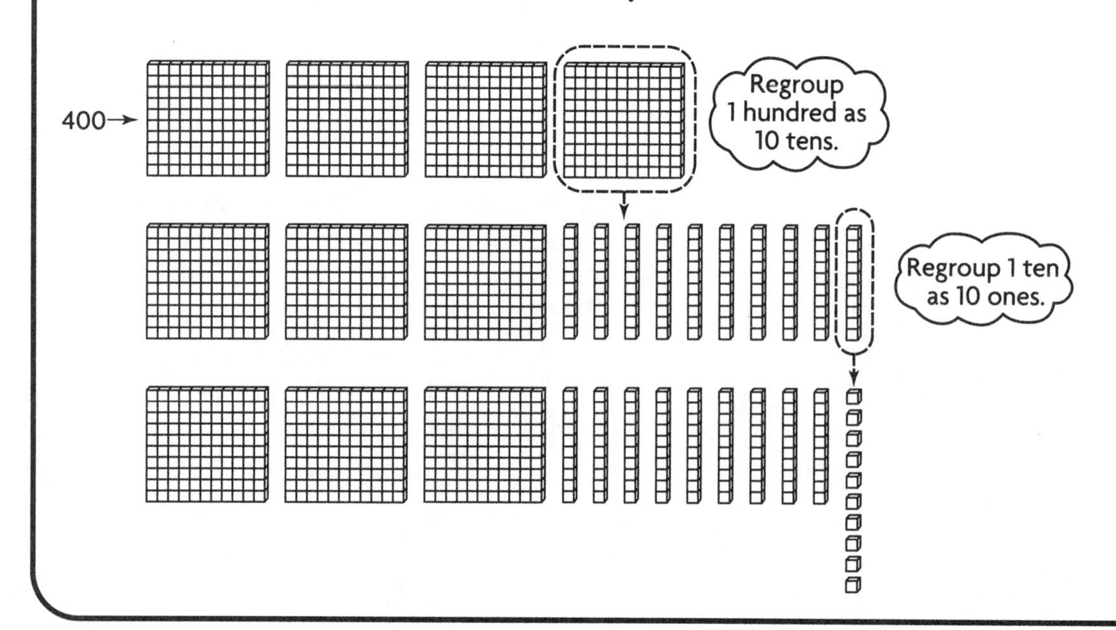

Name ______________________ Skill ______________________

Grade 5
Skill 8

Subtraction Across Zeros

Find 500 – 236.

Step 1
There are no ones or tens. Regroup 1 hundred as 10 tens.

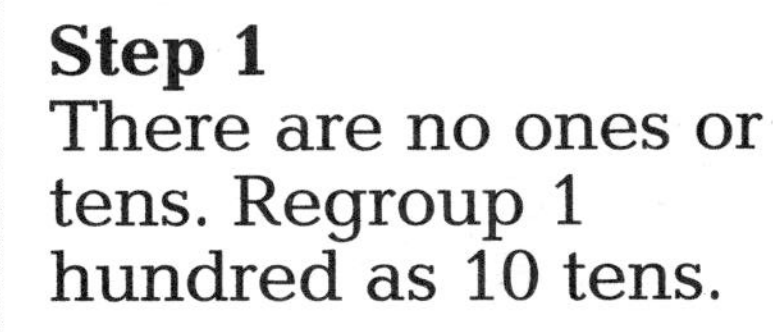

Step 2
Regroup 1 ten as 10 ones. Subtract the ones.

Step 3
Subtract the tens.

Step 4
Subtract the hundreds.

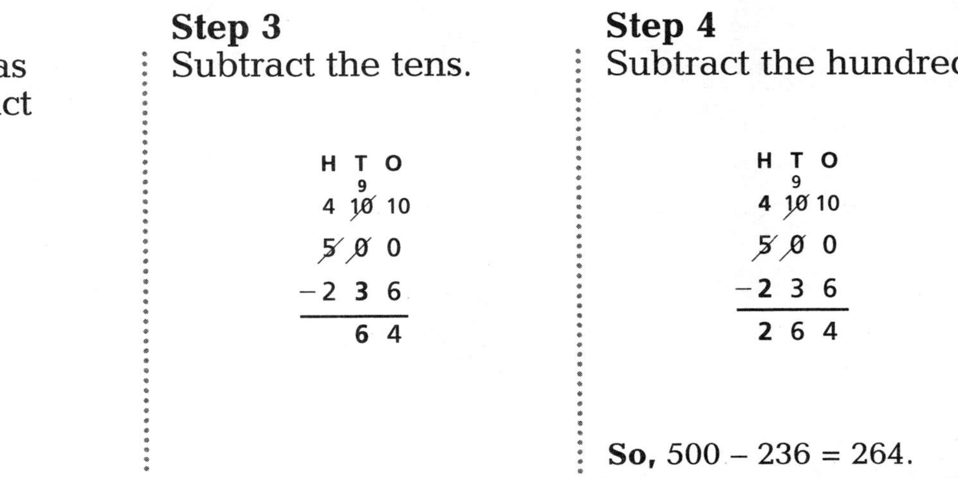

So, 500 – 236 = 264.

Try These

Subtract.

1. H T O
   30 – 17

2. H T O
   250 – 145

3. H T O
   500 – 367

4. H T O
   700 – 628

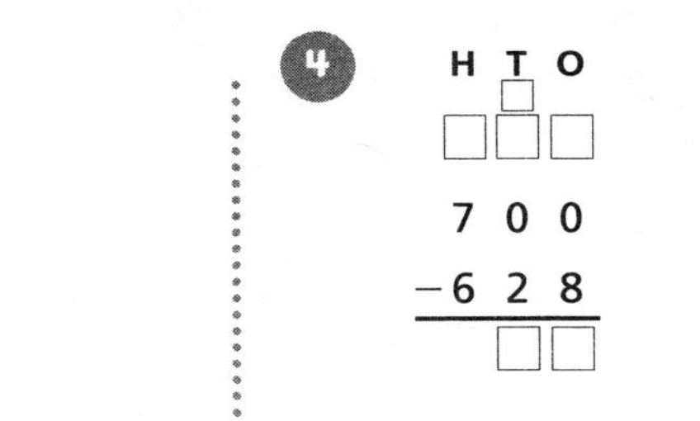

Go to the next side.

Name ______________________ Skill ______________________

Practice on Your Own

Skill 8

```
      H T O
        9
      5 10 10
      6̸ 0̸ 0
    − 2 2 5
      3 7 5
```

There are no ones or tens. So, regroup 1 hundred as 10 tens. Then regroup 1 ten as 10 ones.

Find the difference.

1. H T O 130 − 56	2. H T O 350 − 165	3. H T O 600 − 367	4. H T O 700 − 428
5. H T O 150 − 92	6. H T O 270 − 189	7. H T O 400 − 253	8. H T O 800 − 591
9. H T O 190 − 155	10. H T O 340 − 267	11. H T O 400 − 122	12. H T O 600 − 479
13. 280 − 142	14. 405 − 386	15. 700 − 513	16. 900 − 238

Check

Find the difference.

17. 390 − 265	18. 304 − 129	19. 600 − 418	20. 900 − 752

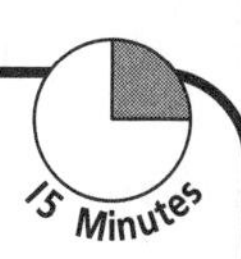

Using Skill 9

OBJECTIVE Use strategies to recall subtraction facts 0 - 9

Begin by recalling for students that they can use strategies to subtract.

Direct students' attention to the strategy Use Addition. Ask: **What operation is the inverse of subtraction?** (addition)

Point out to students that they can think of an addition fact to help them solve 7 − 4 = □. Ask: **What number plus 4 equals 7?** (3) **How are the addition and subtraction sentences related?** (Both have the same three numbers; addition and subtraction are inverse operations)

Direct students' attention to the strategy Use Facts You Know. Point out to students that if they know one subtraction fact, they also know another subtraction fact.

Say: **You know 6 − 4 = 2.**

Ask: **What other subtraction fact can you write using these three numbers?** (6 − 2 = 4)

Next, direct students' attention to the strategy Count Back. Point out to students that they can subtract by using a number line to count back. Remind students that they move left on the number line to subtract. Point to 8 − 3 = □. Ask: **At what number do you start on the number line?** (8) **How many jumps do you make?** (3) **On what number do you stop?** (5)

Finally, direct students' attention to the strategy Subtracting with Zero. Guide students to understand that when they subtract zero from a number, the result is the number itself.

Also, make sure students understand that when they subtract a number from itself, the result is zero.

TRY THESE Exercises 1–8 provide practice using strategies to subtract.

- **Exercises 1 and 5** Use Addition.
- **Exercises 2 and 6** Count Back.
- **Exercises 3 and 7** Use Facts You Know.
- **Exercises 4 and 8** Subtracting with Zero.

PRACTICE ON YOUR OWN Review the examples at the top of the page. Ask students to explain the inverse operations of addition and subtraction.

CHECK Determine if students can recall subtraction facts through 9.

Success is indicated by 3 out of 3 correct responses.

Students who successfully complete the **Practice on Your Own** and **Check** are ready to move to the next skill.

COMMON ERRORS

- Students may subtract incorrectly.
- Students may add instead of subtract.

Students who made more than 5 errors in the **Practice on Your Own**, or who were not successful in the **Check** section, may benefit from the **Alternative Teaching Strategy** on the next page.

Optional

Alternative Teaching Strategy

15 Minutes

Use Counters to Model Subtraction Facts 0–9

OBJECTIVE Use counters to model subtraction facts 0–9

MATERIALS index cards, counters, paper, pencils

You may wish to have students work in pairs. One student writes the addition sentence for a fact through 9 and models it with the counters. The other student then uses the same numbers to write a subtraction fact and model it with the counters.

Prepare an index card with 5 + 3 = 8 on one side and 8 − 5 = 3 on the other side. Distribute the counters, paper and pencils. Display 5 + 3 = 8.

Have one partner model the first addend with the counters. Ask: **How many counters do you show?** (5)

Now have the students model the second addend with counters. Ask: **How many counters do you show for the second addend?** (3)

How many counters do you have in all? (8)

What is the sum? (8)

Have students record the addition number sentence. (5 + 3 = 8)

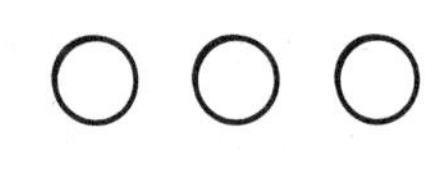

Direct the students' partners to leave the 8 counters showing. Tell the students that they are now going to model the related subtraction fact.

Display 8 − 5 = 3. Ask: **How many counters do you take away?** (5)

Have the students remove 5 counters. Ask:

How many counters do you have left? (3)

What is the difference? (3)

Have students record the subtraction number sentence. (8 − 5 = 3)

Guide students to understand that addition and subtraction are inverse operations, that is, one *undoes* the other.

Repeat the activity with similar examples. When the students show understanding of the meaning of subtraction, remove the counters and have them try an exercise using only paper and pencil. Ask students to explain how addition can be used to check the difference of a given subtraction sentence.

Name ______________________ Skill ______________________

Grade 5 **Skill 9**

Subtraction Facts 0–9

Use strategies to recall subtraction facts.

Use Addition
Subtraction is the inverse of addition.
Use an addition fact to recall subtraction.

$7 - 4 = \square$
$3 + 4 = 7$
So, $7 - 4 = \mathbf{3}$

What number plus 4 equals 7?

Use Facts You Know
Use a fact you know.
$6 - 2 = \square$
If you know $6 - 4 = 2$, then you also know $6 - 2 = 4$.

So, $6 - 2 = 4$.

Count Back
$8 - 3 = \square$

0 1 2 3 4 5 6 7 8 9 10

Say: 8. Count back 3: 7, 6, **5**.
So, $8 - 3 = \mathbf{5}$.

Subtracting with Zero
$4 - 0 = \square$
When you subtract 0 from a number, the difference is that number.
So, $4 - 0 = 4$.

$4 - 4 = \square$
When you subtract a number from itself, the difference is 0.

So, $4 - 4 = 0$.

Try These

Use strategies to find the difference.

1. $3 + $ ______ $= 5$

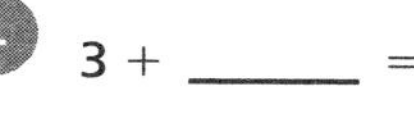

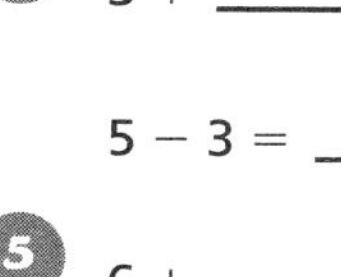

   $5 - 3 =$ ______

2. 0 1 2 3 4 5 6 7 8 9 10

   $6 - 1 =$ ______

3. $9 - 5 = 4$

   $9 - 4 =$ ______

4. $8 - 8 =$ ______

   $8 - 0 =$ ______

5. $6 + $ ______ $= 9$

   $9 - 6 =$ ______

6. 0 1 2 3 4 5 6 7 8 9 10

   $7 - 2 =$ ______

7. $6 - 4 = 2$

   $6 - 2 =$ ______

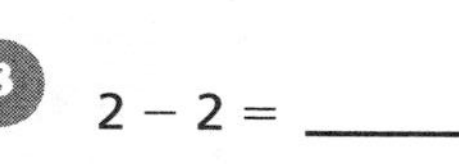

8. $2 - 2 =$ ______

   $2 - 0 =$ ______

Go to the next side.

Name ______________________ Skill ______________________

Practice on Your Own

Skill 9

$9 - 3 = \square$

Count back to subtract.

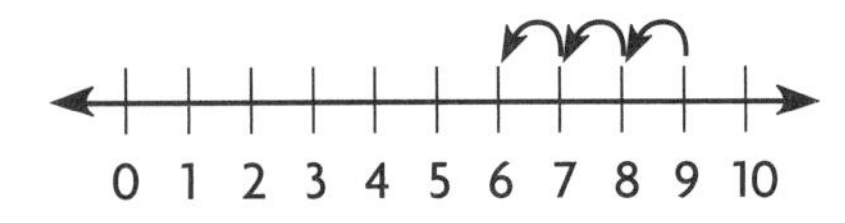

So, 9 – 3 = 6.

$7 - 5 = \square$

Think about addition facts to subtract.

2 + 5 = 7

So, 7 – 5 = 2.

Use strategies to find the difference.

1. 2 + ______ = 5
   5 – 2 = ______
2. 4 + ______ = 6
   6 – 4 = ______
3. 4 + ______ = 9
   9 – 4 = ______
4. 7 – 2 = ______
5. 9 – 1 = ______
6. 6 – 3 = ______

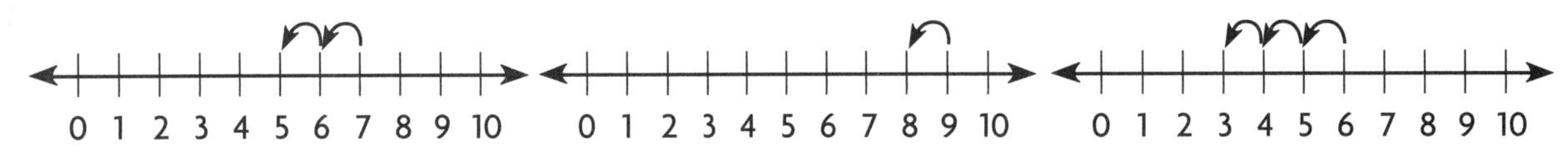

7. 9 – 4 = 5
   9 – 5 = ______
8. 7 – 6 = 1
   7 – 1 = ______
9. 8 – 5 = 3
   8 – 3 = ______
10. 6 – 6 = ______
    6 – 0 = ______
11. 7 – 7 = ______
    7 – 0 = ______
12. 9 – 9 = ______
    9 – 0 = ______
13. 9 – 7 = ______
14. 8 – 5 = ______
15. 7 – 3 = ______
16. 9 – 6 = ______
17. 7 – 0 = ______
18. 5 – 2 = ______

Check

Use strategies to find the difference.

19. 6 – 3 = ______
20. 9 – 5 = ______
21. 8 – 8 = ______

Using Skill 10

OBJECTIVE Use strategies to recall subtraction facts

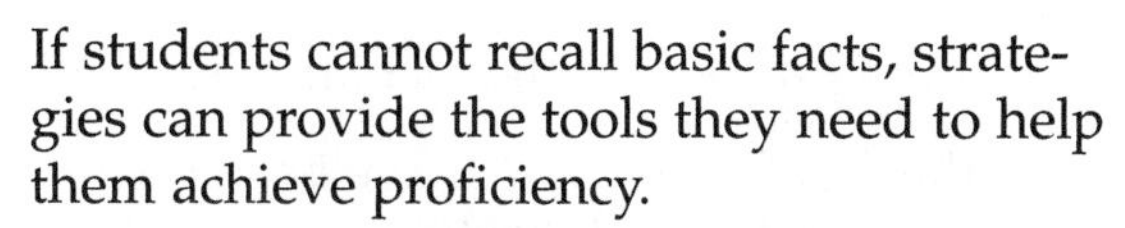

If students cannot recall basic facts, strategies can provide the tools they need to help them achieve proficiency.

You may wish to reintroduce the strategies for subtraction using 'easier' facts (those from 0–9). The first step should be to develop understanding of the strategies. For example, before you discuss the *Use Addition* section, review the strategy using the easier subtraction fact, $7 - 5 = \square$.

Guide the students by saying:

You can think of this subtraction sentence as an addition sentence. Just ask yourself, "What number plus 5 is 7?" Provide other easy facts and have students formulate the addition question.

For the *Count Back* strategy, display a large number line and review how to find $8 - 3 = \square$.

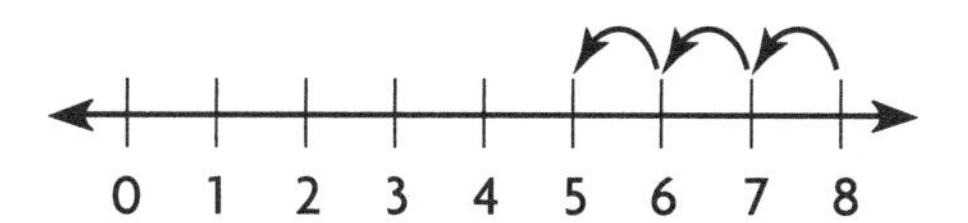

Note that the 8 is the starting number and that students count the spaces between numbers, not the tick marks.

For the *Use Facts You Know*, help students realize that they may have already memorized related subtraction facts that can be used to help them find a difference they think they do not know. Ask:

"What subtraction fact is related to $5 - 3 = \square$? ($5 - 2 = \square$) $7 - 6 = \square$? ($7 - 1 = \square$) $9 - 4 = \square$? ($9 - 5 = \square$)"

TRY THESE Exercises 1–6 provide practice using the strategies introduced.

- **Exercises 1 and 4** Use addition.
- **Exercises 2 and 5** Count back.
- **Exercises 3 and 6** Use facts you know.

PRACTICE ON YOUR OWN Review the strategies at the top of the page.

The exercises that follow provide practice using the subtraction strategies to find the difference.

CHECK Determine if students can use the strategies to find the difference. Success is indicated by 3 out of 4 correct responses.

Students who successfully complete **Practice on Your Own** and **Check** are ready to move to the next skill.

COMMON ERRORS

- Students may use tick marks to count back, thus their answers will be 1 more than the actual difference.
- Students may not know when or how to apply a strategy. For example, students may inappropriately count back 5, 6 or more, lose count, and then record an incorrect answer.

Students who made more than 2 errors in the **Practice on You Own**, or who were not successful in the **Check** section, may benefit from the **Alternative Teaching Strategy** on the next page.

Optional

Alternative Teaching Strategy

Fact Family Cards

OBJECTIVE Make fact family cards to review subtraction facts 10–19

MATERIALS oaktag or poster board, masking tape, magic markers, and rulers

You may wish to have students work in pairs to make the fact family cards. Give each pair 1 inch × 5 inch oaktag strips, a 12 inch × 4 inch oaktag strip, magic markers, ruler, and a piece of masking tape. Have the students divide the larger strip into 3 sections. Next, the shorter strip should be folded to make a sleeve that can easily slide over the larger strip. After the sleeve has been tested to make sure it slides easily, tape the ends overlapping.

To assist the students with the fact families, list them on the board, such as 9, 2, 11; 6, 5, 11; 8, 3, 11; 7, 4, 11 and 1, 10, 11. Have the partners take turns reading the fact families and writing the numerals on the cards. The numerals should be written from left to right with one numeral in each section of the marked strip. Each section should be the same size.

9	2	11

Repeat the process for the fact families to 19.

After the fact family cards have been completed, show the students how to slide the sleeve over the strip and cover a number.

To practice the subtraction facts the partner selects a fact family strip, slides the sleeve to cover one of the numbers, and then shows this to the other student. The student names the missing number in the family. The sleeve is removed to check.

For the card below the student would say: "Eleven minus nine is..."

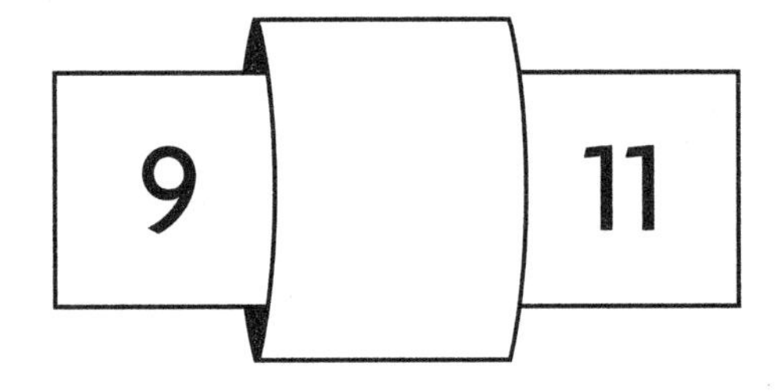

Have the partners take turns. Encourage students to review a few facts every day. Students should also keep a record of the facts they know and the facts they still need to master.

Name ______________________ Skill ______________________

Grade 5
Skill 10

Subtraction Facts 10–19

You can use strategies to recall subtraction facts.

Use Addition
Subtraction is the inverse of addition. Use an addition fact to recall subtraction.

$15 - 7 = \blacksquare$
$7 + \mathbf{8} = 15$
So, $15 - 7 = 8$.

Think: What number plus 7 equals 15?

Count Back
$12 - 3 = \blacksquare$

6 7 8 9 10 11 12 13

Say: 12. Count back 3: 11, 10, **9**.
So, $12 - 3 = 9$.

Use Facts You Know
Use a fact you know.
$13 - 9 = \blacksquare$

If you know $13 - 4 = 9$, then you also know $13 - 9 = 4$.
So, $13 - 9 = 4$.

Try These

Use strategies to find the difference.

1. $9 +$ ____ $= 17$

   $17 - 9 =$ ____

2. 5 6 7 8 9 10 11 12

   $11 - 3 =$ ____

3. $14 - 6 = 8$

   $14 - 8 =$ ____

4. $8 +$ ____ $= 15$

   $15 - 8 =$ ____

5. 9 10 11 12 13 14 15

   $14 - 5 =$ ____

6. $16 - 9 = 7$

   $16 - 7 =$ ____

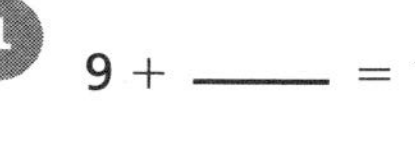

Name ______________________ Skill ______________________

Practice on Your Own

Use Addition
16 − 9 = □

Think about addition facts to help you subtract.

7 + 9 = 16

So, 16 − 9 = 7.

Count Back
11 − 2 = □

Count back to subtract.

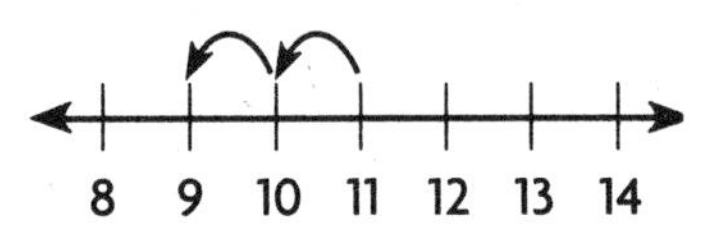

So, 11 − 2 = 9.

Use Facts You Know
15 − 8 = □

If you know 15 − 7 = 8, then 15 − 8 = 7.

Use strategies to find the difference.

1. 4 + ____ = 13
   13 − 4 = ____
2. 5 + ____ = 14
   14 − 5 = ____
3. 9 + ____ = 12
   12 − 9 = ____
4. 8 + ____ = 13
   13 − 8 = ____

5. 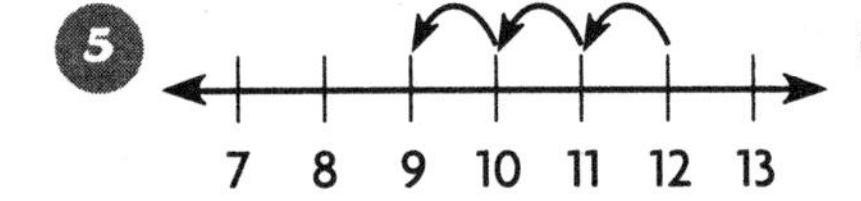

   12 − 3 = ____
6. 

   14 − 5 = ____
7. 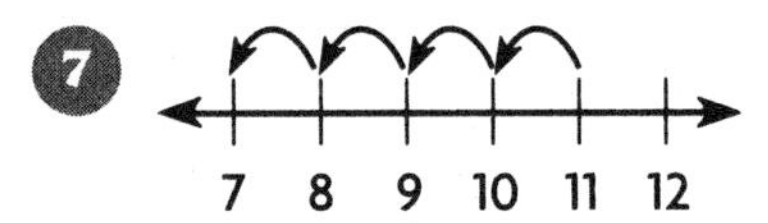

   11 − 4 = ____

8. 10 + 4 = 14
   14 − 4 = ____
9. 8 + 9 = 17
   17 − 8 = ____
10. 8 + 7 = 15
    15 − 8 = ____
11. 9 + 7 = 16
    16 − 9 = ____

12. 14 − 6 = ____
13. 17 − 9 = ____
14. 12 − 4 = ____
15. 11 − 5 = ____

16. 17 − 9 = ____
17. 15 − 6 = ____
18. 12 − 5 = ____
19. 11 − 3 = ____

Check

Use strategies to find the difference.

20. 16 − 9 = ____
21. 12 − 5 = ____
22. 16 − 8 = ____
23. 18 − 9 = ____

Name ______________________ Skill ______________________

SKILL 8

TRY THESE

1. 2, 10, 13
2. 2, 9, 4, 10, 105
3. 4, 9, 10,10, 133
4. 6, 9, 10, 10, 72

PRACTICE

1. 2, 10, 74
2. 2, 9, 4, 10, 185
3. 5, 9, 10,10, 233
4. 6, 9, 10, 10, 272
5. 0, 14, 10, 58
6. 1, 16, 10, 81,
7. 3, 9, 10, 10, 147
8. 7, 9, 10, 10, 209
9. 8, 10, 35
10. 2, 13, 10, 73
11. 3, 9, 10, 10, 278
12. 5, 9, 10, 10, 121
13. 138
14. 19
15. 187
16. 662

CHECK

17. 125
18. 175
19. 182
20. 148

SKILL 9

TRY THESE

1. 2; 2
2. 5
3. 5
4. 0; 8
5. 3; 3
6. 5
7. 4
8. 0; 2

PRACTICE

1. 3; 3
2. 2; 2
3. 5; 5
4. 5
5. 8
6. 3
7. 4
8. 6
9. 5
10. 0; 6
11. 0; 7
12. 0; 9
13. 2
14. 3
15. 4
16. 3
17. 7
18. 3

CHECK

19. 3
20. 4
21. 0

Answer Card

Whole Number Subtraction

Grade 5

Name ________________ Skill ________________

Answer Card

Whole Number Subtraction

Grade 5

SKILL 10

TRY THESE

1. 8; 8
2. 8
3. 6
4. 7; 7
5. 9
6. 9

PRACTICE

1. 9; 9
2. 9; 9
3. 3; 3
4. 5; 5
5. 9
6. 9
7. 7
8. 10
9. 9
10. 7
11. 7
12. 8
13. 8
14. 8
15. 6
16. 8
17. 9
18. 7
19. 8

CHECK

20. 7
21. 7
22. 8
23. 9

Number Sense

Money

Skill 11 Grade 5

Multiply Money

Using Skill 11

OBJECTIVE Multiply amounts of money less than a dollar

Explain that in this lesson students will multiply amounts of money that are less than one dollar. Point out that multiplying money is the same as multiplying whole numbers except that there is a dollar sign and decimal point in the factor and in the product.

In Step 1, remind students that when there are no dollars, a zero is used in the dollars place. Discuss how dollars, dimes, and pennies are like hundreds, tens, and ones. Make sure students know that ten pennies equal one dime and that ten dimes equal one dollar.

Direct students' attention to Step 2. Ask:
Do you regroup the pennies? Explain. (Yes, because there are more than 9 pennies, I can regroup 15 pennies as 1 dime 5 pennies.)
When do you add the regrouped dime? (Possible response: First I multiply the dimes and then I add the regrouped dime to that product.)
How many dimes do you have now? (7)
Must you regroup the dimes? Explain. (No, there are fewer than 10 dimes.)
What do you do after you multiply the dimes? (Write the product with a zero in the dollar place, a dollar sign, and a decimal point between the dollars and dimes.)

TRY THESE In Exercises 1–4 students multiply money, regrouping in different ways.

- **Exercise 1** Regroup pennies.
- **Exercise 2** No regrouping.
- **Exercise 3** Regroup pennies.
- **Exercise 4** Regroup pennies and dimes.

PRACTICE ON YOUR OWN Review the example at the top of the page. Ask students to compare multiplying money amounts with multiplying whole numbers. How are they alike? How are they different?

CHECK Determine if students can regroup correctly, place the dollar sign and decimal point in the product, and place a zero where appropriate. Success is indicated by 3 out of 4 correct responses.

Students who successfully complete the **Practice on Your Own** and **Check** are ready to move to the next skill.

COMMON ERRORS

- Students may not place the decimal point or dollar sign correctly in the product.
- Students may add a regrouped digit to the factor before multiplying, instead of adding it to the product.
- Students may not know multiplication facts.

Students who made more than 4 errors in the **Practice on Your Own**, or who were not successful in the **Check** section, may benefit from the **Alternative Teaching Strategy** on the next page.

Optional

Alternative Teaching Strategy
Use Models to Multiply Money

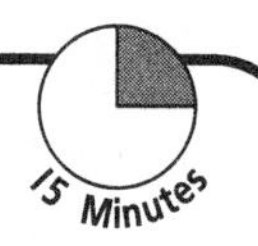

OBJECTIVE Use play money to practice multiplying with money

MATERIALS play money, paper

Distribute the play money and present the example 2 × $0.29. Have students point to the dollar sign and the decimal point. If necessary, review how to draw a dollar sign by making an S with two vertical strike-throughs. Emphasize that the decimal point is always between the dollars and the dimes.

Explain to students that they can model and record the multiplication with money, just as they do with whole numbers.

Have students model 2 × $0.29 in play money by displaying 2 groups of 2 dimes and 9 pennies.

2 groups of 2 dimes and 9 pennies

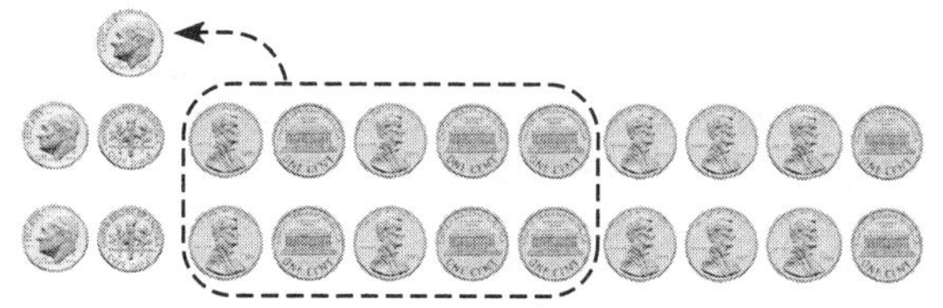

5 dimes 8 pennies
$0.58

First have students find the total number of pennies. Recall that 10 pennies equal 1 dime, and that 10 dimes equal 1 dollar. Guide students through the regrouping procedure.

Record the multiplication on large paper, as students explain each step.

Record the product as $0.58 and explain that when multiplying money, students must write the dollar sign and decimal point in the product.

Refer to the top factor and point out that the decimal point separates the dollars and cents. Say: **When there are no dollars in the product, you write a zero in the dollars place.**

Repeat the activity several times. First have students model without regrouping and then have them model regrouping both pennies and dimes.

When students show an understanding of the regrouping process and placing the decimal point and dollar sign, have them multiply money amounts without models, recording the multiplication on paper only.

Grade 5 **Skill 11**

Multiply Money

Find $0.25 × 3.

Step 1 Think about how to multiply whole numbers.

	dollars	.	dimes	pennies
$	0	.	2	5
×				3

Step 2 Multiply the digits in the pennies place. Regroup 10 pennies as 1 dime.

Think:
3 × 5 = 15
15 pennies = 1 dime, 5 pennies

	dollars	.	dimes	pennies
$	0	.	2	5
×				3
				5

Step 3 Multiply the digits in the dimes place. Add the dime that was regrouped.

Think:
3 × 2 = 6
6 dimes + 1 dime = 7 dimes

	dollars	.	dimes	pennies
			1	
$	0	.	2	5
×				3
			7	5

Step 4 Multiply the dollars. Write the dollar sign and decimal point in the product.

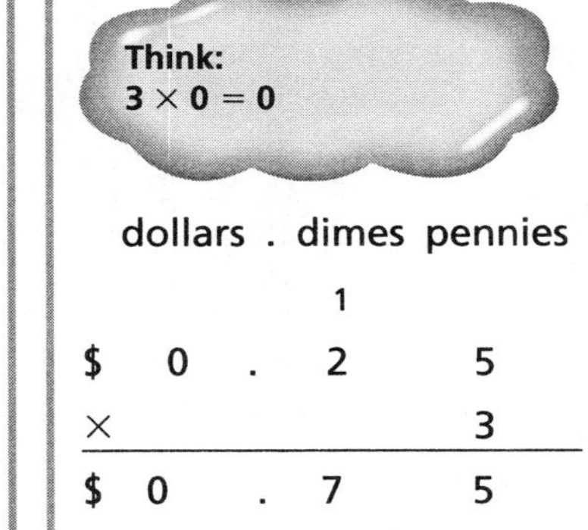

	dollars	.	dimes	pennies
			1	
$	0	.	2	5
×				3
$	0	.	7	5

So, $0.25 × 3 = $0.75

Try These

Multiply.

1.

	dollars	.	dimes	pennies
$	0	.	2	5
×				2
$				

2.

	dollars	.	dimes	pennies
$	0	.	1	0
×				6
$				

3.

	dollars	.	dimes	pennies
$	0	.	0	5
×				4
$				

4.

	dollars	.	dimes	pennies
$	0	.	2	5
×				5
$				

Go to the next page.

Name ______________________ Skill ______________________

Practice on Your Own

Skill 11

Find $0.50 × 3.

	dollars	.	dimes	pennies
$	0	.	5	0
×				3
$	1	.	5	0

3 × 0 pennies = 0 pennies
3 × 5 dimes = 15 dimes.
Regroup as 1 dollar, 5 dimes.
3 × 0 dollars + 1 dollar = 0 dollars + 1 dollar
= 1 dollar

Remember to write a decimal point and dollar sign in the product.

So, $0.50 × 3 = $1.50.

Multiply.

1. dollars . dimes pennies
   $0 . 0 5
   × 5
   $

2. dollars . dimes pennies
   $0 . 1 0
   × 8
   $

3. dollars . dimes pennies
   $0 . 2 5
   × 5
   $

4. dollars . dimes pennies
   $0 . 2 1
   × 9
   $

5. $0 . 5 1
   × 4
   $

6. $0 . 2 5
   × 6
   $

7. $0 . 5 0
   × 5
   $

8. $0 . 1 5
   × 3
   $

9. $0 . 0 5
   × 3
   $

10. $0 . 2 5
    × 4
    $

11. $0 . 1 0
    × 1 0
    $

12. $0 . 5 5
    × 5
    $

13. $0.05 × 4

14. $0.01 × 5

15. $0.50 × 2

16. $0.25 × 8

Multiply.

17. $0.21 × 8

18. $0.50 × 3

19. $0 . 1 0
    × 7
    $

20. $0 . 2 5
    × 5
    $

Name ______________________ Skill ______________________

Answer Card

Money

Grade 5

SKILL 11

TRY THESE

1. $0.50
2. $0.60
3. $0.20
4. $1.25

PRACTICE

1. $0.25
2. $0.80
3. $1.25
4. $1.89
5. $2.04
6. $1.50
7. $2.50
8. $0.45
9. $0.15
10. $1.00
11. $1.00
12. $2.75
13. $0.20
14. $0.05
15. $1.00
16. $2.00

CHECK

17. $1.68
18. $1.50
19. $0.70
20. $1.25

Number Sense

Whole Number Multiplication

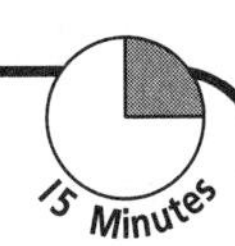

Using Skill 12

OBJECTIVE Recall multiplication facts 1 through 5

Review the definitions for *factors* and *products*. Then have students look at the first example. Suggest that they can use a number line to model multiplication facts.

Direct students to the number line in the first example. Clarify for students that the factor 4 represents (or tells them) the number of spaces to "jump over" on the number line, and that the factor 3 represents the number of times to "jump." Ask: **From which number do you start?** (0)

How many jumps do you make on the number line? (3)

How many spaces do you jump over each time? (4)

Continue: The number you land on after the third jump is the product. What is the product 3 × 4? (12)

Ask similar questions as you work through the second example. Explain that in this example the variable n represents the product, they are finding the value of n.

TRY THESE
In Exercises 1–3 students use the number line to find the product or the value of n.

- **Exercises 1, 2** Find the product.
- **Exercise 3** Find the value of n.

PRACTICE ON YOUR OWN Review the example at the top of the page. Ask students to explain how the number line models the multiplication sentence $5 \times 4 = n$.

CHECK Determine if students know the multiplication facts to 5. Success is indicated by 3 out of 4 correct responses.

Students who successfully complete the **Practice on Your Own** and **Check** are ready to move to the next skill.

COMMON ERRORS

- Students may start with number 1 on the number line.
- Students may not understand the concept of multiplication.

Students who made more than 4 errors in the **Practice on Your Own**, or who were not successful in the **Check** section, may benefit from the **Alternative Teaching Strategy** on the next page.

Optional

Alternative Teaching Strategy

15 Minutes

Use Models to Show Multiplication Facts 1–5

OBJECTIVE Use counters to model multiplication facts from 1 through 5

MATERIALS counters, centimeter grid paper

Distribute counters to each student. Recall that *factors* are the numbers that are multiplied; the *product* is the result, or answer.

Display the following:

3×4

Ask: **What are the factors?** (3 and 4) **Does this expression tell you what the product is?** (no) Continue: You can use counters to find the product.

Ask: **How many counters will you have if there are 3 groups with 4 counters in each group?**

Demonstrate how to use the counters to model the multiplication fact by showing 3 groups of 4 counters each. Let students add all the counters to find 12.

3×4

3 groups of 4 equals?

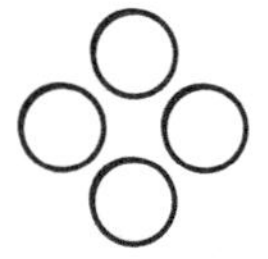
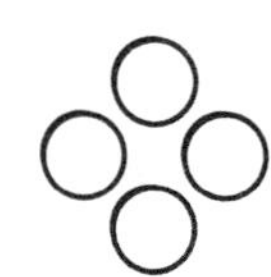
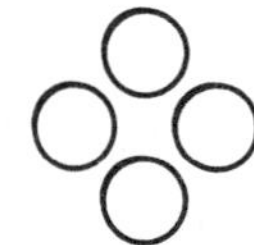

$3 \times 4 = 12$

Display the completed multiplication sentence $3 \times 4 = 12$ and have students point out the product.

Repeat this activity for other facts through 5. Have students model the multiplication with counters, find the product, and then write a complete multiplication sentence for each fact.

When the students show an understanding of how to model multiplication facts, have them complete a multiplication table for facts through 5 without using models.

You could begin by showing students how to record the fact for 3×4 on the multiplication table.

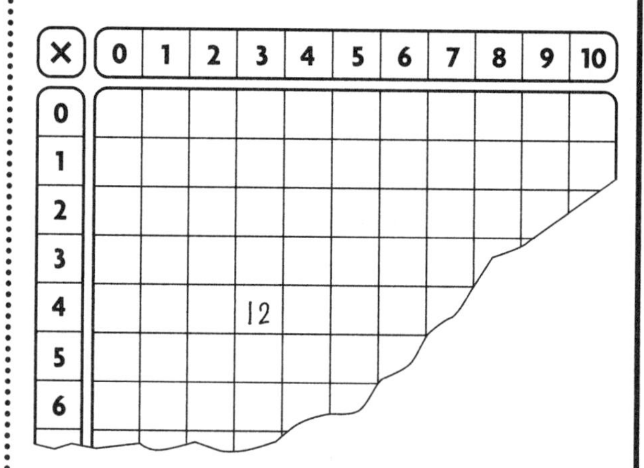

×	0	1	2	3	4	5	6	7	8	9	10
0											
1											
2											
3											
4				12							
5											
6											

Name ______ Skill ______

Grade 5 **Skill 12**

Multiplication Facts 1–5

The numbers you multiply are called **factors**.
The answer is called the **product**.

Use jumps on a number line to find the product.

3×4.

Think: 3 jumps of 4

Start at 0. Make 3 jumps of 4 spaces on the number line.

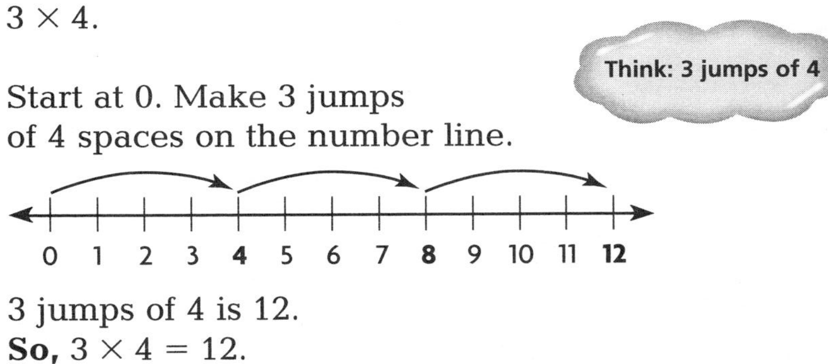

3 jumps of 4 is 12.
So, $3 \times 4 = 12$.

Find the value of n.
$2 \times 5 = n$

Think: 2 jumps of 5

Start at 0. Make 2 jumps of 5 spaces on the number line.

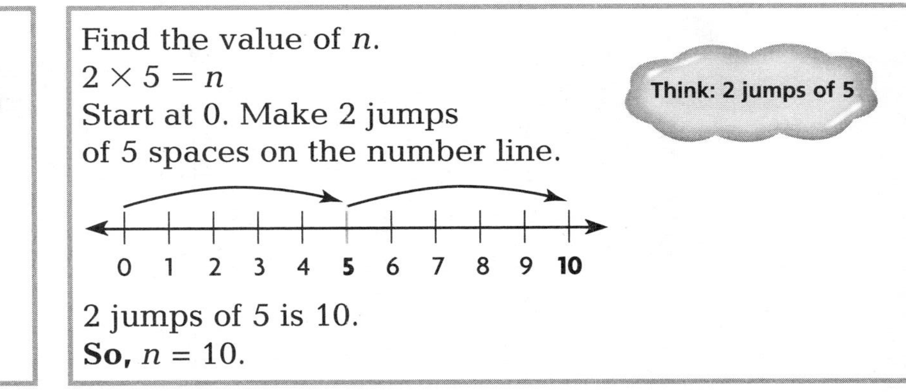

2 jumps of 5 is 10.
So, $n = 10$.

Try These

Find the product or the value of n.

1 2×7

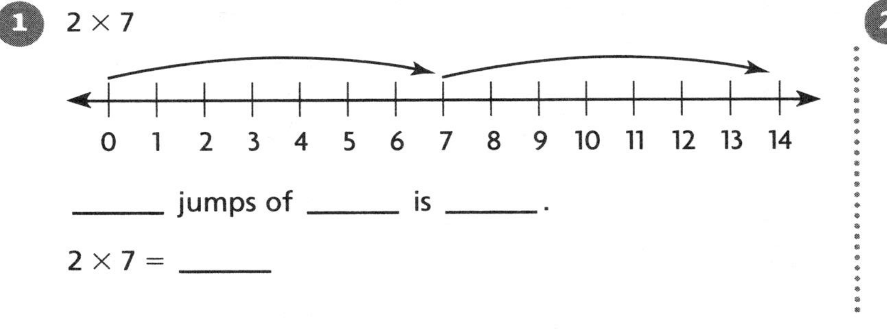

______ jumps of ______ is ______.

$2 \times 7 =$ ______

2 $4 \times 2 = n$

0 1 2 3 4 5 6 7 8

______ jumps of ______ is ______.

$n =$ ______

Go to the next side.

Name ______________________ Skill ______________________

Practice on Your Own

Skill 12

Find the value of *n*.
$5 \times 4 = n$

Think: 5 jumps of 4

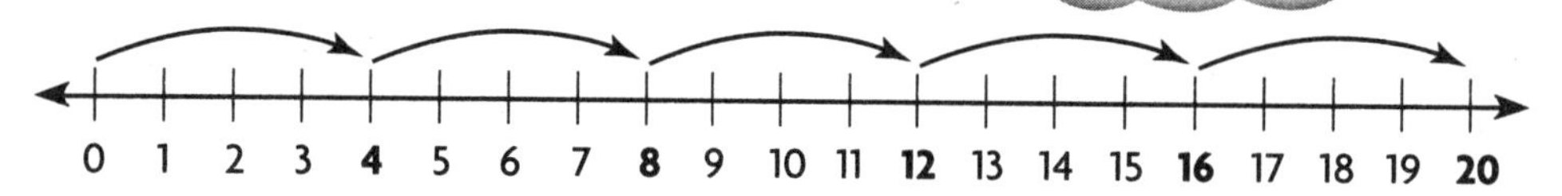

5 jumps of 4 is 20.
So, $n = 20$.

Find the product or the value of *n*.

1 2×6

____ jumps of ____ is ____

$2 \times 6 =$ ____

2 3×3

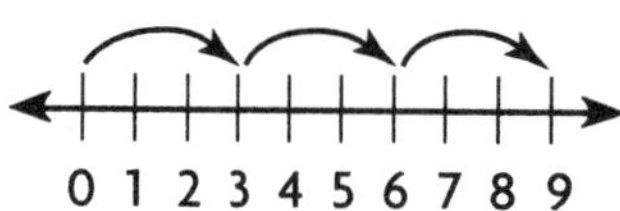

____ jumps of ____ is ____

$3 \times 3 =$ ____

3 $5 \times 2 = n$

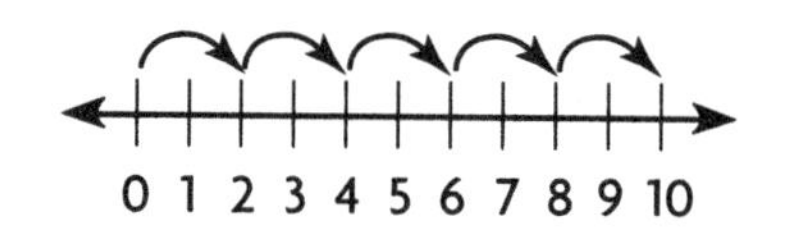

____ jumps of ____ is ____

$n =$ ____

4 2×2

____ jumps of ____ is ____

$2 \times 2 =$ ____

5 $2 \times 8 = n$

____ jumps of ____ is ____

$n =$ ____

6 $3 \times 5 = n$

____ jumps of ____ is ____

$n =$ ____

7 $4 \times 7 =$ ____

8 $5 \times 8 =$ ____

9 $5 \times 7 =$ ____

10 $5 \times 6 =$ ____

11 $5 \times 1 =$ ____

12 $8 \times 3 =$ ____

13 $5 \times 5 =$ ____

14 $8 \times 4 =$ ____

15 $2 \times 3 = n$
$n =$ ____

16 $4 \times 1 = n$
$n =$ ____

17 $5 \times 3 = n$
$n =$ ____

18 $4 \times 9 = n$
$n =$ ____

Check

Find the product or the value of *n*.

19 $3 \times 9 =$ ____

20 $7 \times 2 =$ ____

21 $4 \times 6 = n$
$n =$ ____

22 $5 \times 9 = n$
$n =$ ____

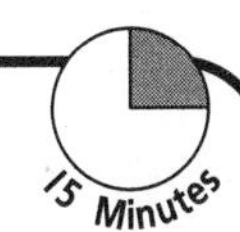

Using Skill 13

OBJECTIVE Recall multiplication facts 6–10

MATERIALS tiles

Have students look at the first example and use tiles to show the array for 6×4.

Ask: **How many rows of tiles did you show?** (6) **How many tiles are in each row?** (4) **How many tiles are there in all?** (24)

Be sure students are familiar with the words used in multiplication.

Ask: **What do you call the numbers you multiply?** (factors) **What do you call the result?** (product)

Have students practice writing the multiplication fact two ways—horizontally and vertically.

For the second example, have students show the array for 7×5 and then tell the meaning of the variable in $7 \times 5 = n$.

Ask: **What does the letter *n* stand for?** (the product, or 35)

Have students lay out more arrays for other multiplication facts. Direct them to write the multiplication fact for each, both horizontally and vertically.

TRY THESE Exercises 1–3 model finding products.

- **Exercise 1** Multiply factors 8 and 4.
- **Exercise 2** Multiply factors 5 and 4.
- **Exercise 3** Multiply factors 6 and 8.

PRACTICE ON YOUR OWN Focus on the example at the top of the page. Have students identify the factors and the product. Relate *n* to the product.

CHECK Be sure students understand that, in these exercises, "find the value of *n*" means "find the product." Success is indicated by 3 out of 3 correct responses.

Students who successfully complete the **Practice on Your Own** and **Check** are ready to move to the next skill.

COMMON ERRORS

- Students may confuse the number of rows with the number of tiles in a row.
- Students may have trouble using the horizontal or vertical format.

Students who made more than 3 errors in the **Practice on Your Own**, or who were not successful in the **Check** section, may benefit from the **Alternative Teaching Strategy** on the next page.

Optional

Alternative Teaching Strategy

Multiplication Facts 6–10

20 Minutes

OBJECTIVE Use arrays to model multiplication facts 6-10

MATERIALS tiles

Lay out an array of tiles. Show 6 rows of 3 tiles. Have students count the tiles.

Ask: **How many tiles are there?** (18) **How many rows?** (6) **How many tiles are in each row?** (3) **Ask a student to write the multiplication sentence for the array.** ($6 \times 3 = 18$)

Have students record $6 \times 3 = 18$ in a multiplication table such as the one below:

×	0	1	2	3	4	5	6	7	8	9	10
0											
1											
2											
3							18				
4											
5											
6											

Have students use the 18 tiles to make another array, but this time to show $9 \times 2 = 18$.

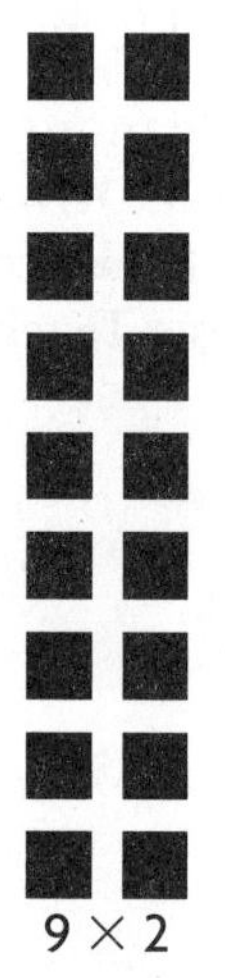

9×2

Ask students to record $9 \times 2 = 18$ in the multiplication table.

Tell students to model $8 \times 4 = n$.

Ask: **What does *n* stand for?** (the total number of tiles) **How many rows will you display?** (8) **How many tiles will you put in each row?** (4)
What is *n*? (32)

Have students repeat the activity several times, making arrays to help them fill in the multiplication table. Encourage them to write two sentences for each fact. For example, for 7×5, students should write $7 \times 5 = n$, and $n = 35$.

Finally, have them complete the table without making arrays.

Grade 5 Skill 13

Multiplication Facts 6–10

Use an array to find products.

Find 6×4.
Count the tiles.

Think: There is a total of 24 tiles.

6 rows of **4** is **24**.

So, $6 \times 4 = 24$.

$$\begin{array}{r} 4 \\ \times 6 \\ \hline 24 \end{array}$$

Find the value of n.
$7 \times 5 = n$
Count the tiles.

Think: There is a total of 35 tiles.

7 rows of **5** is **35**.

So, $n = 35$.

$$\begin{array}{r} 5 \\ \times 7 \\ \hline 35 \end{array}$$

Try These

Find the product or the value of n.

1. 8×4

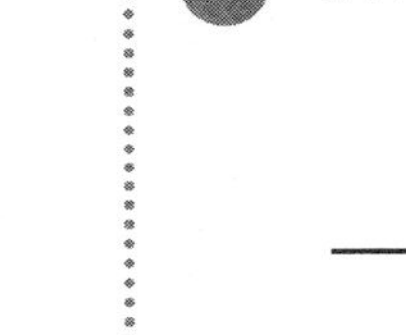

_____ rows of
_____ is _____.
_____ × _____
= _____

2. $\begin{array}{r} 5 \\ \times 4 \\ \hline \end{array}$

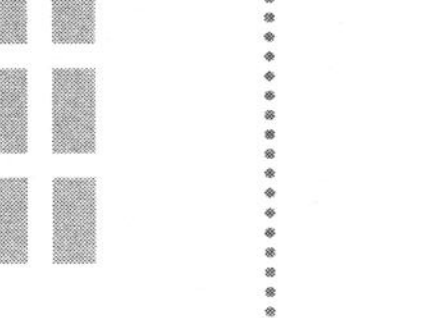

_____ rows of
_____ is _____.
_____ × _____
= _____

3. $6 \times 8 = n$

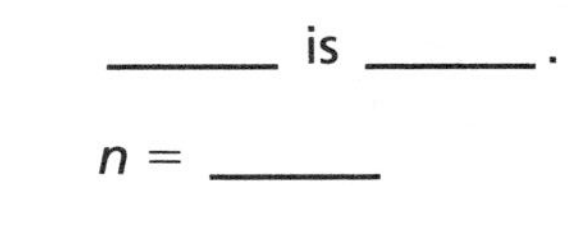
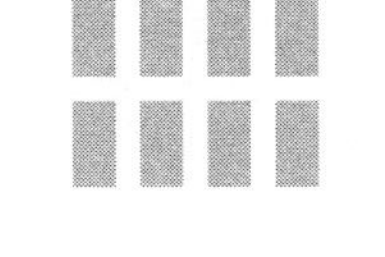
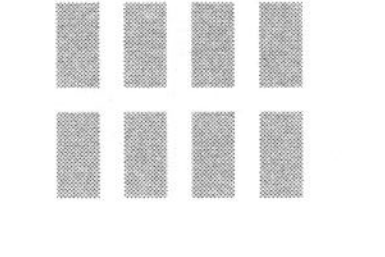

_____ rows of
_____ is _____.
n = _____

Go to the next side.

Name _____ Skill _____

Name ______________________ Skill ______________________

Practice on Your Own

Find the value of n.
$6 \times 6 = n$

Count the tiles.

Think: There is a total of 36 tiles.

Think:
Count to find the total number of tiles.

6 rows of **6** is **36.**
$6 \times 6 = 36$

$$\begin{array}{r} 6 \\ \times 6 \\ \hline 36 \end{array}$$

Find the product or the value of n.

1. 6×5

_____ rows of _____ is _____.

_____ $\times$ _____ $=$ _____

2. $6 \times 10 = n$

_____ rows of _____ is _____.

$n =$ _____

3. $8 \times 5 =$ _____
4. $6 \times 7 =$ _____
5. $9 \times 4 =$ _____
6. $10 \times 5 =$ _____
7. $\begin{array}{r} 3 \\ \times 9 \\ \hline \end{array}$
8. $\begin{array}{r} 6 \\ \times 8 \\ \hline \end{array}$
9. $\begin{array}{r} 7 \\ \times 7 \\ \hline \end{array}$
10. $\begin{array}{r} 5 \\ \times 9 \\ \hline \end{array}$
11. $8 \times 3 = n$ $n =$ _____
12. $6 \times 9 = n$ $n =$ _____
13. $9 \times 7 = n$ $n =$ _____
14. $10 \times 4 = n$ $n =$ _____

Check

Find the product or the value of n.

15. $9 \times 5 =$ _____

16. $\begin{array}{r} 10 \\ \times\ 7 \\ \hline \end{array}$

17. $5 \times 7 = n$ $n =$ _____

Skill 14 Grade 5 Multiplication Properties

Using Skill 14

15 Minutes

OBJECTIVE Use the Commutative and Associative Properties, Property of One, and Zero Property to multiply

You may wish to review the terms used in multiplication. Recall for students that the numbers you multiply are called factors and that the answer is called the product.

Draw students' attention to the Commutative Property. Explain to students that the arrangement of the squares is called an array. Since the overall shape is a rectangle, the arrangement is called a rectangular array. Ask: **What factors are represented by the first rectangular array?** (3, 4) **What factors are represented by the second rectangular array?** (4, 3) **Does the order of the factors result in different products?** (no) **Why?** (Both products equal 12.)

Use the arrays to help students recognize that two factors can be multiplied in any order and the product remains the same.

Direct students' attention to the Associative Property. Be sure students recognize that there are more than two factors. Point out to students that the first group of arrays represents 3 groups of 4×2 and that the second groups of arrays represents 4 groups of 2×3. Ask: **Are the factors the same in each multiplication sentence?** (yes) **What is the product $(4 \times 2) \times 3$?** (24) **What is the product $4 \times (2 \times 3)$?** (24) **Are the products the same?** (yes) **What is different?** (The factors are grouped differently in the multiplication sentences.)

Point out to students that when multiplying three or more factors, grouping factors differently does not change the product.

Continue to ask similar questions to help students recognize the Property of One and the Zero Property.

TRY THESE Exercises 1–4 model four different properties of multiplication.

- **Exercise 1** Commutative Property.
- **Exercise 2** Associative Property.
- **Exercise 3** Property of One.
- **Exercise 4** Zero Property.

PRACTICE ON YOUR OWN Review the example at the top of the page. Be sure students can identify the characteristics of each property.

CHECK Determine if students understand the multiplication property illustrated in each exercise.

Success is indicated by 4 out of 4 correct responses.

Students who successfully complete the **Practice on Your Own** and **Check** are ready to move to the next skill.

COMMON ERRORS

- Students may not understand the meaning of order and grouping.
- Students may not recall multiplication facts.

Students who made more than 5 errors in the **Practice on Your Own**, or who were not successful in the **Check** section, may benefit from the **Alternative Teaching Strategy** on the next page.

Optional

Alternative Teaching Strategy

15 Minutes

Modeling the Commutative and Associative Properties

OBJECTIVE Use counters to model the Commutative and Associative Properties of Multiplication

MATERIALS counters, flip chart

Explain to students that they will be using counters to model the Commutative and Associative Properties of Multiplication. Distribute counters and have students model a 5 by 2 array and a 2 by 5 array. Help students visualize the orientation of a row as moving from left to right.

For the first array, ask students to count the number of counters in each row and then record the numbers on a flip chart.

Repeat for the second array.

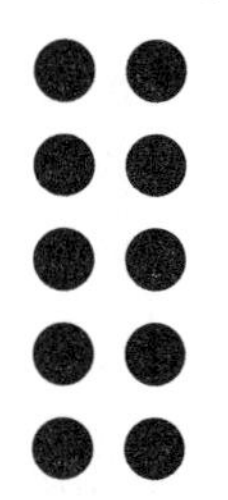

Guide students to see that in both groups the total number of counters equals 10.

On the flip chart, write:

2 rows of 5 counters = 10 counters

↓

$2 \times 5 = 10$

5 rows of 2 counters = 10 counters

↓

$5 \times 2 = 10$

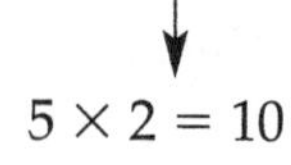

Guide students to verbalize that the order in which two numbers are multiplied does not change the product. Tell them that this is called the Commutative Property of Multiplication.

Next, have students model two 3 by 5 arrays and three 5 by 2 arrays.

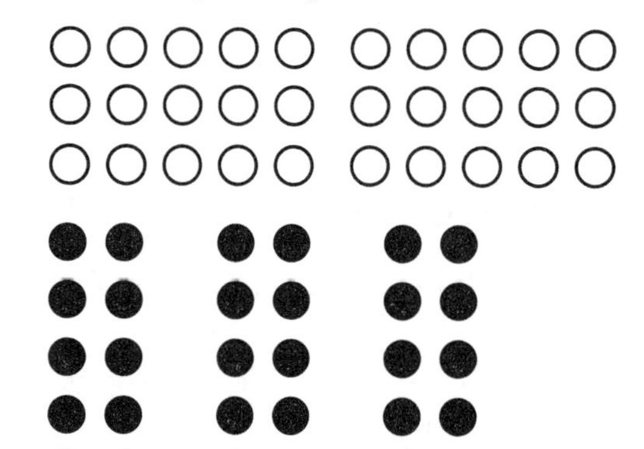

Have students use the same approach as they did with the Commutative Property to illustrate that different groupings of three factors does not change the product. Help students conclude that:

$$(3 \times 5) \times 2 = 3 \times (5 \times 2)$$

Tell them that this is called the Associative Property of Multiplication.

Continue to illustrate both properties with different factors.

Name ______ Skill ______

Grade 5 Skill 14

Multiplication Properties

You can use multiplication properties to help you multiply.

Commutative Property
You can multiply two factors in any order without changing the product.

$3 \times 4 = 12$ $4 \times 3 = 12$

same

$3 \times 4 = 4 \times 3$

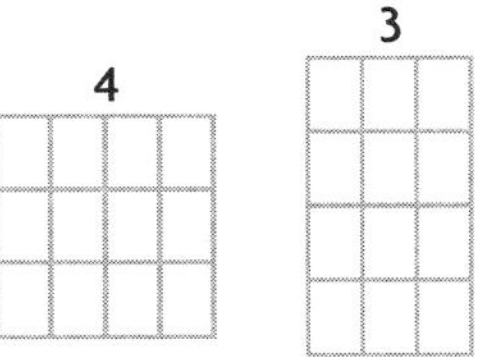
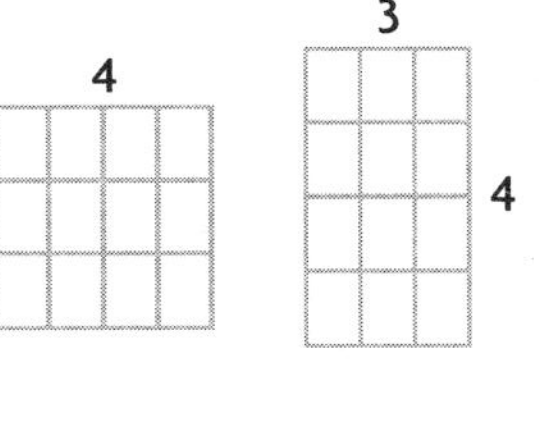

Associative Property
When multiplying three or more factors, any two of the factors can be multiplied, and the remaining factors may then be multiplied without changing the product.

$(4 \times 2) \times 3 = 24$ $4 \times (2 \times 3) = 24$

same

$(4 \times 2) \times 3 = 4 \times (2 \times 3)$

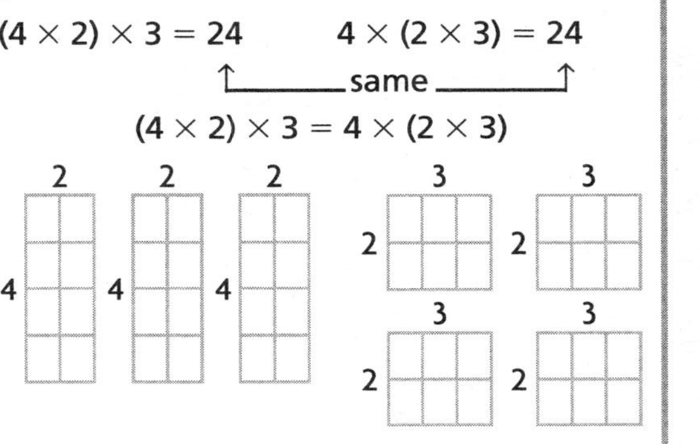

Property of One
The product of any number and 1 equals that number.

$1 \times 9 = 9$
$57 \times 1 = 57$

Zero Property
The product of any number and 0 equals 0.

$72 \times 0 = 0$
$0 \times 26 = 0$

Try These

Complete to show each multiplication property.

1. Commutative Property

$6 \times 5 = 5 \times$ ____

$4 \times$ ____ $= 9 \times 4$

2. Associative Property

$(7 \times 2) \times$ ____ $= 7 \times (2 \times 3)$

$21 \times (4 \times 11) = (21 \times$ ____$) \times 11$

3. Property of One

____ $\times 93 = 93$

$47 =$ ____ $\times 1$

4. Zero Property

$81 \times 0 =$ ____

$52 \times$ ____ $= 0$

Go to the next side.

Name ______________________ Skill ______________________

Practice on Your Own

Skill 14

Think:

In multiplication, changing the order or grouping of the factors does not change the product.

$6 \times 9 = 9 \times 6$ **Commutative Property**

$18 \times (5 \times 3) = (18 \times 5) \times 3$ **Associative Property**

Multiplying any number by 1 equals that number.

$24 \times 1 = 24$ **Property of One**

Multiplying any number by 0 equals 0.

$35 \times 0 = 0$ **Zero Property**

Complete to show the multiplication property.

1. Commutative Property
   $3 \times 5 = 5 \times$ ___
2. Associative Property
   $(6 \times 7) \times$ ___ $= 6 \times (7 \times 3)$
3. Property of One
   $63 \times 1 =$ ___
4. Zero Property
   $18 \times$ ___ $= 0$

Complete. Name the multiplication property.

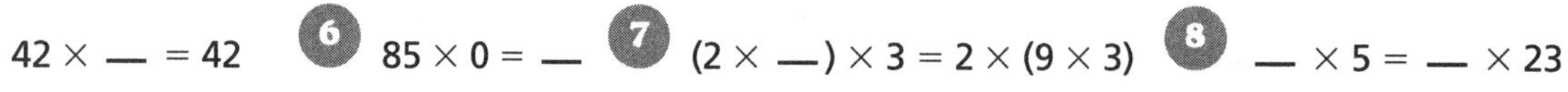

5. $42 \times$ ___ $= 42$
   ____
6. $85 \times 0 =$ ___
   ____
7. $(2 \times$ ___$) \times 3 = 2 \times (9 \times 3)$
   ____
8. ___ $\times 5 =$ ___ $\times 23$
   ____

Complete to show the multiplication property.

9. $(8 \times 9) \times 3 =$ ___________
   Associative
10. $0 \times 42 =$ ___________
    Zero Property
11. $17 \times 1 =$ ___________
    Property of One
12. $3 \times 5 = 5 \times$ ___________
    Commutative
13. $(5 \times 2) \times 3 =$ ___________
    Associative
14. $7 \times 45 = 45 \times$ ___________
    Commutative
15. $0 \times 5 =$ ___________
    Zero Property
16. $1 \times 42 =$ ___________
    Property of One

Check

Complete to show the multiplication property.

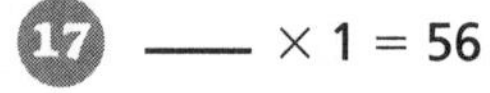

17. ___ $\times 1 = 56$
    Property of One
18. $(8 \times 10) \times 7 =$ ___
    Associative Property
19. $94 \times 41 = 41 \times$ ___
    Commutative Property
20. $0 \times 81 =$ ___
    Zero Property

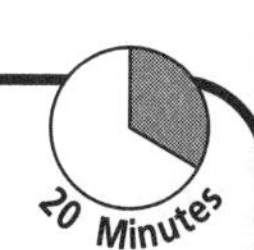

Using Skill 15

OBJECTIVE Use the Distributive Property to multiply numbers

MATERIALS paper

Recall with students that the Distributive Property allows them to "break apart" numbers so the numbers are easier to work with using mental math.

On Skill 15, draw attention to the second factor in the multiplication 3×16. Ask:

The factor 16 is renamed as the sum of which two numbers? (10 + 6)

Why does 3 × 16 represent the same number as 3 × (10 + 6)? (16 = 10 + 6)

Direct students' attention to Step 2.

Say: In the original expression, 16 is multiplied by 3. Then ask:

In the new expression, what should I multiply both 10 and 6 by? (3)

Reconfirm for students that 10 and 6 represent the factor 16 "broken apart."
The number 3 is the other factor and has not changed.

Continue with Step 3. Ask:

Where do the numbers 30 and 18 come from? ($3 \times 10 = 30$, $3 \times 6 = 18$)

If necessary, refer students to the partial products in Step 2. Help students recognize that the final product is the sum of the two partial products. Ask:
What is the final product? (48)

Guide students to recognize that the final sum of the partial products, 48, represents the product 3×16.

TRY THESE Exercises 1–3 model the Distributive Property.

- **Exercise 1** Rewrite 28 as 20 and 8.
- **Exercise 2** Rewrite 61 as 60 and 1.
- **Exercise 3** Rewrite 17 as 10 and 7.

PRACTICE ON YOUR OWN Review the example at the top of the page. Have students identify the sum that is used to represent one factor. Have students name the partial products. Then ask students to name the final product.

CHECK Make sure students can use the Distributive Property when multiplying a 2-digit number by a 1-digit number.

Success is indicated by 3 out of 3 correct responses.

Students who successfully complete the **Practice on Your Own** and **Check** are ready to move to the next skill.

COMMON ERRORS

- Students may "break apart" a factor with addends that do not equal the factor.
- Students may "break apart" one factor correctly, but then forget to use the other factor to compute the partial products.
- Student may add instead of multiply or may fail to regroup when adding to find the final product.

Students who made more than 4 errors in the **Practice on Your Own**, or who were not successful in the **Check** section, may benefit from the **Alternative Teaching Strategy** on the next page.

Optional

Alternative Teaching Strategy

20 Minutes

Area Model and the Distributive Property

OBJECTIVE Use an area model to illustrate the Distributive Property

MATERIALS grid paper

Use the area model to illustrate 3×17.

First show students 17 in expanded form:

$$17 = 10 + 7$$

Outline an area on the grid paper as shown.

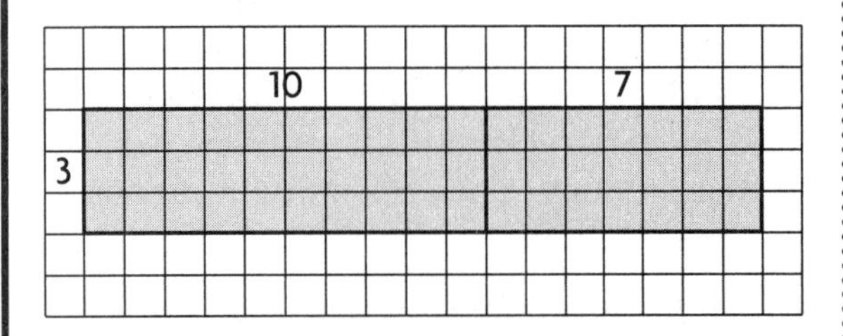

Show students how to use the model to find 3×10. Guide students to see that there are 3 rows of 10 small squares: $10 + 10 + 10 = 30$.

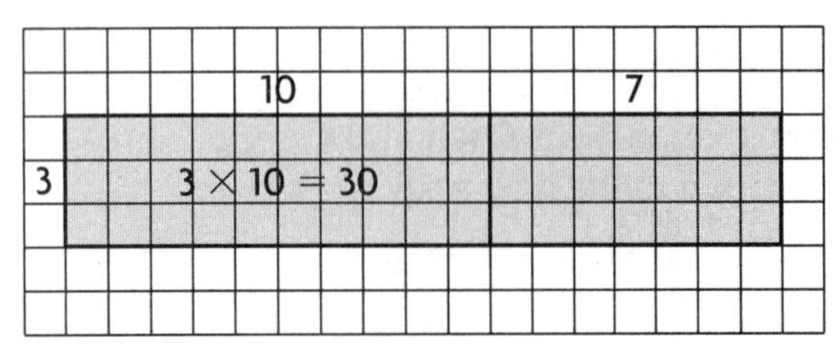

Next, ask students to find 3×7.

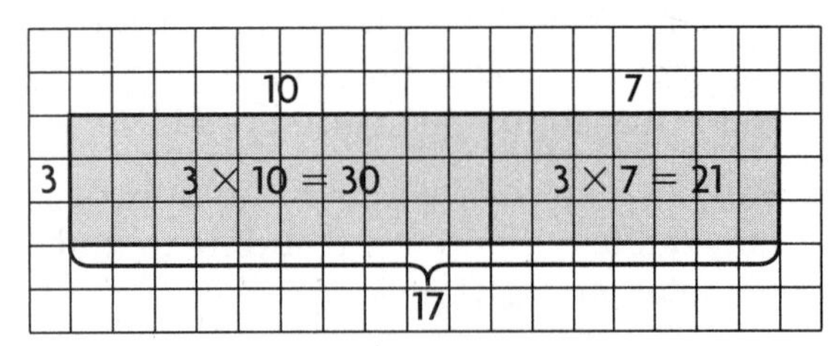

Have students add the two partial products in the area model.

$$30 + 21 = 51$$

Point out to students that the sum 51 is also the product 3×17.

Continue with other similar examples.

Help students realize that by representing one of the factors in expanded form the process of multiplication becomes simpler. You may want to work on the board.
For example:

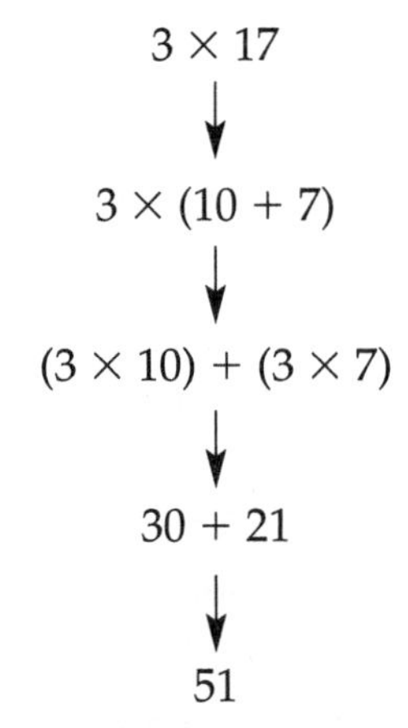

Summarize the steps by pointing out to students that the product of a number and a sum, $3 \times (10 + 7)$, is equal to the sum of the products: $(3 \times 10) + (3 \times 7)$.

Name ______________________ Skill ______________________

Grade 5 Skill 15

Distributive Property

Find $3 \times 16 = \square$. You can use the Distributive Property to help you multiply.

Break apart one factor. Rewrite 16 as 10 + 6.	Multiply the parts.	Add the products.
Think: $16 = 10 + 6$ **So:** $3 \times 16 = 3 \times (10 + 6)$	$3 \times 16 = (3 \times 10) + (3 \times 6)$ ↓ ↓ $30 + 18$	$30 + 18 = 48$ **So,** $3 \times 16 = 48$.

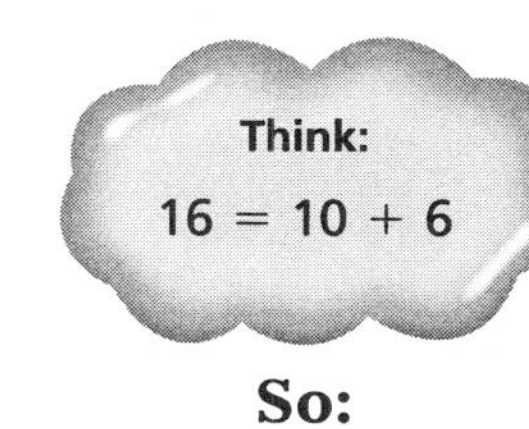

Try These

Find the product.

1. 3×28

   $3 \times 28 = (3 \times 20) + (3 \times \_\_\_)$

   $= \_\_\_ + \_\_\_$

   $= \_\_\_$

2. 2×61

   $2 \times 61 = (2 \times 60) + (2 \times \_\_\_)$

   $= \_\_\_ + \_\_\_$

   $= \_\_\_$

3. 4×17

   $4 \times 17 = (\_\_\_ \times 10) + (\_\_\_ \times 7)$

   $= \_\_\_ + \_\_\_$

   $= \_\_\_$

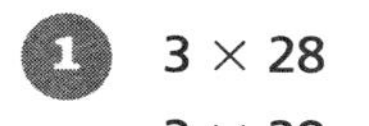
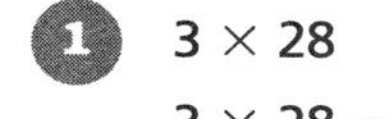

Go to the next side.

Name ______________________ Skill ______________________

Practice on Your Own

Skill 15

Think: Use the Distributive Property to break apart one of the factors. Multiply the parts. Then add the products.

Find $4 \times 18 = \square$.

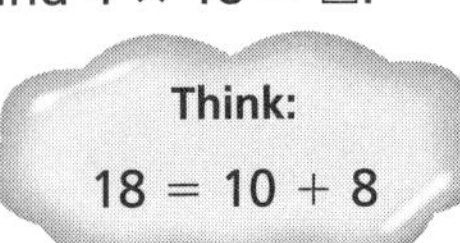

$4 \times 18 = (4 \times 10) + (4 \times 8)$
$= 40 + 32$
$= 72$
$4 \times 18 = 72$

So: $4 \times 18 = 4 \times (10 + 8)$

Find the product.

1. $4 \times 36 =$
   $(4 \times 30) + (4 \times$ ____ $)$
   $=$ ____ $+$ ____
   $=$ ____

2. $5 \times 11 =$
   $($ ____ $\times 10) + ($ ____ $\times 1)$
   $=$ ____ $+$ ____
   $=$ ____

3. $8 \times 32 =$
   $(8 \times$ ____ $) + (8 \times$ ____ $)$
   $=$ ____ $+$ ____
   $=$ ____

4. $5 \times 14 =$
   $(5 \times$ ____ $) + (5 \times$ ____ $)$
   $=$ ____ $+$ ____
   $=$ ____

5. $8 \times 17 =$
   $(8 \times$ ____ $) + (8 \times$ ____ $)$
   $=$ ____ $+$ ____
   $=$ ____

6. $6 \times 24 =$
   $($ ____ $\times 20) + ($ ____ $\times 4)$
   $=$ ____ $+$ ____
   $=$ ____

7. 4×12
   $($ ____ $\times$ ____ $) +$
   $($ ____ $\times$ ____ $) =$ ____

8. 8×21
   $(8 \times 21) =$ ____ $+$ ____
   $=$ ____

9. 7×13
   $(7 \times 13) =$ ____ $+$ ____
   $=$ ____

10. $6 \times 14 =$ ________
    ____

11. $8 \times 12 =$ ________
    ____

12. $3 \times 13 =$ ________
    ____

Check

Find the product.

13. $5 \times 12 =$ ____
    ____

14. $8 \times 16 =$ ____
    ____

15. $4 \times 25 =$ ____
    ____

Skill 16 Grade 5 — Multiply by 10 and 100

Using Skill 16

OBJECTIVE Use patterns to multiply with multiples of 10 and 100

Begin by explaining to the students that in this lesson they multiply by 10 and 100, and by multiples of 10 and 100.

Direct the students' attention to the first example. Suggest that they look at the three multiplication sentences and look for a pattern. Students may notice that the multiplication fact $3 \times 1 = 3$ is repeated in each multiplication sentence.

Reinforce that the product of 10 and any other factor has a zero in the ones place. The product of 100 and any other factor has zeros in both the ones and tens places.

For the second example, guide students to see how the pattern they found is also useful for multiples of 10 and 100.

Ask: **What multiplication fact do you see repeated in each multiplication sentence?** ($5 \times 2 = 10$)
If you follow the pattern, what is 5 × 200? (1,000)

Discuss the pattern for facts that end in zero, for example:

$$8 \times 5 = 40$$

$$8 \times 50 = 400$$

$$8 \times 500 = 4{,}000$$

Point out that they can still use the pattern. Have them note that the extra zero comes from 8×5, the multiplication fact. The pattern tacks zeros onto the product of the fact.

TRY THESE In Exercises 1–3 students multiply by 10 and 100, and by multiples of 10 and 100.

- **Exercises 1–2** Multiply by 10 and 100.
- **Exercise 3** Multiply by 40 and 400.

PRACTICE ON YOUR OWN Review the examples at the top of the page. Remind students first to find the fact, then apply the pattern as they work through the page.

CHECK Determine if students understand the pattern and can apply it when multiplying with 10 and 100, and multiples of 10 and 100.

Success is indicated by 3 out of 3 correct responses.

Students who successfully complete the **Practice on Your Own** and **Check** are ready to move to the next skill.

COMMON ERRORS

- Students may not include the correct number of zeros in the product.
- Students may multiply incorrectly because they do not know the multiplication facts.

Students who made more than 3 errors in the **Practice on Your Own**, or who were not successful in the **Check** section, may benefit from the **Alternative Teaching Strategy** on the next page.

Optional

Alternative Teaching Strategy

Model Multiplying by 10 and 100

15 Minutes

OBJECTIVE Use base-ten blocks to show a pattern for multiplying by 10 and 100, and multiples of 10 and 100.

MATERIALS base-ten blocks, paper

You may wish to have students work in pairs. One student models the multiplication with the blocks, while the other student records each step with paper and pencil.

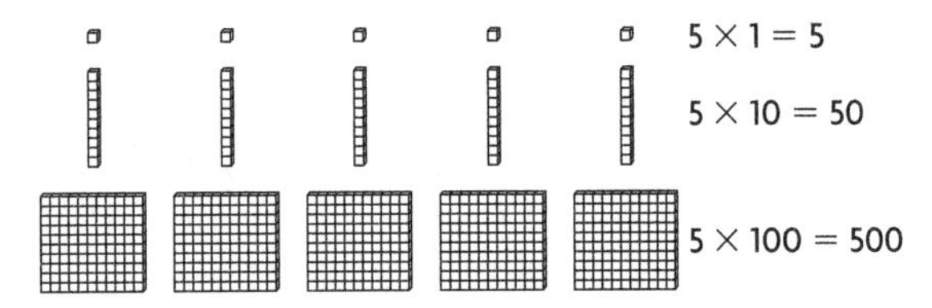

Distribute the base-ten blocks. Have one partner model:

5×1 one = ___ones

5×1 ten = ___ tens

5×1 hundred = ___ hundreds

Have the other partner record:

$5 \times 1 = 5$

$5 \times 10 = 50$

$5 \times 100 = 500$

Suggest that students examine the multiplication sentences they recorded and look for a pattern. Students may note that the multiplication fact $5 \times 1 = 5$ is evident in all three number sentences.

Help students also recognize that tens have a zero in the ones place, and hundreds have zeros in the ones and tens places. Reading a number sentence using place value can help them find all of the digits of the product.

Repeat the modeling and recording until the students are confident they understand the pattern.

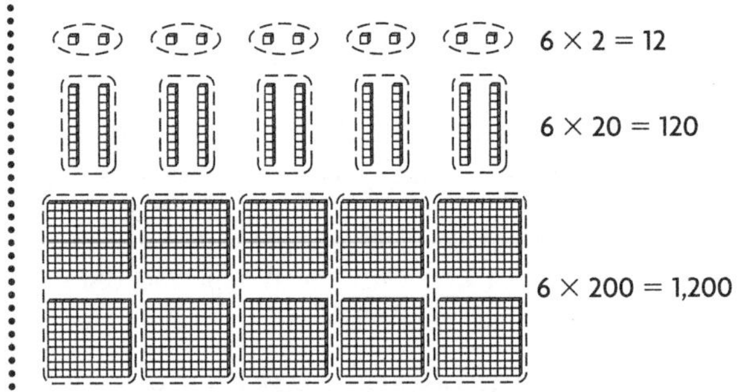

Guide students to see that the pattern also works for multiples of 10 and 100.

6×2 ones = ___ones

$6 \times 2 = 12$

6×2 tens = ___ tens

$6 \times 20 = 120$

6×2 hundreds = ___ hundreds

$6 \times 200 = 1{,}200$

When the students show an understanding of the pattern, have them find the product for these exercises.

$2 \times 100 =$ ___ $5 \times 30 =$ ___ $6 \times 50 =$ ___

$3 \times 900 =$ ___ $4 \times 40 =$ ___ $8 \times 300 =$ ___

Name ____________________ Skill ____________________

Grade 5 Skill 16

Multiply by 10 and 100

You can use a pattern to multiply with tens and hundreds.

Multiply by 10 and 100.
Count to find the number of zeros in each factor.

	Think:
$3 \times 1 = 3$	← fact
$3 \times 10 = 30$	← When you multiply by 10, the ones digit in the product is 0.
$3 \times 100 = 300$	← When you multiply by 100, the final two digits in the product are 0.

Multiply by a multiple of 10 and 100.

	Think:
$5 \times 2 = 10$	The fact has a zero in the product.
$5 \times 20 = 100$	← When you multiply by a multiple of 10, the ones digit in the product is 0.
$5 \times 200 = 1{,}000$	← When you multiply by a multiple of 100, the final two digits in the product are 0.

Try These

Find the product. Use a pattern.

1. $5 \times 1 =$ ______
   $5 \times 10 =$ ______
   $5 \times 100 =$ ______

2. $9 \times 1 =$ ______
   $9 \times 10 =$ ______
   $9 \times 100 =$ ______

3. $6 \times 4 =$ ______
   $6 \times 40 =$ ______
   $6 \times 400 =$ ______

Go to the next side.

Name ______________________ Skill ______________________

Practice on Your Own

Skill 16

Think:
Use the pattern to multiply by 10 and 100.

$7 \times 1 = 7$	$2 \times 6 = 12$	
$7 \times 10 = 70$	$2 \times 60 = 120$	← The ones digits are 0.
$7 \times 100 = 700$	$2 \times 600 = 1,200$	← The ones and tens digits are 0.

Find the product. Use a pattern.

1. $2 \times 1 =$ ______
   $2 \times 10 =$ ______
   $2 \times 100 =$ ______

2. $8 \times 1 =$ ______
   $8 \times 10 =$ ______
   $8 \times 100 =$ ______

3. $4 \times 1 =$ ______
   $4 \times 10 =$ ______
   $4 \times 100 =$ ______

4. $7 \times 2 =$ ______
   $7 \times 20 =$ ______
   $7 \times 200 =$ ______

5. $4 \times 5 =$ ______
   $4 \times 50 =$ ______
   $4 \times 500 =$ ______

6. $8 \times 4 =$ ______
   $8 \times 40 =$ ______
   $8 \times 400 =$ ______

7. $9 \times 40 =$ ______

8. $7 \times 300 =$ ______

9. $4 \times 70 =$ ______

10. $8 \times 60 =$ ______

11. $6 \times 400 =$ ______

12. $7 \times 20 =$ ______

Check

Find the product.

13. $4 \times 100 =$ ______

14. $3 \times 500 =$ ______

15. $5 \times 60 =$ ______

Multiply by 1-Digit Numbers (2-, 3-Digit Factors)

Using Skill 17

OBJECTIVE Multiply by 1-digit numbers

MATERIALS base-ten blocks

Distribute base-ten blocks for students to use to review regrouping ones as tens and ones. Display 14 ones. Ask: **How many ones are there?** (14) Have students show the number another way. (1 ten and 4 ones) Ask: **How many tens?** (1) **How many ones?** (4)

Review regrouping tens as hundreds and tens. Display 15 tens. Ask: **How many tens are there?** (15) Have students show the number another way. (1 hundred and 5 tens) Ask: **How many hundreds are there?** (1) **How many tens?** (5)

Lead students through the steps that show multiplying by a 1-digit number. Focus on the regrouping in Step 1. Ask: **How many ones are there?** (15) **How do you regroup the ones?** (as 1 ten and 5 ones) Point out the placement of the numeral 1 for the regrouped ten.

In Step 2, emphasize how to multiply 6 tens first and then add the 1 regrouped ten. In Step 3, point out that the 1 hundred is multiplied first, then the 1 regrouped hundred is added.

TRY THESE Exercises 1–4 provide practice multiplying by a 1-digit number.

- **Exercise 1** Regroup ones.
- **Exercise 2** Regroup tens.
- **Exercise 3** Regroup ones and tens.
- **Exercise 4** Regroup ones; 0 in tens place.

PRACTICE ON YOUR OWN Focus on the example at the top of the page. Have a student explain the regrouping involved. Encourage students to line up the digits—ones under ones, tens under tens, and so on—as they multiply.

CHECK Determine if student can multiply by 1-digit numbers with regrouping. Success is indicated by 3 out of 4 correct responses.

Students who successfully complete the **Practice on Your Own** and **Check** are ready to move to the next skill.

COMMON ERRORS

- Students may forget to add the regrouped digit or may add it first and then multiply.
- Students may write a digit in the wrong place when regrouping.

Students who made more than 3 errors in the **Practice on Your Own**, or who were not successful in the **Check** section, may benefit from the **Alternative Teaching Strategy** on the next page.

Optional

Alternative Teaching Strategy

20 Minutes

Model Multiplying by 1-Digit Numbers

OBJECTIVE Model multiplying by 1-digit numbers

MATERIALS base-ten blocks

Use base-ten blocks to practice regrouping ones as tens and ones. Give a student 12 ones; the student gives you 1 ten and 2 ones in return. Practice with other numbers of ones. Show the numbers in a place-value chart.

Next practice regrouping tens as hundreds and tens. Give a student 11 tens; the student gives you 1 hundred and 1 ten in return. Practice with other numbers of tens. Show the numbers in a place-value chart.

Use base-ten blocks to show 3 groups of 145.

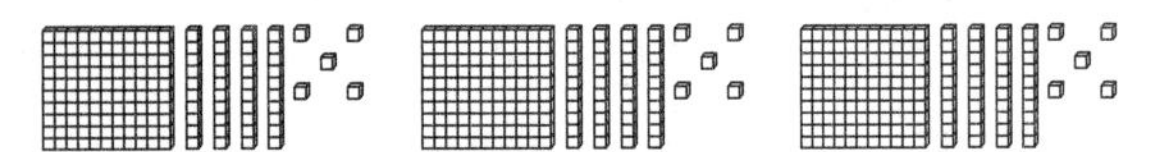

Have students gather up the ones. Ask: **How many ones are there?** (15) Exchange the 15 ones for 1 ten 5 ones. Ask: **Now how many ones do you have?** (5) **How many tens?** (1) Tell students to put the blocks aside.

Have students gather up the tens. Ask: **How many tens are there?** (12) Exchange the 12 tens for 1 hundred 2 tens. **Now how many tens do you have?** (2) **Do you have any other tens to add?** (yes) **How many?** (1) **Then how many tens do you have in all?** (3) Tell students to put the blocks aside.

Have students gather up the hundreds. Ask: **How many hundreds are there?** (3) **Do you have any other hundreds to add?** (yes) **How many?** (1) **Then how many hundreds do you have in all?** (4) **How many tens?** (3) **How many ones?** (5) **What is 3 × 145?** (435)

Lead students through the algorithm for 3 × 145. Then compare the results.

Name ______________________ Skill ______________________

Grade 5 Skill 17

Multiply by 1-Digit Numbers (2-, 3-Digit Factors)

Find 3 × 165

Step 1
Multiply the ones.
Regroup ones as tens and ones.

```
  H T O
    1
  1 6 5
×     3
-------
      5
```

3 × 5 ones = 15 ones
15 ones = 1 ten, 5 ones

Step 2
Multiply the tens. Add the regrouped ten. Regroup tens as **hundreds** and **tens.**

```
  H T O
  1 1
  1 6 5
×     3
-------
  4 9 5
```

3 × 6 tens = 18 tens
18 tens + 1 ten = 19 tens
19 tens = 1 hundred, 9 tens

Step 3
Multiply the hundreds. Add the regrouped hundred.

```
  H T O
  1 1
  1 6 5
×     3
-------
  4 9 5
```

3 × 1 hundred = 3 hundreds
3 hundreds + 1 hundred = 4 hundreds

So, 3 × 165 = 495.

Try These

Multiply.

1.
```
  T O
  □
  2 5
×   2
-----
```

2.
```
  H T O
  □
  1 7 3
×     2
-------
```

3.
```
  H T O
  □ □
  2 5 7
×     3
-------
```

4.
```
  H T O
    □
  2 0 4
×     4
-------
```

Go to the next side.

Name ______________________ Skill ______________________

Practice on Your Own

Skill 17

Th	H	T	O
	1	2	
	4	3	6
×			4
1,	7	4	4

Think: Add the regrouped digits.

4×6 ones = 24 ones. Regroup.
4×3 tens = 12 tens.
12 tens + 2 tens = 14 tens. Regroup.
4×4 hundreds = 16 hundreds.
16 hundreds + 1 hundred = 17 hundreds.
Regroup 17 hundreds as 1 thousand, 7 hundreds.

Multiply.

1. T O — 16×4
2. H T O — 102×7
3. H T O — 122×5
4. H T O — 369×2
5. 20×8
6. 235×3
7. 167×4
8. 342×3
9. 37×4
10. 117×7
11. 269×3
12. 205×3
13. 38×6
14. 230×4
15. 132×5
16. 625×3

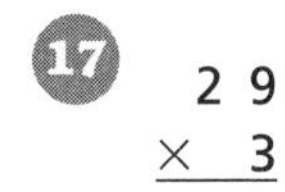

Check

Multiply.

17. 29×3
18. 216×3
19. 478×2
20. 724×3

Skip Count on a Number Line

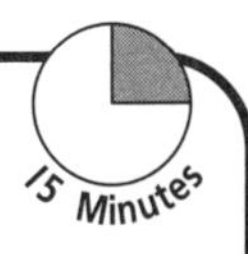

Using Skill 18

OBJECTIVE Skip count on a number line

Direct students' attention to the first number line. Begin by having them locate various points.

Read about skip counting at the top of the page. Look at the first example. Encourage students to trace the arrows that show skip counting by 4.

Ask: **Where do you start?** (at 0) **How many spaces do you skip each time?** (4)

Extend the number line to 36. Ask if anyone can tell you the next three numbers after 24. (28, 32, 36)

Look at the second example. Help students read the labels on the number line—point out that each space separates two numbers but not every number is labeled.

Ask: **How many spaces do you skip each time?** (5)

Extend the number line and have a student tell you the next three numbers after 30. (35, 40, 45)

TRY THESE Exercises 1–4 provide practice in skip counting.

- **Exercise 1** Count by twos.
- **Exercise 2** Count by fives.
- **Exercise 3** Count by tens.
- **Exercise 4** Count by sixes.

PRACTICE ON YOUR OWN Look at the examples at the top of the page. Focus on the two patterns. Ask students to tell you how they skip count to find the missing numbers. (by 4 and by 3) Draw number lines, and have volunteers show skip counting by 4 and then by 3.

CHECK Some students may need to draw number lines to skip count. Success is indicated by 2 out of 2 correct responses.

Students who successfully complete the **Practice on Your Own** and **Check** are ready to move to the next skill.

COMMON ERRORS

- Students may miscount the number of spaces.
- Students may count the launch number as 1 in the skip count.

Students who made more than 1 error in the **Practice on Your Own**, or who were not successful in the **Check** section, may benefit from the **Alternative Teaching Strategy** on the next page.

Optional

Alternative Teaching Strategy

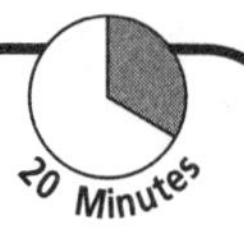

Skip Count on a Number Line

OBJECTIVE Use counters to model skip counting

MATERIALS counters

Use counters to model skip counting by 2. Start with 0 counters.

Ask: **How many counters do you have?** (0)

Give students 2 counters.

Ask: **How many counters do you have?** (2)

Have students draw an arrow from 0 to 2 on a number line. Next, give 2 more counters.

Ask: **How many counters do you have now?** (4)

Have students draw an arrow from 2 to 4 on the number line. Repeat the procedure until the number of counters totals 10.

Ask: **How many counters do you have?** (10) **How many does the number line show?** (10)

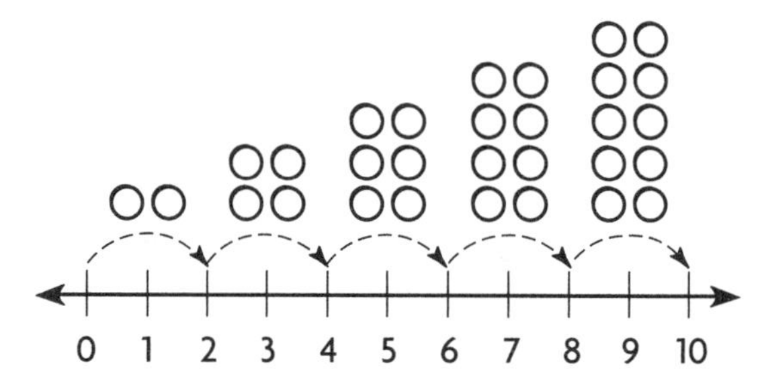

Repeat the activity, this time having students use the counters and number line to find the missing numbers for the following:

3, 6, 9, □, □, □ (12, 15, 18)

Finally, have students work in pairs. They should find the missing numbers, using counters and/or number lines as needed.

5, 10, 15, □, □, □ (20, 25, 30)

4, 8, □, 16, □, □ (12, 20, 24)

10, 20, □, □, 50, □ (30, 40, 60)

Name ______________________ Skill ______________________

Grade 5 Skill 18

Skip Count on a Number Line

Use a number line and skip counting to find the missing numbers in some patterns.

Find the next 3 numbers: 0, 4, 8, 12, ■, ■, ■
Skip count on the number line.

0 1 2 3 4 5 6 7 8 9 10 11 12 13 14 15 16 17 18 19 20 21 22 23 24

The number line shows skip counting by 4.
The next three numbers are: 16, 20, 24.

Find the next 3 numbers: 0, 5, 10, 15, ■, ■, ■
Skip count on the number line.

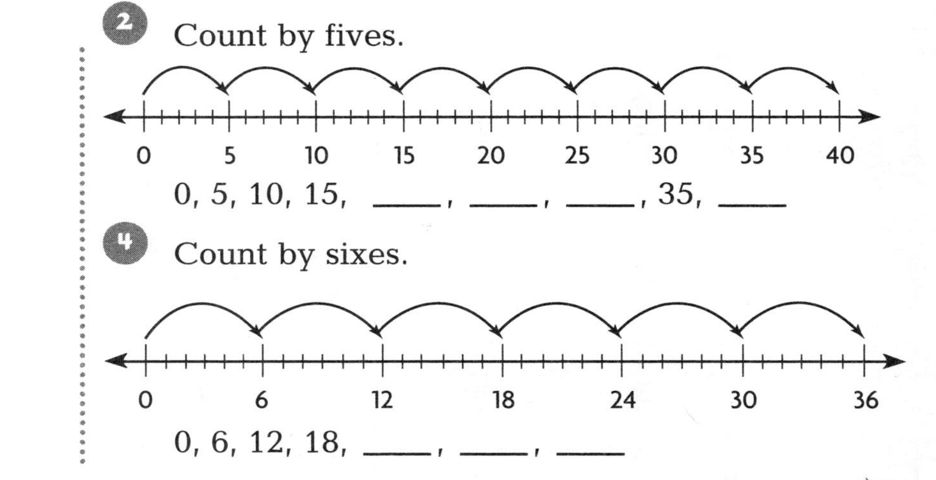

The number line shows skip counting by 5.
The next three numbers are: 20, 25, 30.

Try These

Use the number line to find the missing numbers.

1. Count by twos.

   0 1 2 3 4 5 6 7 8 9 10 11 12 13 14 15 16 17 18 19 20

   0, 2, 4, 6, ____, ____, ____, 14, 16, ____, ____

2. Count by fives.

   0 5 10 15 20 25 30 35 40

   0, 5, 10, 15, ____, ____, ____, 35, ____

3. Count by tens.

   0 10 20 30 40 50 60

   0, 10, 20, ____, ____, ____, ____

4. Count by sixes.

   0 6 12 18 24 30 36

   0, 6, 12, 18, ____, ____, ____

Go to the next side.

Name ________________________ Skill ________________________

Practice on Your Own

Skill 18

Skip count to find the missing numbers.

0, 4, 8, 12, 16, □, □, □.
Count: 4, 8, 12, 16, **20**, **24**, **28**.

The next three numbers are
20, 24, 28.

12, 15, 18, 21, □, □, □.
Count: 12, 15, 18, 21, **24**, **27**, **30**.

The next three numbers are
24, 27, 30.

Skip count to find the missing numbers.

1. Count by fives.

   0, 5, 10, 15, __, 25, __, __, __

2. Count by tens.

   0, 10, 20, 30, __, 50, __, __, __, __

3. Count by twos.

   0, 2, 4, __, __, 10, 12, __, __, __, __

4. Count by threes.

   9, 12, 15, __, __, __, __, __, __

5. Count by fours.

   16, 24, __, 32, __, __, __, __

6. Count by sixes.

   18, 24, 30, __, __, __, __

Check

Skip count to find the missing numbers.

7. Count by fives.

   25, 30, 35, __, __, __, __

8. Count by tens.

   40, __, 60, __, __, 90, __

Skill 19 Grade 5

Multiply by 2-Digit Numbers

Using Skill 19

OBJECTIVE Multiply by 2-digit numbers, with regrouping

Begin the lesson by reviewing the place-value names for 5- and 6-digit numbers. You might also review the definitions for *partial product* and *regrouping*.

Draw attention to Step 1. Help students recognize that the first partial product is the result of multiplying the first factor by the 4 ones in the second factor. Ask: **What factors are used to find the first partial product?** (10,437 and 4) **What are the numbers in the three little boxes above the digits in the thousands, hundreds, and tens places?** (regrouped digits)

Direct students' attention to Step 2. Help students recognize the source of the second partial product. Ask: **When multiplying by tens, what is always in the ones digit of your product?** (0) **What digit was regrouped and where?** (1 ten)

Draw attention to the partial products in Step 3. Ask: **How many partial products are there?** (2) **Which partial product is the result of multiplying by ones?** (the top product) **Which partial product is the result of multiplying by tens?** (the second product) **What do you do with the two partial products?** (Add them.)

TRY THESE Exercises 1–4 model multiplication by 2-digit numbers with regrouping.

- **Exercises 1–2** Multiplication by 2-digit numbers, no regrouping.
- **Exercises 3–4** Multiplication by 2-digit numbers, with regrouping.

PRACTICE ON YOUR OWN Review the example with students. Ask them to identify the place value, partial products, and regrouped digits as they multiply.

CHECK Determine if students multiply by 2-digit numbers, regrouping as appropriate.

Success is indicated by 3 out of 4 correct responses.

Students who successfully complete the **Practice on Your Own** and **Check** are ready to move to the next skill.

COMMON ERRORS

- Students might not recall addition or multiplication facts.
- Students might not regroup correctly when multiplying or when adding partial products.
- Students might confuse place value when writing partial products.

Students who made more than 4 errors in the **Practice on Your Own**, or who were not successful in the **Check** section, may benefit from the **Alternative Teaching Strategy** on the next page.

Optional

Alternative Teaching Strategy

Multiplying with a Model

15 Minutes

OBJECTIVE Use a model of 2-digit multiplication

Use a model to illustrate 28 × 325.

First, help students write 325 in expanded form:

$$325 = 300 + 20 + 5$$

Next, help students write 28 in expanded form.

$$28 = 20 + 8$$

Then, show students how to draw and label a grid like the one below. This model provides a way for students to use mental math to find partial products.

	300	20	5
8			
20			

Once they have labeled the grid, help students complete the multiplication, step by step. Begin by showing how to mentally multiply 8 × 300, then 8 × 20, and so on.

	300	20	5
8	8 x 300 = 2,400		
20			

	325
8	8 x 325 = 2,600
20	20 x 325 = 6,500

When all the multiplication is complete, direct students to add the partial products:

2400 + 160 + 40 + 6,000 + 400 + 100 = 9,100

Explain to students that the sum 9,100 is the product of 28 and 325. You can illustrate this quite dramatically by making the grid proportional to the numbers in expanded form.

Continue with other similar examples. The goal is to have students become comfortable enough with multiplication that they can give up using models and multiply independently at the symbolic level.

Name ______________________ Skill ______________________

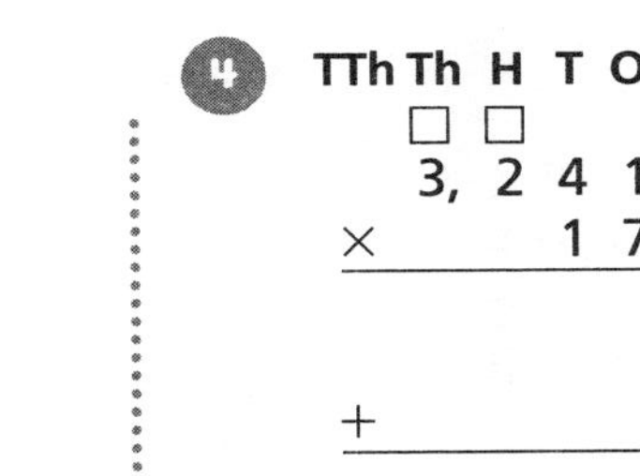

Multiply by 2-Digit Numbers

Find 24 × 10,437.

Step 1 Multiply by the ones.

	TTh	Th	H	T	O	
		1	1	2		
	1	0,	4	3	7	
×				2	4	
	4	1	7	4	8	← 4 × 10,437

Step 2 Multiply by the tens.

	HTh	TTh	Th	H	T	O	
					1		
			1	1	2		
		1	0,	4	3	7	
×					2	4	
		4	1	7	4	8	
	2	0	8	7	4	0	← 20 × 10,437

Step 3 Add the partial products.

	HTh	TTh	Th	H	T	O
					1	
			1	1	2	
		1	0,	4	3	7
×					2	4
		4	1	7	4	8
+	2	0	8	7	4	0
	2	5	0,	4	8	8

So, 24 × 10,437 = 250,488.

Try These

Multiply.

1

	H	T	O
		2	3
×		1	2
+			

2

	Th	H	T	O
		1	2	3
×			2	1
+				

3

	Th	H	T	O
			□	
		2	4	5
×			1	2
+				

4

	TTh	Th	H	T	O
		□	□		
		3,	2	4	1
×				1	7
+					

Go to the next side.

Name ________________________ Skill ________________________

Practice on Your Own

Think: Do you need to regroup?

	HTh	TTh	Th	H	T	O	
			1		2		
		2	0,	3	1	4	
×					2	6	
	1	2	1,	8	8	4	← $6 \times 20{,}314$
+	4	0	6,	2	8	0	← $20 \times 20{,}314$
	5	2	8,	1	6	4	

Multiply.

1. H T O
   34
   × 26
   +

2. Th H T O
   624
   × 27
   +

3. TTh Th H T O
   434
   × 92
   +

4. Th H T O
   408
   × 35
   +

5. TTh Th H T O
   1,773
   × 36
   +

6. TTh Th H T O
   42,176
   × 54
   +

Check

Multiply.

7. 94 × 48 +

8. 837 × 64 +

9. 5,286 × 37 +

10. 29,637 × 84 +

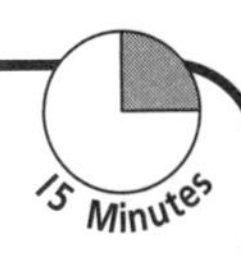

Using Skill 20

OBJECTIVE Write all the factors of a number

You may wish to review multiplication facts with the students.

Draw students' attention to the multiplication table. Tell students that the table shows factors and products. Ask: **What do the numbers in the top row and the numbers in the leftmost column represent?** (Factors of numbers within the table.)

Tell students that the table can help them find the factors of 18. Ask: **How many times does 18 appear in the table?** (4) Tell students that this means they will find 4 sets of factors of 18.

Have students place their fingers on the first 18 in the table. Have them move their finger up the column to the top row. Ask: **What number is in the top row?** (9) Now have them place their finger on the 18 again and move it across the row to the leftmost column. Ask: **What number is in the leftmost column?** (2) Explain to students that 9 and 2 are factors of 18.

Continue in a similar manner. Have students find the other factors of 18 using the table. Point out to students that the factors repeat and that each factor need only be written once. Remind students that 18 also has itself and 1 as factors and that these factors are not found in the table.

Guide students to realize that each number will always have at least 2 factors — the number itself and 1.

TRY THESE In Exercises 1–3 students find the factors of a number.

- **Exercise 1** Factors of 9.
- **Exercise 2** Factors of 14.
- **Exercise 3** Factors of 12.

PRACTICE ON YOUR OWN Review the example at the top of the page. Have students find the four places where 6 appears in the multiplication table. Ask them to name the factors of 6 using the table.

CHECK Determine if students can name all the factors of a number, including the number itself and 1.

Success is indicated by 3 out of 4 correct responses.

Students who successfully complete the **Practice on Your Own** and **Check** are ready to move to the next skill.

COMMON ERRORS

- Students may forget to include the number itself as a factor.
- Students may not give any factors for 1, forgetting that the only factor of 1 is 1 itself.
- Students may include the same factor twice.

Students who made more than 3 errors in the **Practice on Your Own**, or who were not successful in the **Check** section, may benefit from the **Alternative Teaching Strategy** on the next page.

Optional

Alternative Teaching Strategy

Use Arrays to Model Factors

OBJECTIVE Use rectangular arrays to find the factors of a number

MATERIALS graph paper or 40 square tiles

Tell students that they will be making arrays using tiles or graph paper. Have students work in pairs for this activity. Distribute tiles or graph paper to each pair of students. Explain what is meant by a rectangular array of tiles.

Present the two models shown.

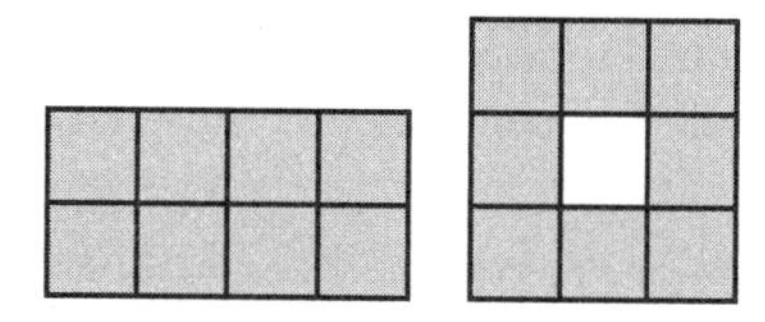

Explain to students that the first model is a rectangular array of tiles. The second model is not a rectangular array of tiles because there is a gap.

Direct each pair to build rectangular arrays using 12 tiles. Students may also draw the rectangular arrays on graph paper. Guide students to build the following arrays.

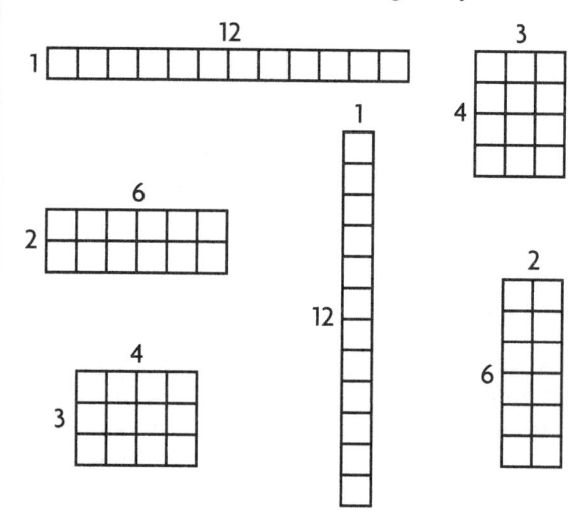

Guide students to recognize that a factor of 12 is one of the dimensions of a rectangular array for 12. Have students write the factors for each array. Point out that the factors repeat.

Next, ask students to name the 6 factors of 12 — that is, the 6 possible dimensions of the 12-tile rectangular array.

Check understanding by directing each pair of students to find the factors of 8. Again, student can use tiles or graph paper. Guide students to build the rectangular arrays below:

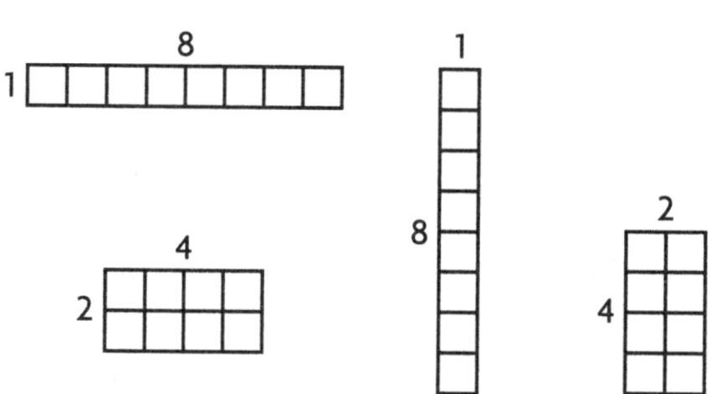

Have students write the factors for each array. Ask students to name the 4 factors of 8.

Continue with similar examples.

Name ______________________ Skill ______________________

Grade 5 **Skill 20**

Factors

You can use a multiplication table to find the factors of a number. Factors are found *across* the top row and *down* the first column in a multiplication table.

Remember: A factor is a number multiplied by another number to find a product. So, factor × factor = product.

Use the table to find the factors of 18.

Step 1 Find 18 in the table.
The number 18 appears four times.

Step 2 Find the factors.
Move *up* the column from 18 to the top row to find one factor. Then move *across* the row from 18 to the first column to find the other factor.

Step 3 Write the factors. Check for factors that repeat.

$1 \times 18 = 18$
$2 \times 9 = 18$
$3 \times 6 = 18$
$6 \times 3 = 18$
$9 \times 2 = 18$
$18 \times 1 = 18$

These factors repeat.

So, the factors of 18 are 1, 2, 3, 6, 9, and 18.

The factors of any number always include 1 and the number itself. The only factor of 1 is 1 itself.

Factors (across) / Factors (down)

×	0	1	2	3	4	5	6	7	8	9
0	0	0	0	0	0	0	0	0	0	0
1	0	1	2	3	4	5	6	7	8	9
2	0	2	4	6	8	10	12	14	16	(18)
3	0	3	6	9	12	15	(18)	21	24	27
4	0	4	8	12	16	20	24	28	32	36
5	0	5	10	15	20	25	30	35	40	45
6	0	6	12	(18)	24	30	36	42	48	54
7	0	7	14	21	28	35	42	49	56	63
8	0	8	16	24	32	40	48	56	64	72
9	0	9	(18)	27	36	45	54	63	72	81

Try These

Use the multiplication table above. Write the factors for each number. Remember to include 1 and the number itself as factors.

1. The factors of 9 are: ____, ____, ____

2. The factors of 14 are: ____, ____, ____, ____

3. The factors of 12 are: ____, ____, ____, ____, ____, ____

Go to the next side.

Name ______________________ Skill ______________________

Practice on Your Own

Skill 20

Find the factors of 6.

Think:

- Find the product 6 in the table.
- Then find the factors by moving across the row and up the column.

	Factor	×	Factor	=	Product
	1	×	6	=	6
	2	×	3	=	6
Repeated	3	×	2	=	6
	6	×	1	=	6

The factors of 6 are: 1, 2, 3, 6.

Factors

×	0	1	2	3	4	5	6	7	8	9
0	0	0	0	0	0	0	0	0	0	0
1	0	1	2	3	4	5	(6)	7	8	9
2	0	2	4	(6)	8	10	12	14	16	18
3	0	3	(6)	9	12	15	18	21	24	27
4	0	4	8	12	16	20	24	28	32	36
5	0	5	10	15	20	25	30	35	40	45
6	0	(6)	12	18	24	30	36	42	48	54
7	0	7	14	21	28	35	42	49	56	63
8	0	8	16	24	32	40	48	56	64	72
9	0	9	18	27	36	45	54	63	72	81

Use the multiplication table above. Write the factors. Remember to include 1 and the number itself.

1. The factors of 8 are ____, ____, ____, ____.
2. The factors of 16 are ____, ____, ____, ____, ____.
3. The factors of 15 are ____, ____, ____, ____.
4. The factors of 27 are ____, ____, ____, ____.
5. The factors of 4 are ________.
6. The factors of 7 are ________.
7. The factors of 10 are ________.

Find the factors. Some products and factors may not be on the multiplication table.

8. The factors of 17 are ________.
9. The factors of 21 are ________.
10. The factors of 25 are ________.
11. The factors of 28 are ________.

Check

Find the factors.

12. The factors of 35: ________
13. The factors of 5: ________
14. The factors of 36: ________
15. The factors of 1: ________

Repeated Factors

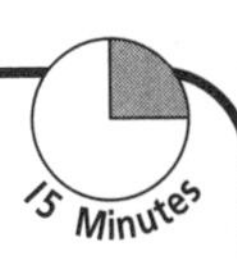

Using Skill 21

OBJECTIVE Use exponents to represent repeated factors

On Skill 21, draw attention to *Using Exponents to Write Expressions*. Be sure students understand the difference between the exponent and the base. Ask: **Which number is the repeated factor?** (6) **How many times is it repeated as a factor?** (4) Help students recognize that when written in exponent form the number 6 is called the *base*. The *exponent* represents the number of times 6 is used as a factor.

Draw attention to the second expression $4 \times 4 \times 5 \times 5 \times 5$. Be sure students recognize that there are two different factors. Ask: **How many times is the factor 4 repeated?** (2) **How many times is the factor 5 repeated?** (3) **In exponent form, what name is given to 4 and 5?** (base) **To 2 and 3? (exponent) How is the expression written in exponent form?** ($4^2 \times 5^3$)

Draw attention to *Finding the Value of Expressions*. Help students recognize that repeated factors can be written in exponent form. Point out that an expression can be evaluated by changing from exponent form to expanded form. Ask: **Is the left side of the equation in exponent or expanded form? (exponent) What is the value of $2^3 \times 5^2$?** (200)

Continue with similar questions as students find the value for the expression $3^5 \times 7^2$.

TRY THESE Exercises 1–2 model writing in exponent form and evaluating an expression.

- **Exercise 1** Write an expression in exponent form.
- **Exercise 2** Find the value of an exponent expression.

PRACTICE ON YOUR OWN Review the example at the top of the page. Have students identify which numbers are the bases and which numbers are the exponents. Point out the use of the expanded form to find the value of the expression.

CHECK Determine if students can write an expression in exponent form and find the value of an exponent expression.

Success is indicated by 4 out of 6 correct responses.

Students who successfully complete the **Practice on Your Own** and **Check** are ready to move to the next skill.

COMMON ERRORS

- Students might not write the correct number of factors when evaluating an exponent expression.
- Students might think of the exponent as a factor instead of the number of factors to be multiplied.
- Students might confuse the meaning of *base* and *exponent*.

Students who made more than 2 errors in each row of the **Practice on Your Own**, or who were not successful in the **Check** section, may benefit from the **Alternative Teaching Strategy** on the next page.

Optional

Alternative Teaching Strategy

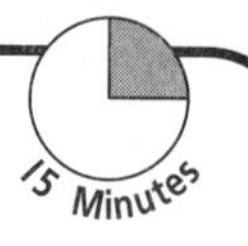

Rolling a Number Cube for Exponents

OBJECTIVE Use expressions with exponents to write repeated factors.

MATERIALS number cube, flip chart

On a flip chart, write:

1st roll (Base)	2nd roll (Exponent)	Expression	Base^{Exponent}

Ask a student to roll a number cube and write the number in the first column. Tell them students that this number is the *base*.

Ask another student to roll the number cube. Explain to students that the second number indicates how many times the number in column 1 is repeated as a factor. It is called an *exponent*.

A roll of 6 followed by a roll of 3 is shown below.

1st roll (Base)	2nd roll (Exponent)	Expression	Base^{Exponent}
6	3	$6 \times 6 \times 6$	

Direct students' attention to the last column. Show students the exponent form.

1st roll (Base)	2nd roll (Exponent)	Expression	Base^{Exponent}
6	3	$6 \times 6 \times 6$	6^3

Have students alternate rolling the number cube and recording the results on the chart.

Help students realize that the expression shown in column 3 represents the repeated multiplication.

After a few tries, you may also guide students to write the final product. Students may wish to use a calculator to find the product of greater numbers.

Name ____________________ Skill ____________________

Grade 5 **Skill** 21

Repeated Factors

Use exponents to represent repeated factors.

An **exponent** shows how many times a number called the *base* is used as a *factor*.

Using Exponents to Write Expressions

- Write the expression $6 \times 6 \times 6 \times 6$ using an exponent.

  4 *repeated* factors

  $6 \times 6 \times 6 \times 6 = 6^4$ (exponent, 4; base, 6)

- Write the expression $4 \times 4 \times 5 \times 5 \times 5$ in exponent form.

  2 *repeated* factors × 3 *repeated* factors

  $4 \times 4 \times 5 \times 5 \times 5 = 4^2 \times 5^3$ (exponents, 2 and 3)

  Exponent Form

Finding the Value of Expressions

- Find the value of the expression $2^3 \times 5^2$.

  3 *repeated* factors × 2 *repeated* factors

  $2^3 \times 5^2 = 2 \times 2 \times 2 \times 5 \times 5$

  $= 8 \times 25 = 200$

  The value of $2^3 \times 5^2$ is 200.

- Find the value of the expression $3^5 \times 7^2$.

  5 *repeated* factors × 2 *repeated* factors

  $3^5 \times 7^2 = 3 \times 3 \times 3 \times 3 \times 3 \times 7 \times 7$

  $= 243 \times 49 = 11{,}907$

  The value of $3^5 \times 7^2$ is 11,907.

Try These

1 Write the expression in exponent form.

$7 \times 7 \times 4 \times 4 \times 4$

Number of times 7 is a factor: 2

Number of times 4 is a factor: ____

$7 \times 7 \times 4 \times 4 \times 4 =$ 7^2 × ____

2 Find the value of the expression.

$2^3 \times 4^2$

$2^3 =$ ____ × ____ × ____ The value of 2^3 is ____ .

$4^2 =$ ____ × ____ The value of 4^2 is ____ .

$2^3 \times 4^2 =$ ____ × ____ = ____

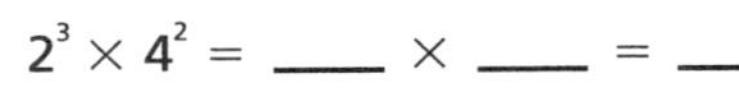

Go to the next side.

Name ____________________ Skill ____________________

Practice on Your Own

Skill 21

Think

An **exponent** shows the number of times a *number* called a **base** is used as a **factor.**

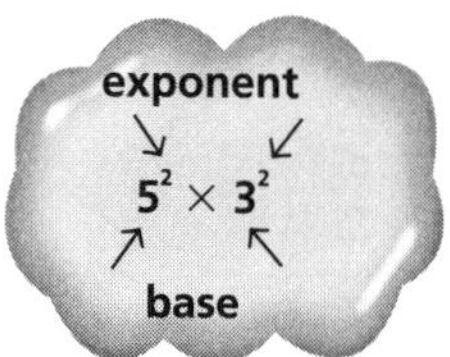

2 repeated factors 2 repeated factors

$5^2 \times 3^2 = 5 \times 5 \times 3 \times 3$

$= 25 \times 9 = 225$

The value of $5^2 \times 3^2$ is 225.

Write in exponent form.

1. $10 \times 10 \times 10 \times 10$
   $10^{\square}$
2. $9 \times 9 \times 9 \times 9 \times 9$
   $9^{\square}$
3. $2 \times 2 \times 2 \times 2 \times 5 \times 5$
   $2^{\square} \times 5^{\square}$
4. $4 \times 4 \times 4 \times 4 \times 4 \times 4$
   $4^{\square}$
5. $7 \times 7 \times 7 \times 7 \times 7$
   $7^{\square}$
6. $2 \times 2 \times 6 \times 6 \times 6$
   $2^{\square} \times 6^{\square}$

Find the value of the expression.

7. $5^2 =$ ___ $\times$ ___
   $=$ ___
8. $2^3 =$ ___ $\times$ ___ $\times$ ___
   $=$ ___
9. $4^2 =$ ___ $\times$ ___
   $=$ ___
10. $2^3 \times 5^2$
    $2^3 \times 5^2 =$ ___ $\times$ ___
    $=$ ___
11. $4^2 \times 2^3$
    $4^2 \times 2^3 =$ ___ $\times$ ___
    $=$ ___
12. $9^2 \times 2^3$
    $9^2 \times 2^3 =$ ___ $\times$ ___
    $=$ ___
13. $6^2 =$ ______
14. $9^2 =$ ______
15. $3^3 =$ ______

Check

Write in exponent form.

16. $6 \times 6 \times 6 \times 6 \times 6 \times 6$
    $6^{\square}$
17. $3 \times 3 \times 3 \times 3 \times 3$
    $3^{\square}$
18. $5 \times 5 \times 9 \times 9 \times 9$
    $5^{\square} \times 9^{\square}$

Find the value of the expression.

19. 8^2
    $8^2 =$ ______
20. 3^4
    $3^4 =$ ______
21. 5^3
    $5^3 =$ ______

Name ______________________ Skill ______________________

Answer Card
Multiplication
Grade 5

Skill 12

TRY THESE

1. 2, 7, 14; 14
2. 4, 2, 8; 8

PRACTICE

1. 2, 6, 12; 12
2. 3, 3, 9; 9
3. 5, 2, 10; 10
4. 2, 2, 4; 4
5. 2, 8, 16; 16
6. 3, 5, 15; 15
7. 36
8. 40
9. 35
10. 44
11. 5
12. 24
13. 25
14. 32
15. 6
16. 4
17. 15
18. 50

CHECK

19. 27
20. 14
21. 24
22. 45

Skill 13

TRY THESE

1. 8, 4, 32; $8 \times 4 = 32$
2. 7, 3, 21; $7 \times 3 = 21$
3. 6, 8, 48, $n = 48$

PRACTICE

1. 6, 5, 30, $6 \times 5 = 30$
2. 6, 10, 60, $n = 60$
3. 40
4. 42
5. 36
6. 50
7. 88
8. 48
9. 49
10. 144
11. $n = 24$
12. $n = 54$
13. $n = 63$
14. $n = 40$

CHECK

15. 45
16. 70
17. $n = 35$

Skill 14

TRY THESE

1. 6; 9
2. 3; 4
3. 1; 47
4. 0; 0

PRACTICE

1. 3
2. 3
3. 63
4. 0
5. 1; Property of One
6. 0; Zero Property
7. 9; Associative Property
8. 23, 5; Commutative Property
9. $8 \times (9 \times 3)$
10. 0
11. 17
12. 3
13. $5 \times (2 \times 3)$
14. 7
15. 0
16. 42

CHECK

17. 56
18. $8 \times (10 \times 7)$
19. 94
20. 0

Name ____________ Skill ____________

Skill 15

TRY THESE

1. 8; 60, 24; 84
2. 1; 120, 2; 122
3. 4, 4; 40, 28; 68

PRACTICE

1. 6; 120, 24, 144
2. 5, 5; 50, 5; 55
3. 30, 2; 240, 16; 256
4. 10, 4, 50, 20, 70
5. 10, 7, 80, 56, 136
6. 6, 6, 120, 24, 144
7. $(4 \times 10) + (4 \times 2)$; 48
8. $(8 \times 20) + (8 \times 1)$; 168
9. $(7 \times 10) + (7 \times 3)$; 91
10. 84
11. 96
12. 39

CHECK

13. 60
14. 128
15. 100

Skill 16

TRY THESE

1. 5, 50, 500
2. 9, 90, 900
3. 24, 240, 2,400

PRACTICE

1. 2, 20, 200
2. 8, 80, 800
3. 4, 40, 400
4. 14, 140, 1,400
5. 20, 200, 2,000
6. 32, 320, 3,200
7. 360
8. 2,100
9. 280
10. 480
11. 2,400
12. 140

CHECK

13. 400
14. 1,500
15. 300

Skill 17

TRY THESE

1. 1; 50
2. 1; 346
3. 1, 2; 771
4. 1; 816

PRACTICE

1. 64
2. 714
3. 610
4. 738
5. 160
6. 705
7. 668
8. 1,026
9. 148
10. 819
11. 807
12. 615
13. 228
14. 920
15. 660
16. 1,875

CHECK

17. 87
18. 648
19. 956
20. 2,172

Answer Card

Multiplication

Grade 5

Name ____________________ Skill ____________________

Answer Card

Multiplication

Grade 5

TRY THESE

1. 8, 10, 12; 18, 20
2. 20, 25, 30; 40
3. 30, 40, 50, 60
4. 24, 30, 36

PRACTICE

1. 20; 30, 35, 40
2. 40; 60, 70, 80, 90
3. 6, 8; 14, 16, 18, 20
4. 18, 21, 24, 27, 30, 33
5. 28; 36, 40, 44, 48
6. 36, 42, 48, 54

CHECK

7. 40, 45, 50, 55
8. 50; 70, 80; 100

TRY THESE

1. 46; 230; 276
2. 123; 2,460; 2,583
3. 1; 490; 2,450; 2,940
4. 1, 2; 22,687; 32,410; 55,097

PRACTICE

1. 204; 680; 884
2. 4,368; 12,480; 16,848
3. 868; 39,060; 39,928
4. 2,040; 12,240; 14,280
5. 10,638; 53,190; 63,828
6. 168,704; 2,108,800; 2,277,504

CHECK

7. 752; 3,760; 4,512
8. 3,348; 50,220; 53,568
9. 37,002; 158,580; 195,582
10. 118,548; 2,370,960; 2,489,508

Name ____________________ Skill ____________________

Answer Card
Multiplication
Grade 5

SKILL 20

TRY THESE

1. 1, 3, 9
2. 1, 2, 7, 14
3. 1, 2, 3, 4, 6, 12

PRACTICE

1. 1, 2, 4, 8
2. 1, 2, 4, 8, 16
3. 1, 3, 5, 15
4. 1, 3, 9, 27
5. 1, 2, 4
6. 1, 7
7. 1, 2, 5,10
8. 1, 17
9. 1, 3, 7, 21
10. 1, 5, 25
11. 1, 2, 4, 7, 14, 28

CHECK

12. 1, 5, 7, 35
13. 1, 5
14. 1, 2, 3, 4, 6, 9, 12, 18, 36
15. 1

SKILL 21

TRY THESE

1. 3; 4^3
2. 2, 2, 2, 8; 4, 4, 16; 8, 16, 128

PRACTICE

1. 4
2. 5
3. 4, 2
4. 6
5. 5
6. 2, 3
7. 5, 5; 25
8. 2, 2, 2; 8
9. 4, 4; 16
10. 8, 25; 200
11. 16, 8; 128
12. 81, 8; 648
13. 36
14. 81
15. 27

CHECK

16. 6
17. 5
18. 2, 3
19. 64
20. 81
21. 125

Number Sense

Whole Number Division

Divide 2-Digit Numbers by 1-Digit Numbers

15 Minutes

Using Skill 22

OBJECTIVE Divide 2-digit by 1-digit numbers, with remainders

You may wish to review the terms used in division and how to set up a division problem:

$$\text{Divisor}\,\overline{)\text{Dividend}}\quad \text{Quotient Remainder (r)}$$

Draw students' attention to Step 1. Point out to students that the place-value chart shows the value of the digits in the dividend and in the quotient. Ask: **In what column in the place-value chart will you put the quotient?** (the ones column)

Point out to students the multiplication facts for 9. Ask: **What is the multiple of 9 nearest to, but not greater than, 56?** (54)

Direct students' attention to Step 2. Focus on the quotient. Ask: **54 is the product of 9 and what other factor?** (6) **Where do you write the number 6?** (in the ones place, in the quotient) Explain that you write the product of 6 and 9, 54, in the division problem. Then you subtract. Ask: **What is 56 minus 54?** (2)

Continue with Step 3. Focus on the remainder. Ask: **What is the remainder?** (2) **The remainder must be less than what number?** (9) **What letter is used to indicate the remainder?** (r) **Where is the remainder written?** (next to the 6)

Guide students to recognize that the remainder cannot be equal to or greater than 9. Explain to students that if the remainder is greater than the divisor, then the number in the quotient should be greater.

TRY THESE Exercises 1–4 provide practice dividing 2-digit numbers by 1-digit numbers.

- **Exercises 1–2** One-digit quotient with remainder.
- **Exercises 3–4** Two-digit quotient with remainder.

PRACTICE ON YOUR OWN Review the example at the top of the page. Remind students to bring down the ones. Exercises 1–4 use prompts and have 1-digit quotients with remainders. Exercises 5–8 have 2-digit quotients and prompts for the division. Exercises 9–12 use prompts for the 1- and 2-digit quotients with remainders. Exercises 13–16 provide no prompts.

CHECK Determine if students can divide 2-digit by 1-digit numbers, with remainders.

Success is indicated by 3 out of 4 correct responses.

Students who successfully complete the **Practice on Your Own** and **Check** section are ready to move to the next skill.

COMMON ERRORS

- Students may not understand the meaning of division and thus, may write quotients that are too large or remainders that are greater than the divisor.
- Students may write digits in the quotient in the wrong places.

Students who made more than 4 errors in the **Practice on Your Own**, or who were not successful in the **Check** section, may benefit from the **Alternative Teaching Strategy** on the next page.

Optional

Alternative Teaching Strategy

15 Minutes

Model Division Using Grouping and Repeated Subtraction

OBJECTIVE Use counters to model grouping and repeated subtraction to model 2-digit division with 1-digit divisors

MATERIALS counters (26), flipchart

If students understand the concept of division but have difficulty placing digits in the correct place in the quotient, you may wish to have them turn their lined paper 90° and use the lines as place-value charts.

If students do not understand the meaning of division, you may wish to use the following strategies.

Grouping Model

Direct students to display 26 counters.

Tell students that they are going to find how many groups of four there are in 26 counters.

Guide students to collect the counters into groups of 4. Ask: **How many groups of 4 do you have?** (6) **Are there any counters left over?** (yes) **How many?** (2)

Explain to students that 26 counters can be divided into 6 groups with 4 counters in each group and one group with only 2 counters. Using the flip chart, point out to students that 26 is the dividend, 4 is the divisor, 6 is the quotient and 2 is the remainder.

Repeated Subtraction Model

Have students display 26 counters again. Tell students that they are going to use subtraction to find how many groups of 4 there are in 26. Have students subtract four counters from the group. Ask:

How many counters are left? (22)

Continue having the students subtract 4 counters and tell how many are left as you write the subtraction on the flip chart. Ask: **How many times did you subtract 4?** (6) **Are there any counters left over?** (yes) **How many?** (2)

$$\begin{array}{r} 26 \\ -\ 4 \\ \hline 22 \\ -\ 4 \\ \hline 18 \\ -\ 4 \\ \hline 14 \\ -\ 4 \\ \hline 10 \\ -\ 4 \\ \hline 6 \\ -\ 4 \\ \hline 2 \end{array}$$

Tell students that there are 6 groups of 4, with a remainder of 2.

Explain to students that the process of subtracting groups of 4 is the same as dividing by 4.

Continue with similar examples. Guide students to see that the division algorithm is a quicker method for answering problems involving repeated grouping or subtraction.

Name ____________________ Skill ____________________

Grade 5 Skill

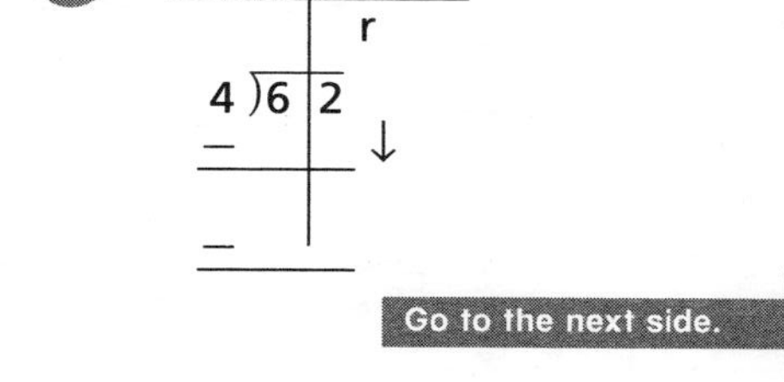

Divide 2-Digit Numbers by 1-Digit Numbers

Find 56 ÷ 9

Tens digit is less than the divisor. $5 < 9 \rightarrow 9\overline{)56}$

Step 1

Think of multiplication facts that have products close to 56.

$9 \times 5 = 45$

$\boxed{9 \times 6 = 54}$

$9 \times 7 = 63$

Step 2

Write 6 in the ones place in the quotient. Multiply and record the product under the dividend. Subtract.

$$\begin{array}{r} 6r2 \\ 9\overline{)56} \\ 9 \times 6 \rightarrow \underline{-54} \\ 2 < 9 \rightarrow \quad 2 \end{array}$$

Step 3

Compare the remainder with the divisor. If the remainder is less than the divisor, write it next to the quotient. Check your work if the remainder is greater than the divisor.

Find 37 ÷ 2

Tens digit is greater than the divisor. $3 > 2 \rightarrow 2\overline{)37}$

Step 1

Divide the tens digit. Write the 1 in the tens place in the quotient.

Step 2

Multiply. Record. Subtract.

Step 3

Compare the remainder with the divisor.

Step 4

Bring down the 7 ones. Repeat step 2.

Step 5

Compare the remainder with the divisor. If the remainder is less than the divisor, write it next to the quotient.

$$\begin{array}{r} 18r1 \\ 2\overline{)37} \\ 2 \times 1 \rightarrow \underline{-2\downarrow} \\ 17 \\ 2 \times 8 \rightarrow \underline{-16} \\ 1 \end{array}$$

Try These

Complete the division.

1. Tens | Ones, r — $3\overline{)17}$, –, 2

Think:
$3 \times 4 = 12$
$\boxed{3 \times 5 = 15}$
$3 \times 6 = 18$

2. Tens | Ones, r — $5\overline{)46}$, –

Think:
$5 \times 7 = 35$
$5 \times 8 = 40$
$\boxed{5 \times 9 = 45}$

3. Tens | Ones, r — $2\overline{)53}$, – ↓, –

4. Tens | Ones, r — $4\overline{)62}$, – ↓, –

Go to the next side.

Name ________________________ Skill ________________________

Practice on Your Own

Think:

- Divide 8 tens by 4.
- Record the 2 in the tens place in the quotient.
- Multiply. Then subtract.
- Bring down the ones.
- Divide 2 ones by 4. Record 0 in the quotient.

Tens	Ones
2	0 r2
4)8	2
−8	↓
0	2
−	0

$2 < 4$ The remainder is less than the divisor.

Divide.

1. Tens | Ones — $4\overline{)3\,4}$ r — 2

2. Tens | Ones — $6\overline{)5\,5}$ r

3. Tens | Ones — $5\overline{)3\,8}$ r

4. Tens | Ones — $9\overline{)3\,8}$ r

5. 2 7 r1 — $3\overline{)8\,2}$ — 2

6. 1 2 r2 — $8\overline{)9\,8}$ — 8

7. 1 4 r4 — $5\overline{)7\,4}$ — 4

8. 2 1 r1 — $2\overline{)4\,3}$ — 3

9. $6\overline{)5\,2}$ r

10. $8\overline{)2\,2}$ r

11. $9\overline{)9\,7}$ r

12. $4\overline{)8\,7}$ r

13. $8\overline{)3\,5}$

14. $5\overline{)4\,9}$

15. $2\overline{)5\,9}$

16. $3\overline{)4\,3}$

Check

Divide.

17. $9\overline{)7\,3}$

18. $8\overline{)5\,8}$

19. $6\overline{)6\,7}$

20. $5\overline{)5\,4}$

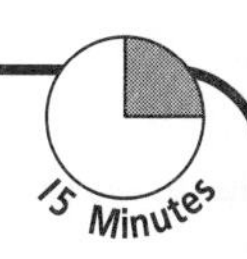

Using Skill 23

OBJECTIVE Use multiplication to check division

You may wish to review the names given to numbers in a division problem (divisor, dividend, quotient, and remainder) as shown at the top of Skill 23.

On Skill 23, draw attention to the multiplication shown in Step 1. Be sure students understand which numbers are being multiplied and why. Note that multiplication is the inverse of division—so multiplication "undoes" division. Ask:

What numbers are being multiplied? (4 and 13) **Say: So, you divide 53 by 4 and find that there are 13 groups of 4 with 1 leftover.**

Direct students' attention to Step 2. Help students focus on the remainder. Ask:

How is the remainder a part of the check step? (You have to add the remainder to the product in order to find the dividend.)

Continue with Step 3. Ask: **How do you know if the answer is correct?** (The sum in Step 2 should equal the dividend, 53.)

Help students recognize that the check step of a quotient with a remainder involves multiplication first, followed by addition. Then a comparison can be made with the dividend.

TRY THESE Exercises 1–3 model the checking process for division.

- **Exercises 1–2** Check 2-digit quotient with a remainder.
- **Exercise 3** Check 3-digit quotient with a remainder.

PRACTICE ON YOUR OWN Review the example at the top of the page. Ask students to name the numbers that are multiplied and added in the checking process. Then ask students to name the number in the original problem to which the result is compared.

CHECK Determine if students can identify which numbers are to be multiplied and which number is to be added.

Success is indicated by 2 out of 3 correct responses.

Students who successfully complete the **Practice on Your Own** and **Check** are ready to move to the next skill.

COMMON ERRORS

- Students may not compare the result of multiplication and addition to the dividend.
- Students may multiply the remainder and divisor, then add the quotient.
- Students may not know the multiplication facts.
- Students may add incorrectly.

Students who made more than 4 errors in the **Practice on Your Own**, or who were not successful in the **Check** section, may benefit from the **Alternative Teaching Strategy** on the next page.

Optional

Alternative Teaching Strategy

15 Minutes

Check Division with Repeated Addition

OBJECTIVE Use repeated addition to check division

MATERIALS counters (27); flip chart

Review the division problem below:

```
            quotient
               ↓
               4  r3 remainder
divisor    6 ) 27     dividend
```

Explain to students that the dividend (27) represents the number of counters in the whole group—it is the number being divided. The divisor (6) represents the number of counters in each group, the quotient (4) represents the number of equal groups and the remainer represents the number of left-over counters.

Tell students that one way to check whether a quotient is correct is to use repeated addition.

To check whether the quotient and remainder 4 r3 is the correct answer to 27 ÷ 6, present students with 27 counters. Ask a volunteer to cluster the counters into groups of 6, and set aside any left-over counters.

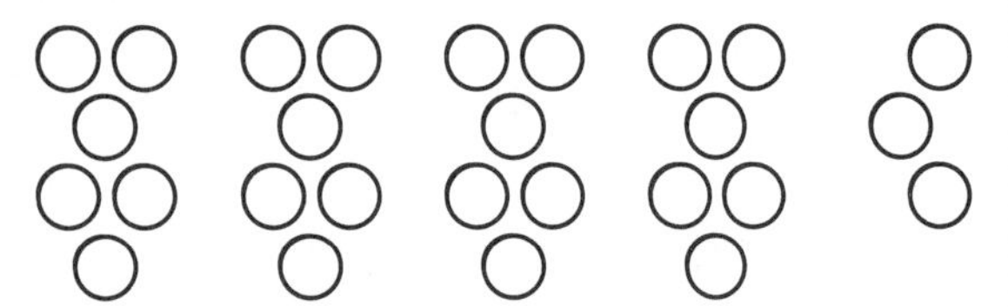

4 groups of 6 and 3 leftover

On a flip chart, help students record the following information.

Number of Groups	Number of Counters in Each Group		
1	6	⟶	6
2	6 + 6	⟶	12
3	6 + 6 + 6	⟶	18
4	6 + 6 + 6 + 6	⟶	24
left-over counters:			+ 3
		TOTAL:	27

Direct students' attention to the total. Help students compare the sum with the dividend. Note that if the sum matches the dividend, then the answer is correct.

Continue with similar examples.

After a few tries, guide students to conclude that a quicker alternative to repeated addition is to multiply. So, using the example above, they can check by multiplication:

$$6 + 6 + 6 + 6 + 3 = 27$$

↓

$$4 \times 6 + 3 = 27$$

Name ____________________ Skill ____________________

Grade 5 Skill 23

Check Division

Check the answer for 53 ÷ 4.

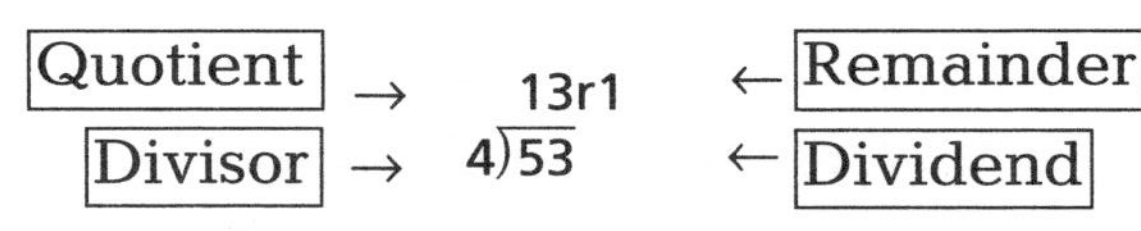

Quotient → 13r1 ← Remainder
Divisor → 4)53 ← Dividend

Step 1
Multiply the quotient by the divisor.

13r1
4)53

13 ←quotient
× 4 ←divisor
52

Step 2
Add the remainder.

13 ← quotient
× 4 ← divisor
52
+ 1 ← remainder
53

Step 3
Compare the sum with the dividend.

Sum		Dividend
53	=	53

If the sum and the dividend are equal, then the division is correct.

So, the answer 13 r1 is correct.

Try These

Check the answer.

1. 13 r3
5)68

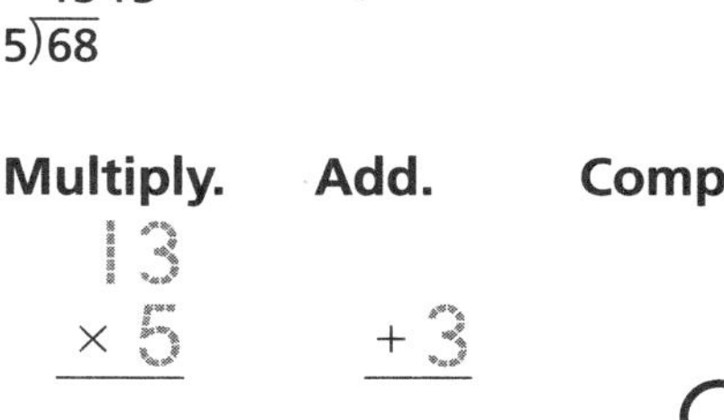

Multiply. 13 × 5 **Add.** + 3 **Compare.** ___ ◯ 68

Is the answer correct? ________

2. 26 r1
2)53

Multiply. × ___ **Add.** + ___ **Compare.** ___ ◯ 53

Is the answer correct? ________

3. 134 r2
7)940

Multiply. × ___ **Add.** + ___ **Compare.** ___ ◯ 940

Is the answer correct? ________

Go to the next side.

Name ______________________ Skill ______________________

Practice on Your Own

Skill 23

Is the division correct?

	Multiply.	**Add.**	**Compare.**
158 r1 3)475	158 quotient × 3 divisor 474	474 +1 remainder 475 dividend	475 = 475

Think: Are the sum and the dividend equal?

The division is correct.

Check the division.

1. 15 r3 / 5)78

   Multiply. Add. Compare.

   15 × 5 ___ +3 ___ ___ ○ 78

   Is the division correct? ______

2. 15 r1 / 6)91

   Multiply. Add. Compare.

   15 × 6 ___ +1 ___ ___ ○ 91

   Is the division correct? ______

3. 257 r2 / 3)773

   Multiply. Add. Compare.

   257 × 3 ___ +2 ___ ___ ○ 773

   Is the division correct? ______

4. 24 r1 / 4)97

   Multiply. Add. Compare.

   × ___ + ___ ___ ○ ___

   Is the division correct? ______

5. 396 r1 / 2)793

   Multiply. Add. Compare.

   × ___ + ___ ___ ○ ___

   Is the division correct? ______

6. 278 r1 / 3)835

   Multiply. Add. Compare.

   × ___ + ___ ___ ○ ___

   Is the division correct? ______

7. 28 r1 / 3)85

   Check:

8. 37 r1 / 2)75

   Check:

9. 127 r3 / 7)892

   Check:

Check

Check the division.

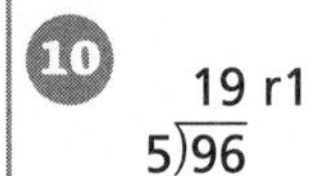

10. 19 r1 / 5)96

11. 304 r2 / 3)914

12. 195 r3 / 4)783

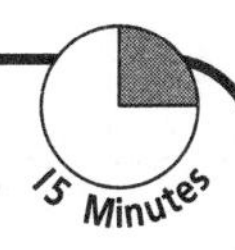

Using Skill 24

OBJECTIVE Use multiplication facts or a pattern to divide by 10

Direct students' attention to the first example. Recall with students that multiplication and division are inverse operations. Point to 50 ÷ 10. Tell students that they can think of a multiplication fact for 10. Ask:

What number multiplied by 10 equals 50? (5)

Direct students' attention to the second example. Review the patterns shown. Ask:

What number is the same in each division sentence? (10) **What is 60 divided by 10?** (6) **What is 600 divided by 10?** (60)

Say:

When you divide by ten, the quotient has one fewer zero than the dividend.

Tell students that there is another pattern. Say: **As you increase the dividend by a factor of 10, the quotient also increases by a factor of 10. Each quotient has one more zero than the previous quotient.**

Direct students' attention to the third example. Point out that they can find the quotient by using a pattern. Ask:

How many zeros are in the dividend, 4,400? (2) **How many zeros will there be in the quotient?** (1) **What is the quotient?** (440)

TRY THESE Exercises 1–3 provide practice with using multiplication facts or a pattern to divide by 10.

- **Exercises 1–2** Use multiplication facts.
- **Exercises 3–4** Use a pattern.

PRACTICE ON YOUR OWN Review the examples at the top of the page. Have students identify the multiplication fact and the pattern used. In Exercises 1–6 students are prompted to use multiplication facts or patterns to divide by 10. In Exercises 7–12 students divide by 10 and tell what strategy they used. Exercises 7–12 use two ways of writing division.

CHECK Determine if students can divide by 10. Success is indicated by 2 out of 3 correct responses.

Students who successfully complete the **Practice on Your Own** and **Check** are ready to move to the next skill.

COMMON ERRORS

- Students may write an incorrect number of zeros in the quotient.
- Students may not see the patterns.

Students who made more than 1 error in the **Practice on Your Own**, or who were not successful in the **Check** section, may benefit from the **Alternative Teaching Strategy** on the next page.

Optional

Alternative Teaching Strategy

Use Place Value to Divide by 10

25 Minutes

OBJECTIVE Use place value to demonstrate patterns of dividing by 10

MATERIALS place-value pocket charts, number cards 0–9

Begin by reviewing the pattern of multiplying by ten. Use place-value word names as well as numbers. Display the following:

$10 \times 5 = 50$ 1 ten × 5 ones = 5 tens

$10 \times 50 = 500$ 1 ten × 5 tens = 5 hundreds

$10 \times 500 = 5{,}000$ 1 ten × 5 hundreds = 5 thousands

Ask:

What operation is the inverse of multiplication? (division)

Tell students that you will repeat the pattern using division. Display the following:

$50 \div 10 = 5$ 5 tens ÷ 1 ten = 5 ones

$500 \div 10 = 50$ 5 hundreds ÷ 1 ten = 5 tens

$5{,}000 \div 10 = 500$ 5 thousands ÷ 1 ten = 5 hundreds

Guide students to understand the patterns shown.

Have students work in pairs for this activity. One partner puts a number in the place-value pocket chart, while the other student writes the division expression. You may wish to present the division in both formats, for example, $50 \div 10$ and $10\overline{)50}$.

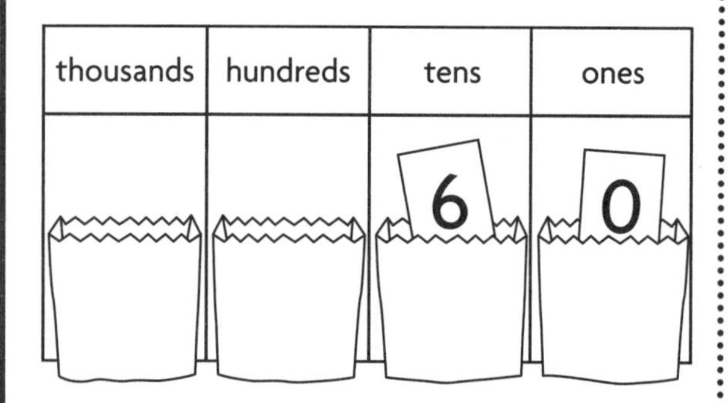

Distribute the place-value pocket charts and the number cards. Present 6×10. Have one partner place the 6 in the place-value pocket chart while the other partner records the multiplication. Ask:

What is 6 times 10? (60)

Have the partner move the 6 in the place-value pocket chart. Ask: **In what direction on the place-value chart does the 6 move?** (to the left) Ask the student to place a number card for 0 in the ones place to show 60.

Continue the activity for 60×10 and 600×10. Guide students to realize that the digit 6 moves one place to the left in each successive exercise.

Now repeat the activity using division. Present $6{,}000 \div 10$. Have one partner place 6,000 in the place-value pocket chart while the other partner records the division. Ask:

What is 6,000 ÷ 10? (600)

Have the partner move the number cards (and remove one 0) in the place-value pocket chart to show 600. Ask:

In what direction on the place-value chart does the 6 move? (to the right)

Continue the activity for $600 \div 10$ and $60 \div 10$. Lead students to understand that the digit 6 moves one place to the right in each successive exercise.

Name ______________________ Skill ______________________

Grade 5 **Skill 24**

Divide by 10

Use multiplication facts you know or a pattern to divide by 10.

Use a multiplication fact for 10 to divide by 10.

$50 \div 10 =$ _____

Think: $10 \times 5 = 50$

So, $50 \div 10 = 5$

Find a pattern to divide greater numbers by 10.

dividend	quotient
$60 \div 10 = 6$	
$600 \div 10 = 60$	
$6{,}000 \div 10 = 600$	
$60{,}000 \div 10 = 6{,}000$	

One pattern is: The quotient has one fewer zero than the dividend.
Another pattern is: Each quotient has one more zero than the previous quotient.

Use one of the patterns to find the quotient.

$4{,}400 \div 10 = \square$

Think: The quotient will have one fewer zero than 4,400.

So, $4{,}400 \div 10 = 440$.

Try These

Divide by 10.

1. $80 \div 10 = \square$
   Think: $10 \times$ __ $= 80$
   $80 \div 10 =$ _____

2. $40 \div 10 = \square$
   Think: $10 \times$ __ $= 40$
   $40 \div 10 =$ _____

3. $30 \div 10 = 3$
   $300 \div 10 = 30$
   $3{,}000 \div 10 =$ _____

4. $70 \div 10 = 7$
   $700 \div 10 = 70$
   $7{,}000 \div 10 =$ _____

Go to the next side.

Name ______________________________ Skill ______________________________

Practice on Your Own

Skill 24

Think:

Use a multiplication fact for 10 or a pattern to help you divide by 10.

Use a multiplication fact.

60 ÷ 10 = □

Think: 10 × 6 = 60

So, 60 ÷ 10 = 6

Use a pattern.

9,000 ÷ 10 = □

90 ÷ 10 = 9

900 ÷ 10 = 90

9,000 ÷ 10 = 900

Think: The quotient has one fewer zero than the dividend.

Divide.

1. 60 ÷ 10 = □
   Think: 10 x _____ = 60
   60 ÷ 10 = _____

2. 30 ÷ 10 = □
   Think: 10 x _____ = 30
   30 ÷ 10 = _____

3. 100 ÷ 10 = □
   Think: 10 x _____ = 100
   100 ÷ 10 = _____

4. 20 ÷ 10 = 2
   200 ÷ 10 = 20
   2,000 ÷ 10 = _____

5. 50 ÷ 10 = 5
   500 ÷ 10 = 50
   5,000 ÷ 10 = _____

6. 80 ÷ 10 = 8
   800 ÷ 10 = 80
   8,000 ÷ 10 = _____

Divide by 10. Tell whether you used a multiplication fact for 10 or a pattern.

7. 70 ÷ 10 _____

8. 420 ÷ 10 _____

9. 5,500 ÷ 10 _____

10. $10\overline{)400}$ _____

11. 
    $10\overline{)750}$ _____

12. $10\overline{)9{,}200}$ _____

Check

Divide by 10.

13. $10\overline{)40}$

14. $10\overline{)800}$

15. 8,200 ÷ 10 = _____

Divide by 1-Digit Numbers

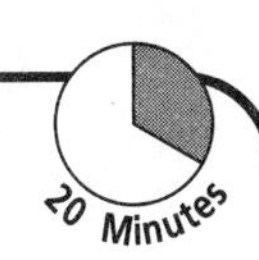

Using Skill 25

OBJECTIVE Divide up to a 4-digit number by a 1-digit number with no remainder

You may wish to copy the example onto the board, and work through the division with the students.

For Step 1, ask: **What digit in the dividend do you consider first to decide where to place the first digit in the quotient?** (the digit in the greatest place value position; the 2) **Can you divide 2 by 7?** (no)

Help students understand that when the divisor is greater than the digit in the greatest place-value position, they should consider the first two digits in the dividend. If the divisor can divide those two digits, then the first digit in the quotient is placed above the second digit. Ask: **What does the check mark in the quotient tell you?** (where to place the first digit in the quotient; in the hundreds place)

For Step 2, ask: **What division fact is close to 29 ÷ 7?** (28 ÷ 7 = 4) **What digit should you write in the hundreds place in the quotient?** (4)

Work through Steps 3 and 4, having students supply the basic fact for finding the next digit in the quotient. Point out the "multiply, subtract, bring down, and compare" steps that are used until there is nothing left to divide.

TRY THESE Exercises 1–3 provide the grid lines for the place value chart to help students align the digit as they divide. The prompts for the subtraction and bring down steps are provided.

- **Exercise 1** Two-digit quotient.
- **Exercises 2–3** Three-digit quotient.

PRACTICE ON YOUR OWN Discuss the **Think** question and the thought cloud suggestions given for the example at the top of the page. You may wish to have the students work through each step. Exercises 1–3 are similar to the **Try These** exercises. In Exercises 4–6, prompts are given to help the students decide where to put the first digit in the quotient. The place-value grid lines have been removed to allow students to practice aligning digits.

CHECK Determine if students can divide 3-digit and 4-digit numbers by 1-digit numbers with no remainders. Success is indicated by 2 out of 3 correct responses.

Students who successfully complete the **Practice on Your Own** and **Check** are ready to move to the next skill.

COMMON ERRORS

- Students may not know multiplication or division facts and thus may use an incorrect fact each time they place a digit in the quotient and multiply.
- Students may subtract incorrectly.
- Students may forget to bring down digits in any place and thus stop dividing before the division is complete.
- Students may write the first digit of the quotient in the wrong position.

Students who made more than 1 error in the **Practice on Your Own**, or who were not successful in the **Check** section, may benefit from the **Alternative Teaching Strategy** on the next page.

Optional

Alternative Teaching Strategy

Use Models to Divide by 1-Digit Numbers

20 Minutes

OBJECTIVE Divide hundreds by 1-digit numbers with no remainder

MATERIALS base-ten blocks, lined notebook paper

Students work in pairs in this activity. One partner models the step, while the other writes the digits in the quotient for the division example. Partners take turns modeling and writing the digits in the quotient.

Distribute the hundreds, tens, and ones blocks, and the lined notebook paper.

First tell the students to turn the lined notebook paper so that the lines are vertical. Write 2)246 on the board. Then tell the students to write the division problem on the lined paper. Each digit should be between 2 lines.

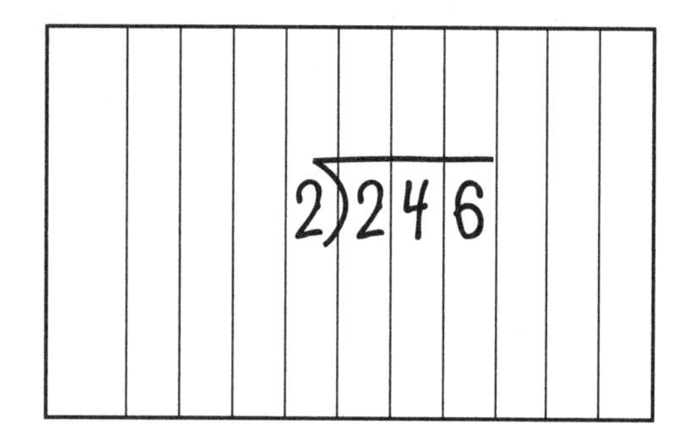

Have the students show the dividend using the hundreds, tens and ones.

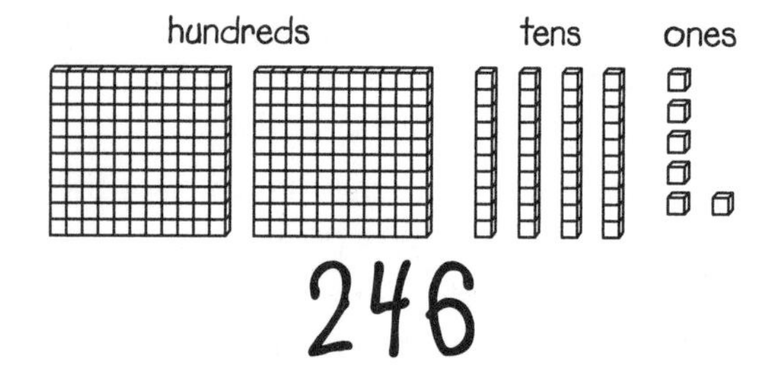

The student doing the writing should label the columns, H, T, O, to correspond with the model. Ask the students to divide the 2 hundreds into 2 equal parts. **Where will you write the first digit in the quotient?** (hundreds place)

The students should record the first digit. (1) Next, ask the students to divide the 4 tens into 2 equal parts and record the result as the next digit in the quotient.

(4 tens ÷ 2 = 2 tens; next digit: 2) Finally, students should divide the 6 ones into 2 equal parts and record the result in the quotient. (6 ones ÷ 2 = 3 ones; next digit: 3) **What is the quotient?** (123)

Before you begin the next part of the activity, you may wish to review renaming hundreds as tens, and tens as ones. Use this example for the students to model:

4)156

After the students have written the example on the vertical lined paper, have the students model the dividend using the hundred, tens, and ones blocks. Ask the students to decide if they can divide the 1 hundred. (No, 1 hundred cannot be divided by 4.)

What should you do to the hundreds so that you can divide? (Regroup the hundreds as 10 tens, and place the 10 tens with the 5 tens.) **Where will you write the first digit of the quotient?** (tens place)

Have students divide the tens, record the result in the quotient, and then regroup tens to divide the ones. (156 ÷ 4 = 39)

You may wish to have students try one more modeling example, before they work independently using paper and pencil.

Name ______________________ Skill ______________________

Grade 5 Skill 25

Divide by 1-Digit Numbers

Find 2,996 ÷ 7.

Step 1 Decide where to place the first digit of the quotient.
Think
7 is greater than 2.
So, the first digit can't be in the thousands place.
The first digit will be in the hundreds place.

```
  Th H T O
      ✓
7 )2, 9 9 6
     7<29
```

Step 2 Divide 29 hundreds. Multiply. Subtract. Compare. Bring down the tens.

```
  Th H T O
     4 2 8
7 )2, 9 9 6
  -2 8 ↓
     1 9
```

Think
$4 \times 7 = 28$
$29 - 28 = 1$
$7 < 19$

Step 3 Divide 19 tens. Multiply. Subtract. Compare. Bring down the ones.

```
  Th H T O
     4 2 8
7 )2, 9 9 6
  -2 8 ↓
     1 9
   - 1 4 ↓
       5 6
```

Think
$2 \times 7 = 14$
$19 - 14 = 5$
$7 < 56$

Step 4 Divide 56 ones. Multiply. Subtract. Compare.

```
     4 2 8
7 )2,9 9 6
  - 2 8 ↓
     1 9
   - 1 4 ↓
       5 6
     - 5 6
         0
```

Think
$8 \times 7 = 56$
$56 - 56 = 0$

So, 2,996 ÷ 7 = 428

Try These

Divide.

1. 265 ÷ 5 = □

```
  H T O
5 )2 6 5
  -__↓
   -__
```

2. 3,776 ÷ 4 = □

```
  Th H T O
4 )3, 7 7 6
  -__↓
    -__↓
      -__
```

3. 456 ÷ 4 =

```
  H T O
4 )4 5 6
  -__↓
   -__↓
    -__
```

Go to the next side.

Name ____________________ Skill ____________________

Practice on Your Own

Skill 25

Example 2,030 ÷ 5 = □

5 > 2, but 5 < 20
So, the first digit of the quotient is in the hundreds place.

Think:
Where will you place the first digit of the quotient?

```
 Th H T O
     4 0 6
 5 )2, 0 3 0
   − 2 0↓↓
       0 3 0
     − 3 0
         0
```

Bring down tens.
5 < 3, so there is not enough to divide. So, write a zero in the tens place in the quotient. Then bring down the ones. Divide.

Divide.

1. H T O
   6)2 8 2

2. Th H T O
   9)1, 3 9 5

3. Th H T O
   8)1, 6 6 4

4. 5)2 9 5

5. 5)3, 2 2 5

6. 3)1, 5 0 9

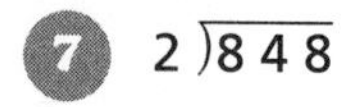

Check

Divide.

7. 2)8 4 8

8. 4)3, 6 8 4

9. 5)3, 5 0 5

Division Patterns (Divide by Multiples of 10)

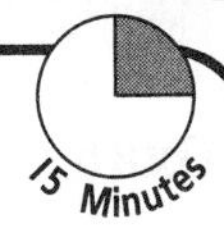

Using Skill 26

OBJECTIVE Use a pattern to divide by multiples of 10

Call on students to state some multiples of 10. Then have them look at the example for 6,000 ÷ 30.

Ask: **What is the dividend?** (6,000) **What is the divisor?** (30) **Are 6,000 and 30 multiples of 10?** (yes)

Guide students through the steps of the pattern for dividing with multiples of 10.

Ask: **What basic fact can you use?** (6 ÷ 3 = 2) **How do you get the first step in the pattern?** (Multiply both 6 and 3 by 10 to get 60 ÷ 30 = 2.)

Continue working through the pattern, relating 60, 600, and 6,000 with 2, 20, and 200. Ask: **What happens to the number of zeros in the dividends 60, 600, and 6,000?** (It increases by 1.) **What happens to the number of zeros in the quotients 2, 20, and 200?** (It increases by 1.) **What happens to the divisor, 30?** (It stays the same.)

TRY THESE Exercises 1–4 provide practice with patterns for dividing.

- **Exercises 1 and 2** Divide through ten thousands.
- **Exercises 3 and 4** Divide through hundred thousands.

PRACTICE ON YOUR OWN Focus on the example at the top of the page. Ask: **What basic fact can you use?** (25 ÷ 5) **How do you get the first step in the pattern?** (Multiply both 25 and 5 by 10 to get 250 ÷ 50 = 5.)

Have a student show two ways of writing the division.

250,000 ÷ 50 and 50 $\overline{)250{,}000}$

CHECK Determine if students can use basic facts and identify patterns in multiples of 10 to divide. Success is indicated by 3 out of 4 correct responses.

Students who successfully complete the **Practice on Your Own** and **Check** are ready to move to the next skill.

COMMON ERRORS

- Students may not remember basic division facts.
- Students may have trouble setting up the patterns.
- Students may write incorrect numbers of zeros.

Students who made more than 3 errors in the **Practice on Your Own**, or who were not successful in the **Check** section, may benefit from the **Alternative Teaching Strategy** on the next page.

Optional

Alternative Teaching Strategy

Division Patterns

20 Minutes

OBJECTIVE Use a pattern to divide by multiples of 10

MATERIALS tiles

Students can practice isolating the basic fact they will use to begin the pattern. Show them this division problem.

$$210 \div 30$$

Say: **Name the basic fact you will use to begin the pattern.** (21 ÷ 3)

Continue: **What is a basic fact for 25,000 ÷ 50?** (25 ÷ 5) **2,800 ÷ 40?** (28 ÷ 4) **630,000 ÷ 90?** (63 ÷ 9)

Review the parts of a division example. Students should be able to identify the dividend, the divisor, and the quotient in both division formats.

$210 \div 30 = 7$ and $30\overline{)210}$ with quotient 7

Check to be sure that students understand they will be dividing by multiples of 10. Use all the examples above to identify the divisors and dividends as multiples of 10.

Have students practice making a pattern for dividing by a multiple of 10.

$$4 \div 2 = 2$$

$$40 \div 20 = 2$$

$$400 \div 20 = 20$$

Ask: **Why do the first two steps result in the same quotient?** (Help students use tiles to show that 4 ÷ 2 and 40 ÷ 20 have the same quotient.)

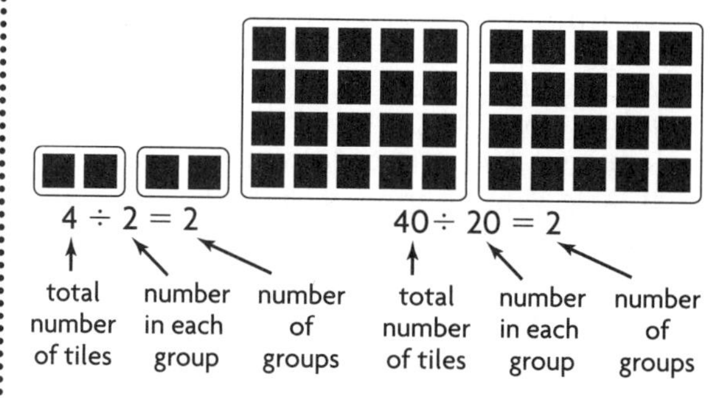

Focus on the second and third steps. Use the example to show students that when they multiply a multiple of 10 by 10, and divide by the same multiple of 10, the next quotient in the pattern will have one more zero.

$$40 \div 20 = 2$$

$$400 \div 20 = 20$$

Repeat the activity with 25,000 ÷ 50, 2,800 ÷ 40, and 630,000 ÷ 90.

Name ______________________ Skill ______________________

Grade 5 **Skill 26**

Division Patterns (Divide by Multiples of 10)

Some multiples of 10:
$1 \times 10 = 10$ $2 \times 10 = 20$ $3 \times 10 = 30$

When you divide by multiples of 10, use a pattern to find the quotient.

Find $6{,}000 \div 30$.

If you know the basic fact, $6 \div 3 = 2$, then you can use a pattern to find $6{,}000 \div 30$.

Think: $6 \div 3 = 2$
So, $60 \div 30 = 2$
$600 \div 30 = 20$
$\mathbf{6{,}000 \div 30 = 200}$

So, $6{,}000 \div 30 = 200$.

Find $240{,}000 \div 60$.

Use the basic fact $24 \div 6 = 4$.
Write a pattern.

$24 \div 6 = 4$
$240 \div 60 = 4$
$2{,}400 \div 60 = 40$
$24{,}000 \div 60 = 400$
$\mathbf{240{,}000 \div 60 = 4{,}000}$

Notice that each quotient has 1 more zero than the quotient before it.

So, $240{,}000 \div 60 = 4{,}000$.

Try These

Complete.

1.
$20 \div 5 =$ ______
$200 \div 50 =$ ______
$2{,}000 \div 50 =$ ______
$20{,}000 \div 50 =$ ______

2.
$35 \div 7 =$ ______
$350 \div 70 =$ ______
$3{,}500 \div 70 =$ ______
$35{,}000 \div 70 =$ ______

3.
$36 \div 4 =$ ______
$360 \div 40 =$ ______
$3{,}600 \div 40 =$ ______
$36{,}000 \div 40 =$ ______
$360{,}000 \div 40 =$ ______

4.
$45 \div 9 =$ ______
$450 \div 90 =$ ______
$4{,}500 \div 90 =$ ______
$45{,}000 \div 90 =$ ______
$450{,}000 \div 90 =$ ______

Go to the next side.

Name ____________________ Skill ____________________

Practice on Your Own

Skill 26

Find 250,000 ÷ 50.

To divide by multiples of 10, use a pattern:

250 ÷ 50 = 5

2,500 ÷ 50 = 50 **Think:** 25 ÷ 5 = 5

25,000 ÷ 50 = 500

250,000 ÷ 50 = 5,000

Each quotient has 1 more zero than the quotient before it.

So, 250,000 ÷ 50 = 5,000.

Complete.

1. 27 ÷ 3 = ______
   270 ÷ 30 = ______
   2,700 ÷ 30 = ______
   27,000 ÷ 30 = ______

2. 42 ÷ 6 = ______
   420 ÷ 60 = ______
   4,200 ÷ 60 = ______
   42,000 ÷ 60 = ______

3. 56 ÷ 8 = ______
   560 ÷ 80 = ______
   5,600 ÷ 80 = ______
   56,000 ÷ 80 = ______
   560,000 ÷ 80 = ______

4. 50)400 50)40,000

5. 70)350 70)3,500

6. 40)320 40)3,200

7. 60)480 60)4,800

8. 30)150 30)1,500

9. 50)350 50)3,500

10. 2,100 ÷ 70 = ______

11. 40)28,000

12. 450,000 ÷ 90 = ______

Check

Find the quotient.

13. 60)3,600

14. 80)48,000

15. 72,000 ÷ 90 = ______

16. 490,000 ÷ 70 = ______

Skill 27 Grade 5

Divide by 2-Digit Numbers

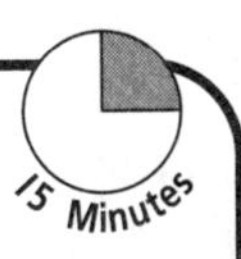

Using Skill 27

OBJECTIVE Divide up to a 4-digit number by a 2-digit number

Display the following pairs of numbers: 50 and 250; 70 and 2,100. Point to each pair.

Ask: **Do these numbers divide evenly?** (yes) Verify the student answers by completing each division. Point out the basic fact embedded in each pair.

Read about compatible numbers at the top of the page. Then work through Step 1. Point out that the compatible numbers divide evenly and are close to the actual numbers. Students should see that the actual dividend, 2,226, is between 2,000 and 2,400—both multiples of 40. So, the estimated quotient is between 50 and 60.

Ask: **How do you know that the first digit of the quotient will be in the tens place?** (The estimated quotient is between 50 and 60, or between 5 *tens* and 6 *tens*.)

Lead students through the actual division in Steps 2 and 3.

Ask: **What is the actual quotient?** (53) **Is it close to the estimate?** (Yes, 53 is between 50 and 60.)

Encourage students to compare the actual quotient with the estimated quotient.

TRY THESE Exercises 1–2 provide practice in estimating quotients using compatible numbers, and dividing.

- **Exercise 1** Divide hundreds by tens.
- **Exercise 2** Divide thousands by tens.

PRACTICE ON YOUR OWN Review the example at the top of the page. Point out that the actual dividend, 1,428, is between 1,200 and 1,500—both multiples of 30. So, the estimated quotient is between 40 and 50.

CHECK Determine if students can correctly divide by 2-digit numbers. Success is indicated by 3 out of 4 correct responses.

Students who successfully complete the **Practice on Your Own** and **Check** are ready to move to the next skill.

COMMON ERRORS

- Students' estimates may be too low or too high causing them to misplace the first digit of the quotient.
- Students may forget to subtract and/or bring down, causing them to incorrectly divide the ones.
- Students may not compare the actual quotient with the estimate to make sure the quotient makes sense.

Students who made more than 2 errors in the **Practice on Your Own**, or who were not successful in the **Check** section, may benefit from the **Alternative Teaching Strategy** on the next page.

Optional

Alternative Teaching Strategy

Use a Multiplication Table

20 Minutes

OBJECTIVE Use a multiplication table to identify compatible numbers.

MATERIALS completed multiplication table

Recalling multiplication facts is important in recognizing compatible numbers. Have students practice recalling some facts, using a completed multiplication table if needed.

Students may need practice recognizing compatible numbers. Give them the following pairs of numbers and have them state compatible numbers. Practice ones and tens together first. For example:

6, 22 (6, 24 or 7, 21) and 3, 13 (3 and 12)

Any students having trouble recalling facts may find it helpful to locate numbers close to 22 and 13 in the multiplication table and then find the factors.

Next, practice with tens and hundreds together, such as 18 and 815 (20 and 800).

For the following pairs of tens and hundreds, have students state two pairs of compatible numbers.

54 and 489 (50 and 500, or 50 and 450)

37 and 171 (30 and 150, or 40 and 200)

Ask: **When you divide, why is it helpful to find compatible numbers first?** (to estimate the quotient)

Why is it helpful to estimate the quotient? (to know where to write the first digit of the actual quotient)

Have students use compatible numbers to find an estimated quotient for 3,276 ÷ 52. (3,000 ÷ 50 = 60 and 3,500 ÷ 50 = 70, so the quotient is about 60 or 70.)

Ask: **Where do you place the first digit of the quotient?** (in the tens place)

Then lead students through the rest of the steps for dividing:

$$\begin{array}{r} 63 \\ 52\overline{)3276} \\ \underline{312} \\ 156 \\ \underline{156} \\ 0 \end{array}$$

312 → $\underline{6} \times \underline{52} = \underline{312}$

156 → $\underline{3} \times \underline{52} = \underline{156}$

Emphasize comparing the estimated quotient and the actual quotient to be sure the actual quotient is reasonable.

Name ______________________ Skill ______________________

Grade 5 Skill 27

Divide by 2-Digit Numbers

Find 2,226 ÷ 42.

Use compatible numbers to help you estimate a quotient. Compatible numbers are numbers that are easy to compute mentally.

Step 1 Estimate the quotient. Decide where to place the first digit in the quotient.

42)2,226

Think: 2,000 and 40 are compatible numbers. 2,400 and 40 are compatible numbers.

40)2,000 = 50 or 40)2,400 = 60

The quotient is probably between 50 and 60. The first digit of the quotient will be in the tens place.

Step 2 Divide 222 tens. Multiply. Subtract. Bring down the ones.

```
   Th H T O
          5
42 )2, 2 2 6
    −2 1 0 ↓
       1 2 6
```

5 × 42 = 210
222 − 210 = 12

Step 3 Divide 126 ones. Multiply. Subtract.

```
   Th H T O
         5 3
42 )2, 2 2 6
    − 2 1 0 ↓
        1 2 6
      − 1 2 6
            0
```

3 × 42 = 126
126 − 126 = 0

So, 2,226 ÷ 42 = 53.

Try These

Complete.

1. 672 ÷ 21 =

```
    H T O
21 )6 7 2
  −______
   −_____
```

Compatible numbers: ________
Estimated quotient: ________
Write the first digit of the quotient in the ________ place.

2. 1,749 ÷ 53 =

```
    Th H T O
53 )1, 7 4 9
  −_________
      −_____
```

Compatible numbers: ________
Estimated quotient: ________
Write the first digit of the quotient in the ________ place.

Go to the next side.

Name ______________________ Skill ______________________

Practice on Your Own

Skill 27

Find 1,428 ÷ 28 = □

Think: 1,200 and 30 are compatible numbers. 1,500 and 30 are compatible numbers.

The quotient will probably be between 40 and 50.

```
   Th H T O
         5 1
  28 )1, 4 2 8
     - 1 4 0↓
         2 8
       - 2 8
           0
```

So, 1, 428 ÷ 28 = 51.

Complete.

1. 27)945 (H T O) Compatible numbers: ________ Estimated quotient: ________

2. 53)4,558 (Th H T O) Compatible numbers: ________ Estimated quotient: ________

3. 48)624 Compatible numbers: ________ Estimated quotient: ________

4. 51)1,887 Compatible numbers: ________ Estimated quotient: ________

5. 21)966

6. 38)722

7. 42)1,512

8. 57)4,104

Check

Complete.

9. 23)897

10. 42)714

11. 48)3,024

12. 81)4,131

Related Facts

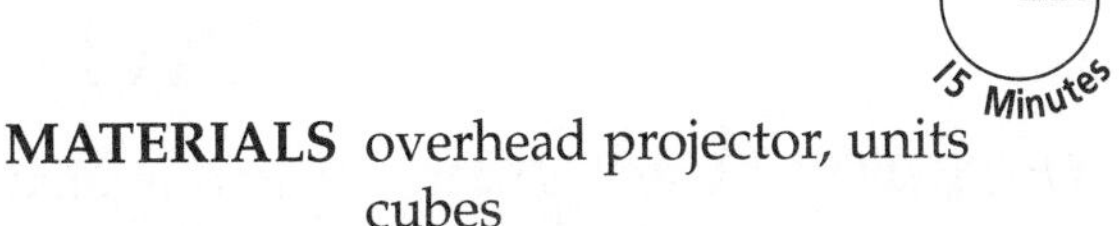

Using Skill 28

OBJECTIVE Use the relationship between multiplication and division to find a related fact

MATERIALS overhead projector, units cubes

Draw attention to the first example. Say: **The array shows the multiplication fact, $3 \times 4 = 12$.**

Ask: **How many rows are there in the array?** (3) **How many columns are there?** (4)

Refer to the second part. Say: **When the array is divided into 4 columns, how many rows are in each column?** (3)

Discuss how the array can show both multiplication and division. Note that since multiplication and division are opposite, or *inverse* operations, one undoes the other.

You may wish to demonstrate on an overhead projector how division undoes multiplication. Arrange 12 cubes in a 3 x 4 array and have students give the multiplication sentence. Then push apart the 4 columns of 3 and describe the result. Say: **I divided the 12 cubes into 4 equal groups. Each group has 3 cubes, so $12 \div 4 = 3$.**

Push the cubes together again. Note that there are now 3 rows of 4 and 3×4 is 12. Observe how, at the start, the cubes represented multiplication; when you divided them into equal groups you "undid" multiplication using division. When you pushed them back together, you "undid" division using multiplication.

Continue with the second example. Connect the array with each of the multiplication facts. Ask the students to tell how many rows and how many columns there are in each array.

TRY THESE Exercises 1–2 provide practice using the concept that multiplication and division are *inverse* operations and that division undoes multiplication. Students are to write the multiplication fact for the array; then they are to write the related division fact.

- **Exercise 1** 4×5 array.
- **Exercise 2** 6×3 array.

PRACTICE ON YOUR OWN Review the array patterns at the top of the page. Ask the students to tell how many rows and how many columns are in each.

CHECK Determine if students can write a related division fact when the multiplication fact is given. Since there are two related division facts, accept either one. Success is indicated by 2 out of 3 correct responses.

Students who successfully complete the **Practice on Your Own** and **Check** are ready to move to the next skill.

COMMON ERRORS

- Students may not understand the inverse relationship between multiplication and division.
- Students may use an incorrect multiplication fact to find a related division fact or vice versa.

Students who made more than 2 errors in the **Practice on Your Own**, or who were not successful in the **Check** section, may benefit from the **Alternative Teaching Strategy** on the next page.

Optional

Alternative Teaching Strategy

20 Minutes

Triangle Flash Cards for Multiplication and Division Facts

OBJECTIVE Use flash cards to find related multiplication and division facts

MATERIALS 3-inch equilateral triangles cut from construction paper or index cards, fine-point markers

Students work in pairs in this activity. Partners write the numbers for fact families on the triangle cards. The product is written in one corner of the triangle and the factors are written in the other two corners.

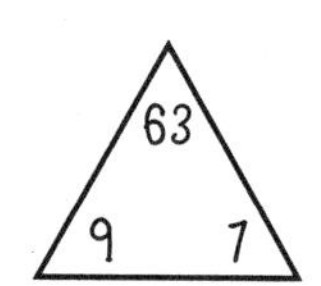

Write the multiplication and division facts.

$9 \times 7 = 63$ $\quad$ $63 \div 7 = 9$

$7 \times 9 = 63$ $\quad$ $63 \div 9 = 7$

List the numbers for different fact families on the board. You may wish to start with the most often missed facts, such as 7, 8, 56; 9, 6, 54; 9, 7, 63; 6, 7, 42; 6, 8, 48 and 9, 8, 72.

After the pairs have made the triangular flashcards, have the students take turns asking each other to give the related multiplication and division facts.

Name ______________________ Skill ______________________

Grade 5 Skill 28

Related Facts

Multiplication and division are **inverse**, or opposite, operations.
Multiplication undoes division. Division undoes multiplication.

Use models to show related facts.

- Use an array to show:
  $3 \times 4 = 12$

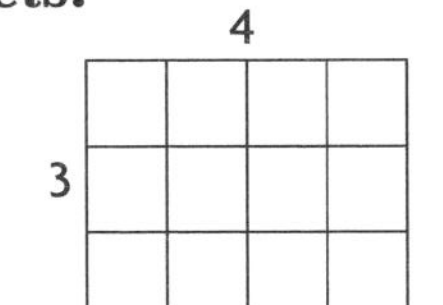

- Divide the array into 4 columns to show:
  $12 \div 4 = 3$

$12 \div 4 = 3$

So $3 \times 4 = 12$ and $12 \div 4 = 3$ are related facts.

Write fact families to show the related facts for 3, 5, and 15.
If you know a multiplication fact, you can write a fact family.

multiplication

$3 \times 5 = 15$
$5 \times 3 = 15$

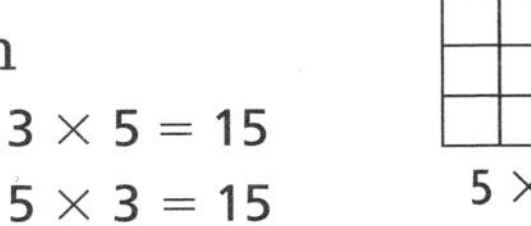

$5 \times 3 = 15$

$3 \times 5 = 15$

division

$15 \div 5 = 3$
$15 \div 3 = 5$

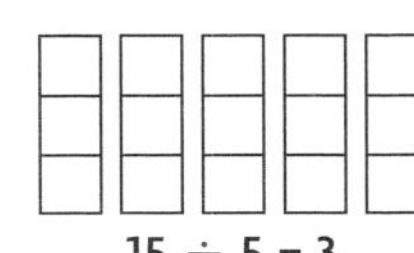

$15 \div 5 = 3$

$15 \div 3 = 5$

Try These

Write two facts for each array.

1. ____ $\times$ ____ $= 20$

   $20 \div$ ____ $=$ ____

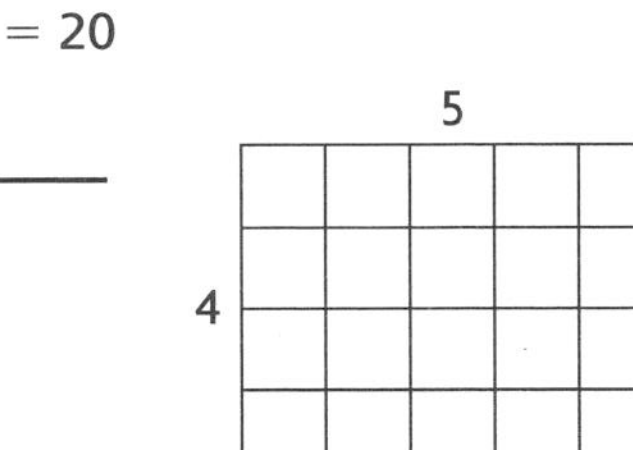
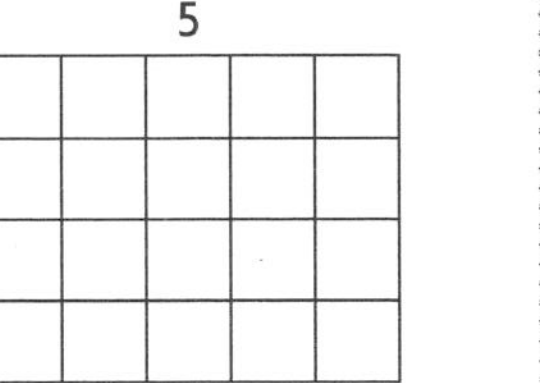

2. ____ $\times$ ____ $= 18$

   $18 \div$ ____ $=$ ____

3

6

Go to the next side.

Name ______________________ Skill ______________________

Practice on Your Own

Skill 28

This fact family shows the related facts for 4, 7, and 28.

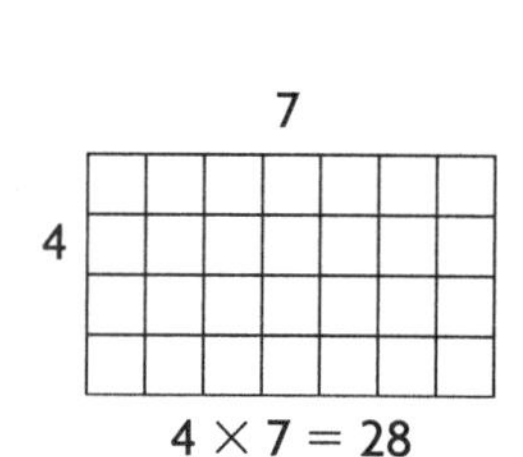

$4 \times 7 = 28$

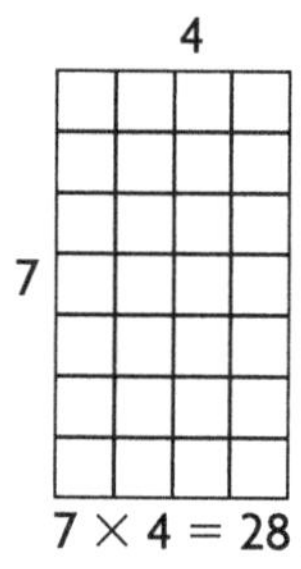

$7 \times 4 = 28$

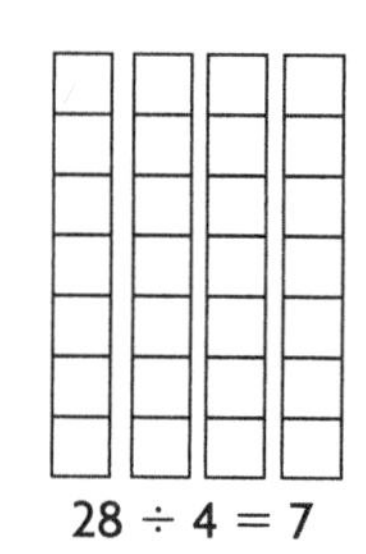

$28 \div 4 = 7$

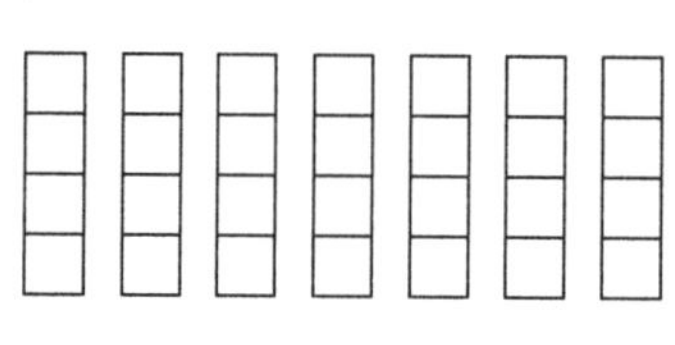

$28 \div 7 = 4$

Write the fact family for each array.

1

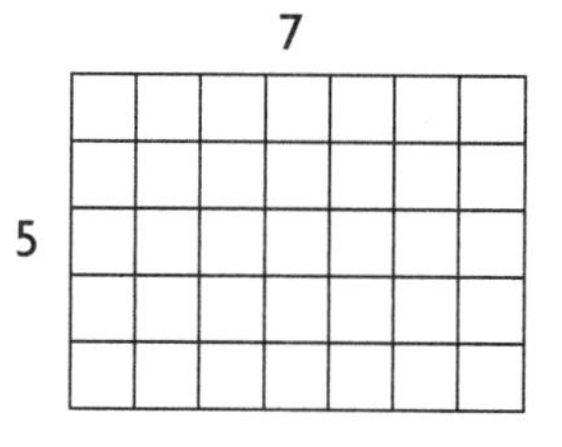

____ × ____ = 35

____ × ____ = 35

35 ÷ ____ = ____

35 ÷ ____ = ____

2

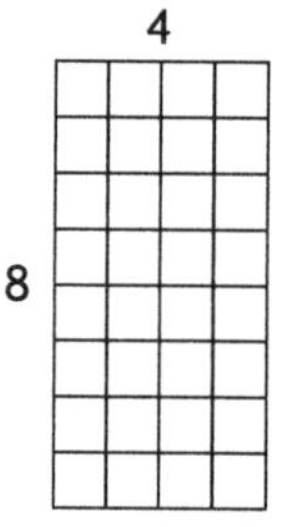

____ × ____ = 32

____ × ____ = 32

32 ÷ ____ = ____

32 ÷ ____ = ____

3

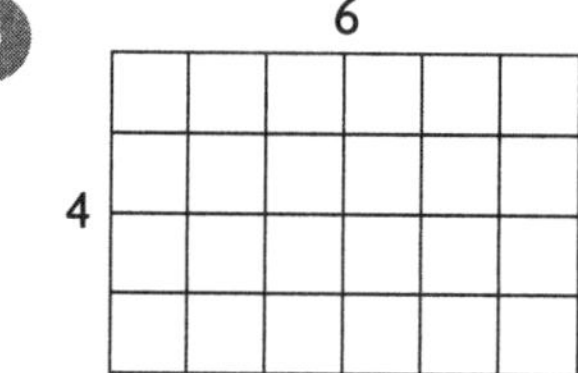

____ × ____ = 24

____ × ____ = 24

24 ÷ ____ = ____

24 ÷ ____ = ____

Complete.

4 $9 \times 6 =$ ____
$54 \div 6 =$ ____

5 $7 \times 6 =$ ____
$42 \div 6 =$ ____

6 $8 \times 5 =$ ____
$40 \div 5 =$ ____

Write a related division fact for each multiplication fact.

7 $5 \times 9 = 45$
____

8 $7 \times 3 = 21$
____

9 $10 \times 6 = 60$
____

10 $5 \times 10 = 50$
____

11 $4 \times 9 = 36$
____

12 $7 \times 8 = 56$
____

Check

Write a related division fact for each multiplication fact.

13 $8 \times 2 = 16$
____

14 $7 \times 9 = 63$
____

15 $6 \times 8 = 48$
____

Skill 29 Grade 5

Division Facts 1–6

Using Skill 29

OBJECTIVE Use strategies to recall division facts

To introduce the *Dividing with Zero and One* section, you may wish to start with the related multiplication fact or 0 and 1 as factors. Display these factors.

Ask: **What is the product when 1 is a factor?** (The product is the other factor.) **What is the product when 0 is a factor?** (The product is zero.)

As you discuss the first part of this section, ask: **What are the related facts for the 6, 1, 6 fact family?** ($6 \times 1 = 6$, $1 \times 6 = 6$, $6 \div 1 = 6$, and $6 \div 6 = 1$)

When discussing the properties of zero in division, ask: **What number times 5 equals 0?** (0)

In the *Use Multiplication Facts* section review the concept that division is the inverse of multiplication, that is, division undoes multiplication. Encourage students to think of a related multiplication fact when they do not know a division fact.

TRY THESE Exercises 1–4 provide practice on the strategies introduced.

- **Exercise 1** Divide with zero and one.
- **Exercises 2–4** Use multiplication facts to divide.

PRACTICE ON YOUR OWN Review the examples at the top of the page. In Exercises 1–2, students divide with zero and one. In Exercises 3–12, students use multiplication facts. In Exercises 13–16, students find the quotient for each division fact.

CHECK Determine if students can find the quotient for each exercise. Students should use the strategies taught.

Success is indicated by a 3 out of 4 correct responses.

Students who successfully complete the **Practice on Your Own** and **Check** are ready to move to the next skill.

COMMON ERRORS

- Students may confuse the role of zero in addition and subtraction with the role of zero in multiplication and division.
- Students may add or subtract instead of divide.
- Students may be unable to apply strategies for recalling basic division facts.

Students who made more than 2 errors in the **Practice on Your Own**, or who were not successful in the **Check** section, may benefit from the **Alternative Teaching Strategy** on the next page.

Optional

Alternative Teaching Strategy

15 Minutes

Oral Practice for Division Facts 1–6

OBJECTIVE Memorize and recall the division facts 1–6

MATERIALS craft stick and fine-point magic markers

Students who have difficulty with division basic facts usually either do not understand the concept of division or have had trouble memorizing the basic facts. After activities have been done to help students understand division, then have the student begin working on memorizing the facts. Students need ample repetition to memorize the basic facts. The practice should be varied daily, and immediate feedback should be provided. These activities can be for a short duration, such as 10 minutes daily.

Students work in pairs in this activity.

Have students prepare the materials to be used by writing just the numbers for each division fact on a craft stick as shown.

12 6 2

Moving from left to right, the dividend is written first, then the divisor. The last number is the quotient. This same division fact stick can be used for the $12 \div 2 = 6$ division fact. Have the students use the related multiplication facts to complete the fact sticks.

Next, students use the completed fact sticks to practice with each other so that they can memorize the division facts.

One partner shows the fact sticks one at a time with the quotient covered. The other student provides the quotient for each. The partners take turns.

Repeat the activity having students cover the middle number for each fact.

Name ____________________ Skill ____________________

Grade 5 **Skill 29**

Division Facts 1–6

Use strategies to recall division facts.

Dividing with Zero and One
Any number divided by 1 is that number.

$6 \div 1 = 6$

Any number divided by itself is 1.

$4 \div 4 = 1$

Zero divided by any number is 0.

$0 \div 5 = 0$

Division by 0 is not possible.

~~2 ÷ 0~~

Use multiplication facts.
Division is the *inverse* of multiplication.

$2 \times 3 = 6$ $6 \div 3 = 2$

6 = **dividend**, 3 = **divisor**, 2 = **quotient**

$12 \div 3 = \square$

Think: $3 \times \square = 12$

$3 \times 4 = 12$

So, $12 \div 3 = 4$.

Try These

Find the quotient.

1. $6 \div 1 =$ ____
   $6 \div 6 =$ ____
   $0 \div 4 =$ ____
   $0 \div 3 =$ ____

2. $6 \times 2 = 12$
   $12 \div 2 =$ ____
   $5 \times 3 = 15$
   $15 \div 3 =$ ____

3. $4 \times 3 = 12$
   $12 \div 3 =$ ____
   $6 \times 4 = 24$
   $24 \div 4 =$ ____

4. $4 \times 5 = 20$
   $20 \div 5 =$ ____
   $7 \times 6 = 42$
   $42 \div 6 =$ ____

Go to the next side.

Name ____________________ Skill ____________________

Practice on Your Own

Skill 29

Divide with Zero and One

$4 \div 1 = 4$ Any number divided by 1 is that number.

$6 \div 6 = 1$ Any number divided by itself is 1.

$0 \div 9 = 0$ 0 divided by any number is 0.

~~$8 \div 0$~~ = Division by 0 is not possible.

Use Multiplication Facts

Multiplication and division are inverse operations.

Find $18 \div 3 = \square$

So, $18 \div 3 = 6$

Think: $3 \times \square = 18$.

Find the quotient.

1. $10 \div 1 =$ ____
   $10 \div 10 =$ ____
2. $0 \div 6 =$ ____
   $0 \div 5 =$ ____
3. $3 \times 5 = 15$
   $15 \div 3 =$ ____
4. $4 \times 2 = 8$
   $8 \div 4 =$ ____

Multiply. Then divide.

5. $2 \times 8 =$ ____
   $16 \div 2 =$ ____
6. $5 \times 7 =$ ____
   $35 \div 5 =$ ____
7. $3 \times 9 =$ ____
   $27 \div 3 =$ ____
8. $4 \times 5 =$ ____
   $20 \div 4 =$ ____
9. $4 \times 9 =$ ____
   $36 \div 4 =$ ____
10. $6 \times 7 =$ ____
    $42 \div 6 =$ ____
11. $5 \times 8 =$ ____
    $40 \div 5 =$ ____
12. $3 \times 8 =$ ____
    $24 \div 3 =$ ____

Divide.

13. $36 \div 6 =$ ____
14. $32 \div 4 =$ ____
15. $7 \div 1 =$ ____
16. $18 \div 3 =$ ____

Check

Divide.

17. $0 \div 5 =$ ____
18. $18 \div 2 =$ ____
19. $45 \div 5 =$ ____
20. $30 \div 6 =$ ____

Skill 30 Grade 5 — Division Facts 7–10

Using Skill 30

OBJECTIVE Use strategies to recall division facts

Begin the lesson by reminding students that, just as they can use strategies to help them recall multiplication facts, they can use strategies to help recall division facts.

Direct students' attention to Skill 30. Review the words used in division.

Ask: **What do you call the number that is divided?** (dividend) **What do you call the number you divide by?** (divisor) **What do you call the result?** (quotient)

Focus on each strategy. For *Use Multiplication Facts,* say: **9 times what number is 63?** Write $9 \times \square = 63$. Have students give the missing factor.

For *Use a Model*, help students draw a 4×8 array of circles. Then relate the total number of circles, 32, to the dividend, the number in each row to the divisor, and the number of rows to the quotient.

For *Use Patterns*, help students complete the chart through $90 \div 10$ to provide more practice. Then have students underline the quotients and the tens digits in the dividends.

TRY THESE Exercises 1–3 provide practice using strategies for division.

- **Exercise 1** Use multiplication facts.
- **Exercise 2** Use a model.
- **Exercise 3** Use patterns.

PRACTICE ON YOUR OWN Focus on the examples at the top of the page. Be sure students understand when the various strategies can be helpful.

CHECK Determine if students can recall division facts.

Success is determined by 4 out of 4 correct responses.

Students who successfully complete **Practice on Your Own** and **Check** are ready to move to the next skill.

COMMON ERRORS

- When drawing a model, students may confuse the number of rows with the number of tiles in a row.
- When using multiplication facts they know, students may recall facts incorrectly.

Students who made more than 4 errors in the **Practice on Your Own**, or who were not successful in the **Check** section, may benefit from the **Alternative Teaching Strategy** on the next page.

Optional

Alternative Teaching Strategy

20 Minutes

Use Arrays to Model Division Facts 7–10

OBJECTIVE Use arrays to model strategies for recalling division facts 7–10

MATERIALS multiplication tables for students to complete, tiles

Explain to students that they will be using models to help them learn strategies for recalling division facts.

For *Use Multiplication Facts,* distribute multiplication tables and have students complete them.

×	0	1	2	3	4	5	6	7	8	9	10
0	0	0	0	0	0	0	0	0	0	0	0
1	0	1	2	3	4	5	6	7	8	9	10
2	0	2	4	6	8	10	12	14	16	18	20
3	0	3	6	9	12	15	18	21	24	27	30
4	0	4	8	12	16	20	24	28	32		
5	0	5	10	15	20	25					
6	0	6	12	18							
7	0	7	14								
8	0	8									

Call out division facts, such as 54 ÷ 9. Have students point to the related multiplication fact (6 × 9) and say the quotient.

Repeat for all the facts. Then have students find quotients without looking at their tables.

For *Use a Model,* distribute tiles. Ask students to find 36 ÷ 9.

Say: **Ask yourself, *How many 9s are in 36?*** Have them use tiles to model the division.

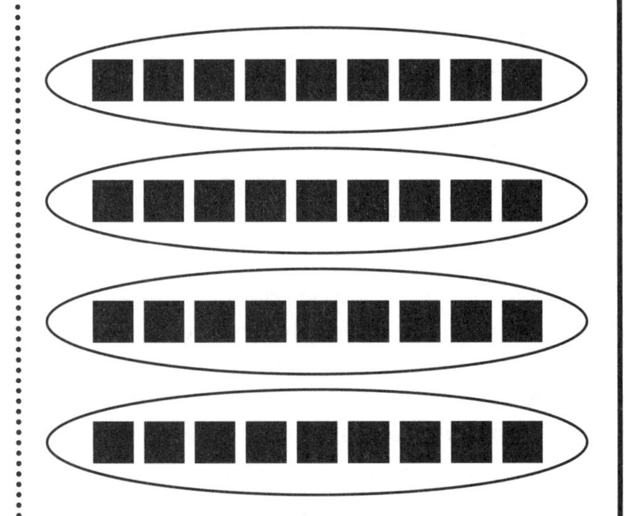

Ask: **What *multiplication fact* does your model show?** (4 × 9)

Lead students to see that this is the related multiplication fact. Repeat the activity several times with other arrays.

Name ______________________ Skill ______________________

Grade 5 **Skill 30**

Division Facts 7–10

Use strategies to recall division facts.

Use Multiplication Facts
Find $63 \div 9$.
Use a multiplication fact you know.

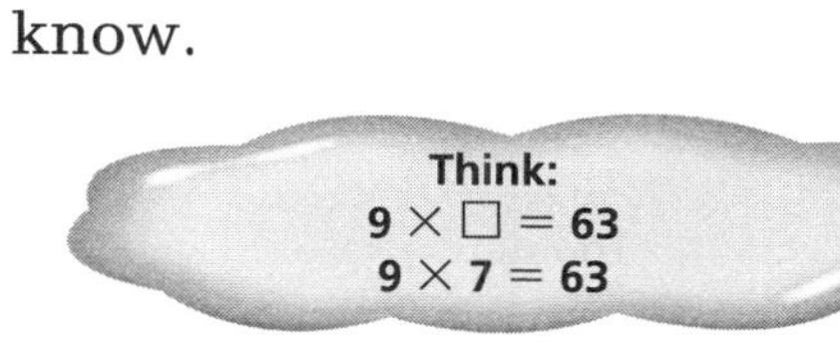

So, $63 \div 9 = 7$.

Use a Model
Find $32 \div 8$.
If you cannot remember the multiplication fact, draw a model.

Think: How many groups of 8 are in 32?

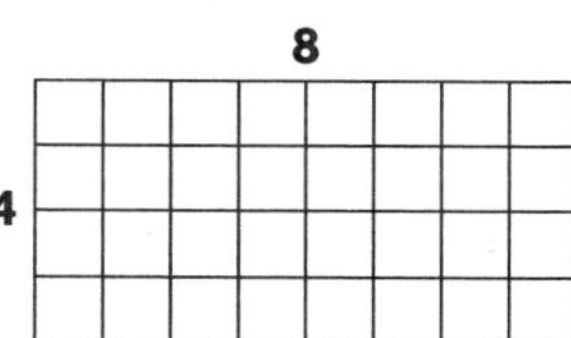

There are four groups of 8 in 32.
So, $32 \div 8 = 4$.

Use Patterns
When you divide by 10, you can see a pattern.

Dividend		Divisor		Quotient
20	÷	10	=	2
30	÷	10	=	3
40	÷	10	=	4
50	÷	10	=	5
60	÷	10	=	6

Notice that for these problems, the quotient is the same as the tens digit of the dividend.

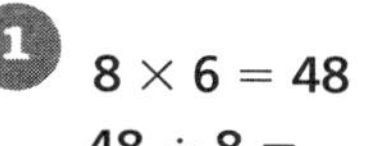

Try These

Find the quotient.

1. $8 \times 6 = 48$
   $48 \div 8 =$ ____

   $8 \times 7 = 56$
   $56 \div 7 =$ ____

2. $54 \div 9 =$ ____

   9
   6

3. $40 \div 10 =$ ____
   $30 \div 10 =$ ____
   $20 \div 10 =$ ____
   $10 \div 10 =$ ____

Go to the next side.

Name ______________________ Skill ______________________

Practice on Your Own

Skill 30

Use a Model

Find $45 \div 9$.

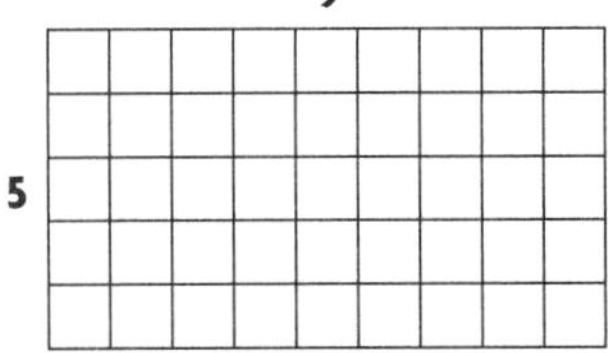

Think: $9 \times \square = 45$.

So, $45 \div 9 = 5$.

Use Multiplication Facts

Multiplication and division are inverse operations. Division undoes multiplication. Find $72 \div 8$.

Think: $8 \times \square = 72$

So, $72 \div 8 = 9$.

Use a Pattern

Dividend		Divisor		Quotient
90	÷	10	=	9
80	÷	10	=	8
70	÷	10	=	7
60	÷	10	=	6

Think: The quotient is the same as the tens digit of the dividend.

Find the quotient.

1. $8 \times 4 = 32$
   $32 \div 8 =$ ____
2. $9 \times 4 = 36$
   $36 \div 9 =$ ____
3. $7 \times 5 = 35$
   $35 \div 7 =$ ____
4. $7 \times 9 = 63$
   $63 \div 7 =$ ____
5. 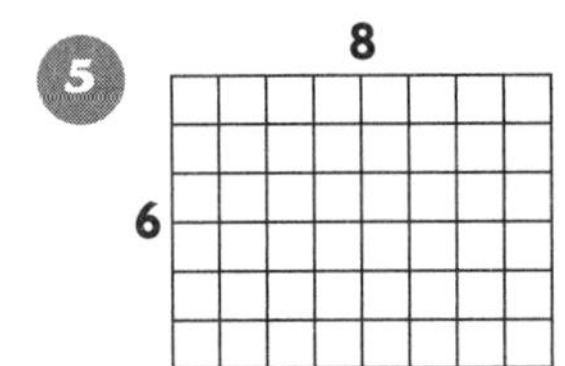

   $48 \div 8 =$ ____
6. 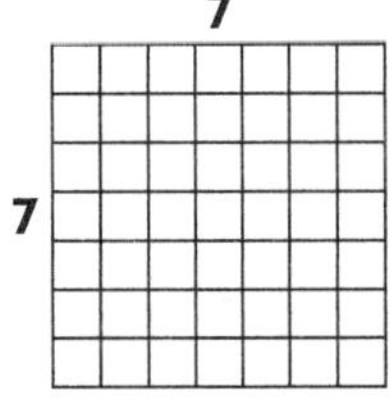

   $49 \div 7 =$ ____
7. $70 \div 10 =$ ____
8. $40 \div 10 =$ ____
9. $20 \div 10 =$ ____
10. $90 \div 10 =$ ____
11. $54 \div 9 =$ ____
12. $56 \div 8 =$ ____
13. $63 \div 9 =$ ____
14. $35 \div 7 =$ ____
15. $24 \div 8 =$ ____
16. $72 \div 9 =$ ____
17. $64 \div 8 =$ ____
18. $28 \div 7 =$ ____

Check

Find the quotient.

19. $50 \div 10 =$ ____
20. $40 \div 8 =$ ____
21. $81 \div 9 =$ ____
22. $42 \div 7 =$ ____

Name ______________________ Skill ______________________

Answer Card

Whole Number Division

Grade 5

TRY THESE

1. 15, 5 r 2
2. 45, 1, 9 r 1
3. 4, 13, 12, 1, 26 r 1
4. 4, 22, 20, 2, 15 r 2

PRACTICE

1. 32, 8 r 2
2. 54, 1, 9 r 1
3. 35, 3, 7 r 3
4. 36, 2, 4 r 2
5. 6, 2, 21, 1
6. 8, 1, 16, 2
7. 5, 2, 20, 4
8. 4, 0, 2, 1
9. 48, 4, 8 r 4
10. 16, 6, 2 r 6
11. 9, 7, 0, 7, 10 r 7
12. 8, 7, 4, 3, 21 r 3
13. 4 r 3
14. 9 r 4
15. 29 r 1
16. 14 r 1

CHECK

17. 8 r 1
18. 7 r 2
19. 11 r 1
20. 10 r 4

TRY THESE

1. 65, 65, 68, 68, =, yes
2. 26, 2, 52, 52, 1, 53, 53, =, yes
3. 134, 7, 938, 938, 2, 940, 940, =, yes

PRACTICE

1. 75, 75, 78, 78, =, yes
2. 90, 90, 91, 91, =, yes
3. 771, 771, 773, 773, =, yes
4. (24 × 4) + 1 = 96 + 1 = 97; yes
5. (396 × 2) + 1 = 792 + 1 = 793; yes
6. (278 × 3) + 1 = 834 + 1 = 835; yes
7. 28 × 3 = 84; 84 + 1 = 85
8. 37 × 2 = 74; 74 + 1 = 75
9. 127 × 7 = 889; 889 + 3 = 892

SKILL 23

CHECK

10. 19 × 5 = 95; 95 + 1 = 96
11. 304 × 3 = 912; 912 + 2 = 914
12. 195 × 4 = 780; 780 + 3 = 783

Name ______________________ Skill ______________________

SKILL 24

TRY THESE

1. 8, 8, 8
2. 4, 4, 4
3. 300
4. 700

PRACTICE

1. 6, 6, 6
2. 3, 3, 3
3. 10, 10, 10
4. 200
5. 500
6. 800
7. 7; facts
8. 42; facts
9. 550; pattern
10. 40; pattern
11. 75; facts
12. 920; pattern

CHECK

13. 4
14. 80
15. 820

SKILL 25

TRY THESE

1. 53
2. 944
3. 114

PRACTICE

1. 47
2. 155
3. 208
4. 59
5. 645
6. 503

CHECK

7. 424
8. 921
9. 701

SKILL 26

TRY THESE

1. 4; 4; 40; 400
2. 5; 5; 50; 500
3. 9; 9; 90; 900; 9,000
4. 5; 5; 50; 500; 5,000

PRACTICE

1. 9; 9; 90; 900
2. 7; 7; 70; 700
3. 7; 7; 70; 700; 7,000
4. 8; 800
5. 5; 50
6. 8; 80
7. 8; 80
8. 5; 50
9. 7; 70
10. 30
11. 700
12. 5,000

CHECK

13. 60
14. 600
15. 800
16. 7,000

Answer Card

Whole Number Division

Grade 5

Name ______________________ Skill ______________________

Skill 27

TRY THESE

1. Compatible numbers: 600 and 20, or 700 and 20; Estimated quotient: 30 or 35; first digit in tens place; 672 ÷ 21 = 32
2. Compatible numbers: 1500 and 50, or 2000 and 50; Estimated quotient: 30 or 40; first digit in tens place; 1749 ÷ 53 = 33

PRACTICE

1. 900 and 30; 30; 945 ÷ 27 = 35
2. 4,500 and 50; 90; 4,558 ÷ 53 = 86
3. 600 and 60; 10; 624 ÷ 48 = 13
4. 1,500 and 50 (or 2,000 and 50); 30 (or 40); 1,887 ÷ 51 = 37
5. 46
6. 19
7. 36
8. 72

CHECK

9. 39
10. 17
11. 63
12. 51

Skill 28

TRY THESE

1. 4, 5; 5, 4
2. 6, 3; 3, 6

PRACTICE

1. 5, 7; 7, 5; 7, 5; 5, 7
2. 8, 4; 4, 8; 4, 8; 8, 4
3. 4, 6; 6, 4; 6, 4; 4, 6
4. 54; 9
5. 42; 7
6. 40; 8
7. 45 ÷ 9 = 5, or 45 ÷ 5 = 9
8. 21 ÷ 3 = 7 or 21 ÷ 7 = 3
9. 60 ÷ 10 = 6 or 60 ÷ 6 = 10
10. 50 ÷ 5 = 10 or 50 ÷ 10 = 5
11. 36 ÷ 4 = 9 or 36 ÷ 9 = 4
12. 56 ÷ 7 = 8 or 56 ÷ 8 = 7

CHECK

13. 16 ÷ 2 = 8 or 16 ÷ 8 = 2
14. 63 ÷ 7 = 9 or 63 ÷ 9 = 7
15. 48 ÷ 6 = 8 or 48 ÷ 8 = 6

Answer Card

Whole Number Division

Grade 5

Name ____________ Skill ____________

Answer Card

Whole Number Division

Grade 5

Skill 29

TRY THESE

1. 6, 1, 0, 0
2. 6, 5
3. 4, 6
4. 4, 7

PRACTICE

1. 10, 1
2. 0, 0
3. 5
4. 2
5. 16; 8
6. 35; 7
7. 27; 9
8. 20; 5
9. 36; 9
10. 42; 7
11. 40; 8
12. 24; 8
13. 6
14. 8
15. 7
16. 6

CHECK

17. 0
18. 9
19. 9
20. 5

Skill 30

TRY THESE

1. 6; 8
2. 6
3. 4; 3; 2; 1

PRACTICE

1. 4
2. 4
3. 5
4. 9
5. 6
6. 7
7. 7
8. 4
9. 2
10. 9
11. 6
12. 7
13. 7
14. 5
15. 3
16. 8
17. 8
18. 4

CHECK

19. 5
20. 5
21. 9
22. 6

Number Sense

Fractions

Skill 31 Grade 5 Understand Fractions

Using Skill 31

OBJECTIVE Write and read the equal parts of a whole as a fraction

Begin by reminding students that a fraction can name a part of a whole or a part of a group. Recall for students that the top number of a fraction is the numerator; it tells how many parts of the whole are being considered. Recall that the bottom number is the denominator; it tells the total number of equal parts.

Direct students' attention to the first example. Point to the square. Ask: **Into how many equal parts is the square divided?** (4) **How many of the equal parts are shaded?** (3)

Point out to students that they can write a fraction to represent the shaded part of the square. Say: **The number of shaded parts is the numerator of the fraction. The total number of equal parts is the denominator.** Ask: **What fraction can you write to represent the shaded part of the square?** ($\frac{3}{4}$)

Tell students that they can read the fraction in different ways. Say:

You can read the fraction as three fourths, or three out of four, or three divided by four.

You may wish to have students read aloud the three ways to read the fraction.

Continue in a similar way with the second example.

TRY THESE In Exercises 1–3 students use models to write a fraction and complete the three ways a fraction can be read.

- **Exercise 1** Four sixths.
- **Exercise 2** Two thirds.
- **Exercise 3** Seven eighths.

PRACTICE ON YOUR OWN Review the example at the top of the page. Have students identify the numerator and denominator of the fraction. Emphasize the three ways a fraction can be read. Exercises 1–9 provide practice writing and reading fractions using prompts.

CHECK Determine if students can write a fraction and provide the three ways the fraction can be read.

Success is indicated by 3 out of 3 correct responses.

Students who successfully complete the **Practice on Your Own** and **Check** are ready to move to the next skill.

COMMON ERRORS

- Students may write the numerator correctly but may write the denominator as the unshaded parts of the model.

Students who made more than 4 errors in the **Practice on Your Own**, or who were not successful in the **Check** section, may benefit from the **Alternative Teaching Strategy** on the next page.

Optional

Alternative Teaching Strategy

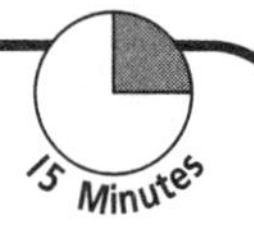

Shade Models to Understand Fractions

OBJECTIVE Shade models to understand fractions

MATERIALS 5 × 8 index cards, each showing figures divided into equal parts, blank 5 x 8 cards, colored markers

Prior to beginning the lesson, prepare 5 × 8 cards showing figures divided into equal parts.

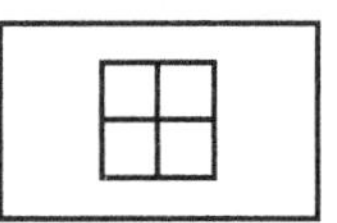 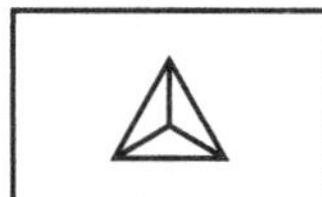 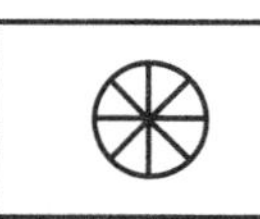

Explain to students that they will shade a figure to represent a fraction. Then, on separate cards they will write the fraction and the three ways the fraction can be read.

Have students work in pairs for this activity. Have one partner shade the figure and write the fraction on a card, while the other partner writes the three ways to read the fraction on three separate cards. There will be a set of 5 cards.

Distribute the card showing a square divided into fourths and 4 blank index cards to each pair of students.

Ask:

How can you show one fourth? (Shade one of the parts.)

Say: **Write the fraction on a separate card.** Check to see that students wrote $\frac{1}{4}$.

Have each partner write the three ways to read the fraction on three separate index cards.

Repeat this activity for each card showing a figure.

When the students are shading the figure, you may wish to encourage them to shade the parts non-consecutively.

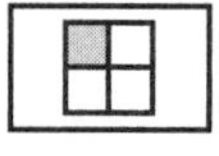 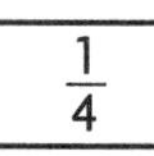

one fourth	one out of four	one divided by four

You may wish to have the students use the cards on another day to match the models to all the correct expressions of their fractions.

Name ______________________ Skill ______________________

Grade 5 **Skill 31**

Understand Fractions

A **fraction** is a number that names a part of a whole.
The top number, the *numerator*, tells how many parts are being used.
The bottom number, the *denominator*, tells the total number of equal parts.

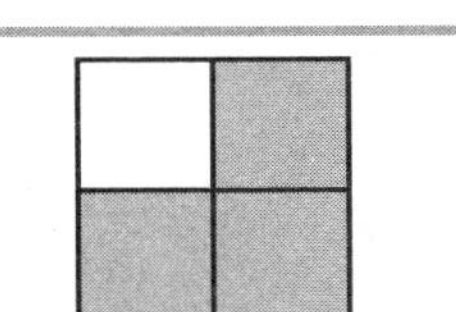

Write: $\frac{3}{4}$

The whole is divided into 4 equal parts.
3 out of **4** parts are shaded.

←equal parts shaded
←total number of equal parts

Read: three fourths
three out of four
three divided by four

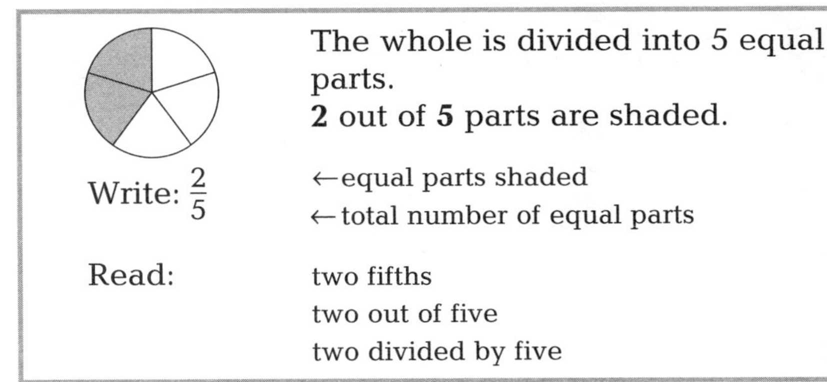

Write: $\frac{2}{5}$

The whole is divided into 5 equal parts.
2 out of **5** parts are shaded.

←equal parts shaded
←total number of equal parts

Read: two fifths
two out of five
two divided by five

Try These

Complete.

1

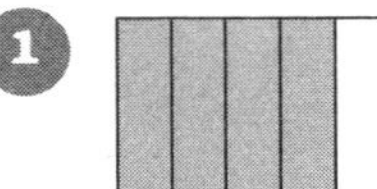

Write:
□ ← equal parts shaded
□ ← total equal parts

Read: ___ sixths
___ out of ___
___ divided by ___

2

Write:
□ ← equal parts shaded
□ ← total equal parts

Read: ___ thirds
___ out of ___
___ divided by ___

3

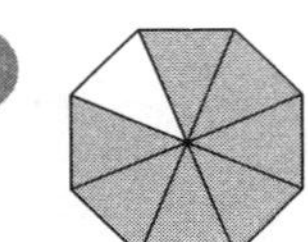

Write:
□ ← equal parts shaded
□ ← total equal parts

Read: ___ eighths
___ out of ___
___ divided by ___

Go to the next side.

Name ______________________ Skill ______________________

Practice on Your Own

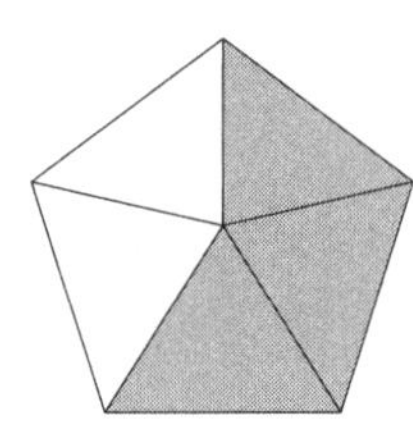

Write: $\frac{3}{5}$ ← equal parts shaded
← total equal parts

The whole is divided into 5 equal parts.
3 out of 5 parts are shaded.

Read: three fifths
three out of five
three divided by five

Complete.

1. Write: $\frac{\square}{\square}$
Read: ____ sevenths
____ out of ____
____ divided by ____

2. Write: $\frac{\square}{\square}$
Read: ____ thirds
____ out of ____
____ divided by ____

3. Write: $\frac{\square}{\square}$
Read: ____ tenths
____ out of ____
____ divided by ____

4. Write: $\frac{\square}{\square}$
Read: ____ ninths

5. Write: $\frac{\square}{\square}$
Read: ____ out of ____

6. Write: $\frac{\square}{\square}$
Read:
____ divided by ____

7. Write: ____
Read: ____ twelfths
____ out of ____
____ divided by ____

8. Write: ____
Read: ____________
____ out of ____
____ divided by ____

9. Write: ____
Read: ____________
____ out of ____
____ divided by ____

Check

Complete.

10. Write: ____
Read: ____________
____ out of ____
____ divided by ____

11. Write: ____
Read: ____________
____ out of ____
____ divided by ____

12. Write: ____
Read: ____________
____ out of ____
____ divided by ____

Skill 32 Grade 5

Compare Fractions

Using Skill 32

OBJECTIVE Compare like and unlike fractions

MATERIALS fraction strips

Distribute fraction strips. Read with students the example for comparing like fractions.

Ask: **What are like fractions?** (fractions with the same denominators)

Have students use the strips to show the fractions $\frac{1}{3}$ and $\frac{2}{3}$. Help them line up the bars so they can compare the parts.

Ask: **How many parts are shaded in $\frac{1}{3}$?** (1) **How many parts are shaded in $\frac{2}{3}$?** (2) **Which is less?** (the fraction with 1 part shaded)

Then have students write the inequality $\frac{1}{3} < \frac{2}{3}$. Check to be sure that they point the inequality symbol in the correct direction. If necessary, practice using the inequality symbol with a few whole numbers, such as $5 < 6$ and $8 > 3$. Remind students that the symbol points to the fraction that "is less than".

Read the example for comparing unlike fractions. Ask: **What are unlike fractions?** (fractions with different denominators)

Have students use strips to show and compare the fractions $\frac{1}{3}$ and $\frac{1}{4}$. Ask: **Which is greater?** ($\frac{1}{3}$)

Some students may notice that when they compared $\frac{1}{3}$ and $\frac{2}{3}$, fractions with like denominators, they compared the numerators: $1 < 2$, so $\frac{1}{3} < \frac{2}{3}$

But, when they compare fractions with unlike denominators they must take into consideration the size of the parts.

TRY THESE Exercises 1–3 model comparing fractions.

- **Exercise 1** Compare fractions with like denominators.
- **Exercises 2–3** Compare fractions with unlike denominators.

PRACTICE ON YOUR OWN Focus on the equivalent fractions at the top of the page. Students should recognize that the amount of shading is equal, but the size of the parts is different.

CHECK Determine if students can compare fractions using the inequality symbol correctly. Success is indicated by 3 out of 3 correct responses.

Students who successfully complete **Practice on Your Own** and **Check** are ready to move on to the next skill.

COMMON ERRORS

- Students may reverse the inequality symbol.
- For equivalent fractions, students may think that the fraction with the greater digits is greater.

Students who made more than 3 errors in the **Practice on Your Own**, or who were not successful in the **Check** section, may benefit from the **Alternative Teaching Strategy** on the next page.

Optional

Alternative Teaching Strategy

Compare Fractions

20 Minutes

OBJECTIVE Use models to compare fractions

MATERIALS counters, congruent paper circles

Give 1 student 1 counter, give another student 3 counters. Ask if they can write a number sentence to compare the numbers of counters. Help them write $1 < 3$.

Repeat with other numbers of counters, and continue to write number sentences using $<$, $>$, and $=$.

Recall with students that number sentences can be written to compare fractions, too.

Model comparing like fractions. Fold a circle into 4 equal parts. Have a volunteer shade 1 part. Ask: **How many equal parts are there in the whole?** (4) **How many parts are shaded?** (1) **What is the fraction for the shaded part?** ($\frac{1}{4}$)

Shade 3 parts of another circle folded into 4 equal parts. Ask: **How many equal parts are there in the whole?** (4) **How many parts are shaded?** (3) **What is the fraction for the shaded part?** ($\frac{3}{4}$)

Write the fractions $\frac{1}{4}$ and $\frac{3}{4}$. Have students look at and compare the shaded parts of the circles. Help students write the number sentence $\frac{1}{4} < \frac{3}{4}$

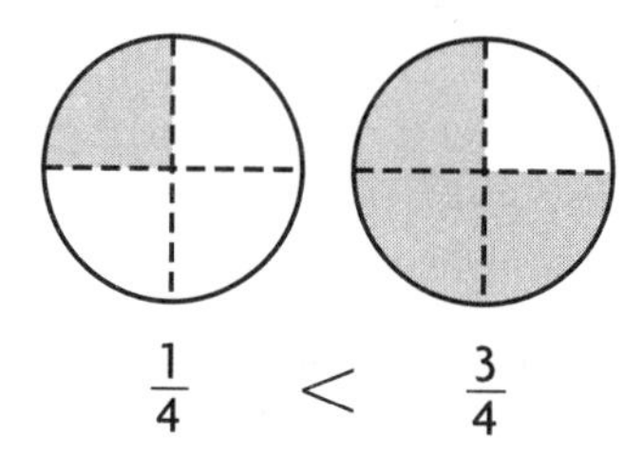

$\frac{1}{4} < \frac{3}{4}$

Repeat the activity for unlike fractions. Have a volunteer shade $\frac{1}{8}$ of a circle folded into 8 parts. Ask:

How many parts are in the whole? (8) **How many parts are shaded?** (1)

Hold up the circle that shows $\frac{1}{4}$ to compare. Ask: **Which is greater?** (Explain that $\frac{1}{4}$ is greater because it is a larger part of the whole—it has more shading than $\frac{1}{8}$.) Help students write $\frac{1}{4} > \frac{1}{8}$.

Repeat for equivalent fractions, such as $\frac{1}{4}$ and $\frac{2}{8}$. Help students write: $\frac{2}{8} = \frac{1}{4}$.

Have students continue to practice using models to compare fractions and to use $<$, $>$, and $=$ in number sentences.

Name ____________________ Skill ____________________

Compare Fractions

Grade 5 **Skill 32**

Use fractions bars to compare fractions. Use > or <.

Remember: > means *is greater than.*
< means *is less than.*

Like Fractions

Compare $\frac{1}{3} \bigcirc \frac{2}{3}$

Think:

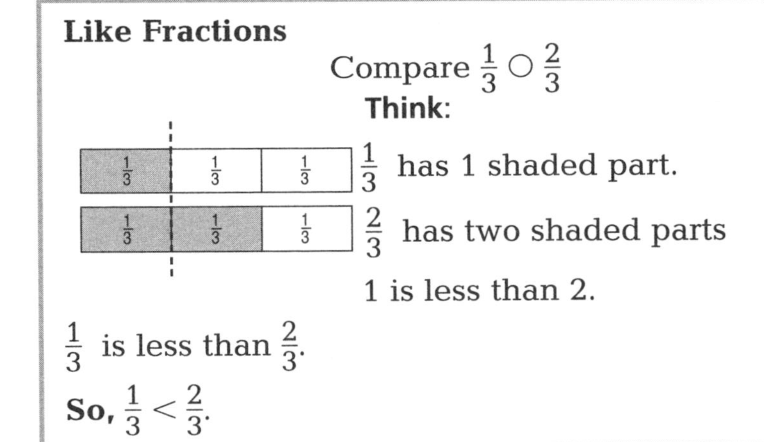

$\frac{1}{3}$ has 1 shaded part.

$\frac{2}{3}$ has two shaded parts

1 is less than 2.

$\frac{1}{3}$ is less than $\frac{2}{3}$.

So, $\frac{1}{3} < \frac{2}{3}$.

Unlike Fractions

Compare $\frac{1}{3} \bigcirc \frac{1}{4}$

Think:

The bar for $\frac{1}{3}$ is larger than the bar for $\frac{1}{4}$.

$\frac{1}{3}$ is greater than $\frac{1}{4}$.

So, $\frac{1}{3} > \frac{1}{4}$.

Try These

Compare. Write *greater than* or *less than*. Then use <, >, or =.

1

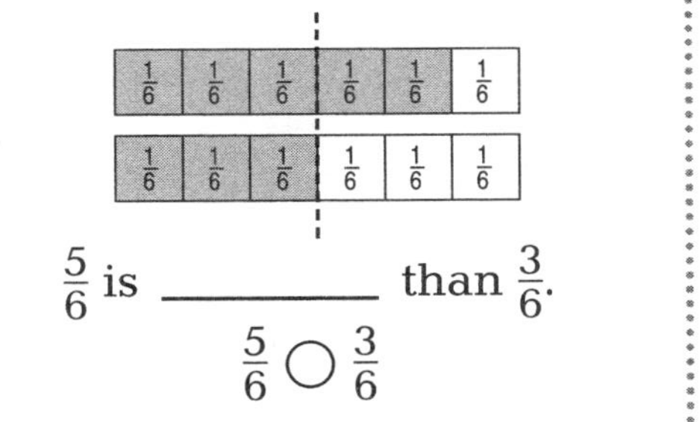

$\frac{5}{6}$ is ______ than $\frac{3}{6}$.

$\frac{5}{6} \bigcirc \frac{3}{6}$

2

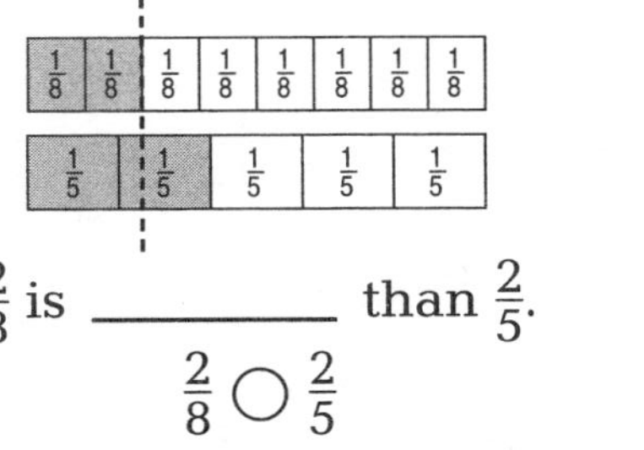

$\frac{2}{8}$ is ______ than $\frac{2}{5}$.

$\frac{2}{8} \bigcirc \frac{2}{5}$

3

$\frac{3}{3}$ is ______ than $\frac{3}{4}$.

$\frac{3}{3} \bigcirc \frac{3}{4}$

Go to the next side.

Name ______________________ Skill ______________________

Practice on Your Own

Skill 32

Remember:
> means *is greater than.*
< means *is less than.*
= means *is equal to.*

Compare $\frac{1}{3} \bigcirc \frac{2}{6}$

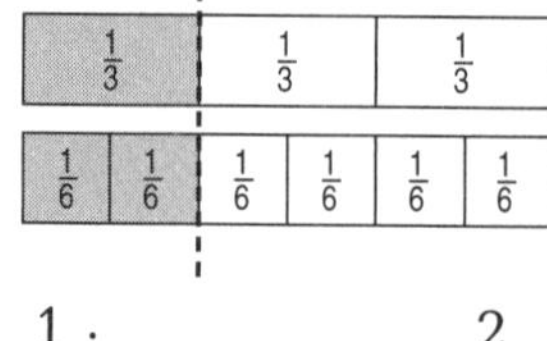

$\frac{1}{3}$ is ________ $\frac{2}{6}$.

So, $\frac{1}{3} = \frac{2}{6}$.

Compare. Write *greater than, less than* or *equal to.* Then use <, >, or =.

1

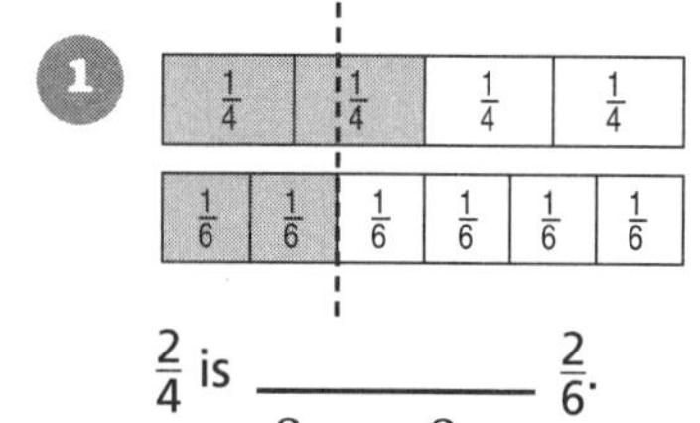

$\frac{2}{4}$ is ________ $\frac{2}{6}$.

So, $\frac{2}{4} \bigcirc \frac{2}{6}$.

2

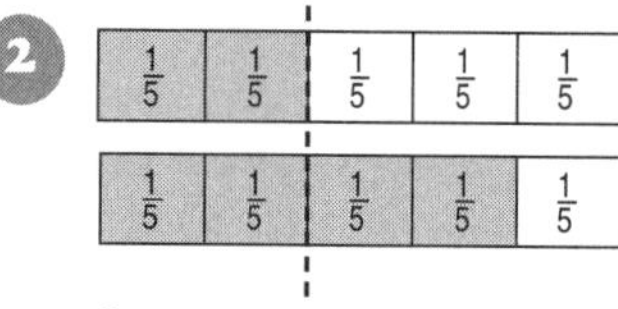

$\frac{2}{5}$ is ________ $\frac{4}{5}$.

So, $\frac{2}{5} \bigcirc \frac{4}{5}$.

3

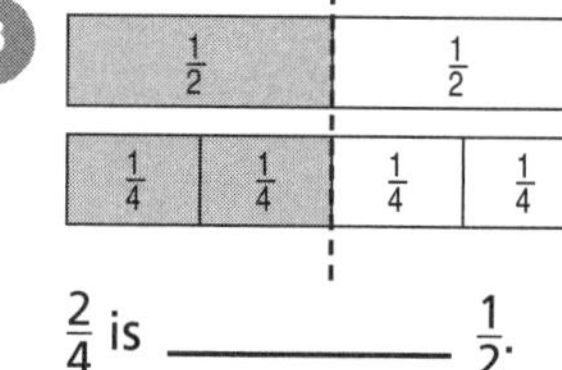

$\frac{2}{4}$ is ________ $\frac{1}{2}$.

So, $\frac{2}{4} \bigcirc \frac{1}{2}$.

4

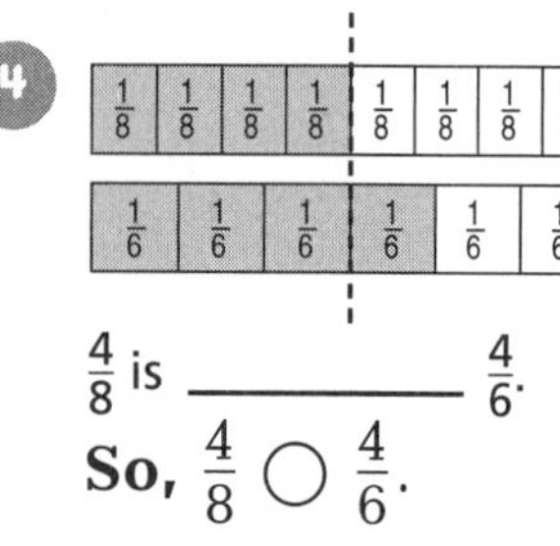

$\frac{4}{8}$ is ________ $\frac{4}{6}$.

So, $\frac{4}{8} \bigcirc \frac{4}{6}$.

5

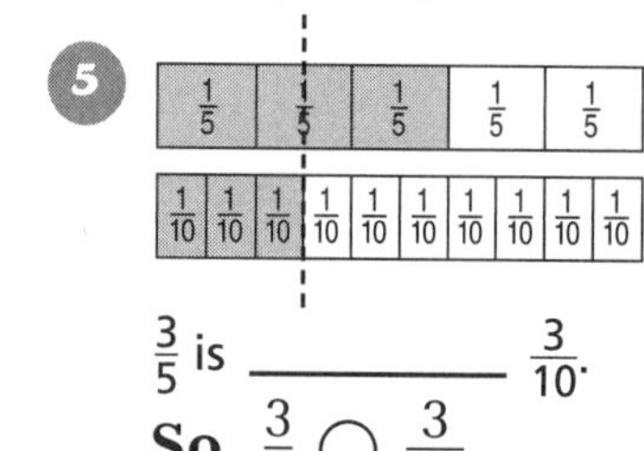

$\frac{3}{5}$ is ________ $\frac{3}{10}$.

So, $\frac{3}{5} \bigcirc \frac{3}{10}$.

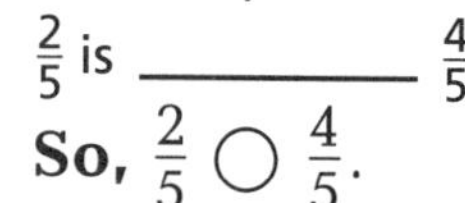

6

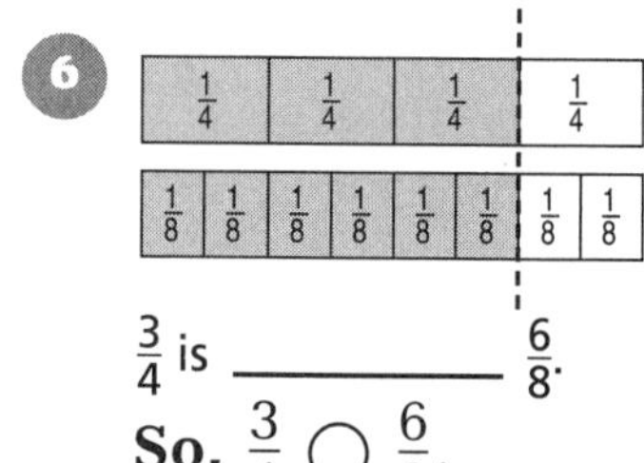

$\frac{3}{4}$ is ________ $\frac{6}{8}$.

So, $\frac{3}{4} \bigcirc \frac{6}{8}$.

Compare fractions. Use <, >, or =.

7

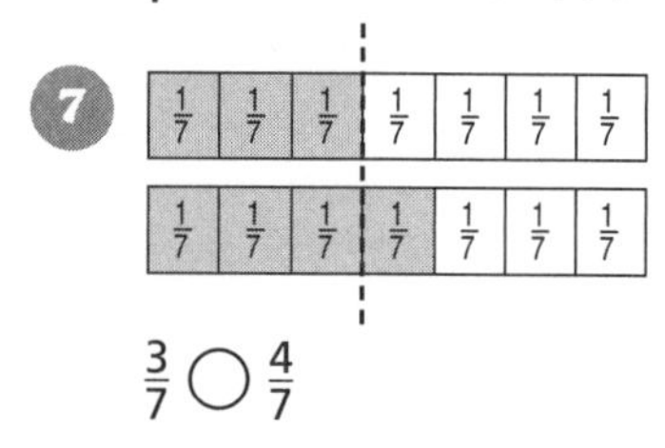

$\frac{3}{7} \bigcirc \frac{4}{7}$

8

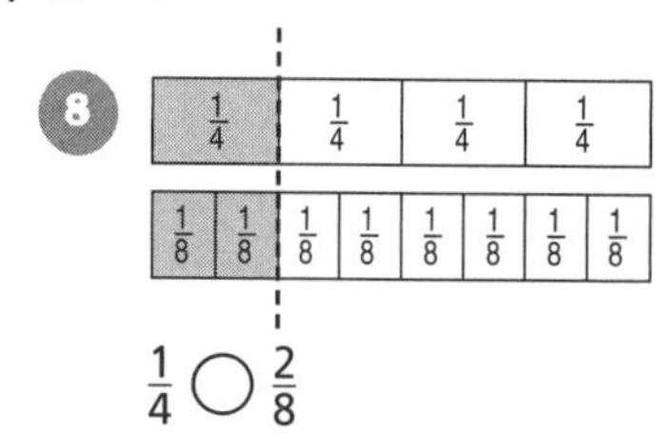

$\frac{1}{4} \bigcirc \frac{2}{8}$

9

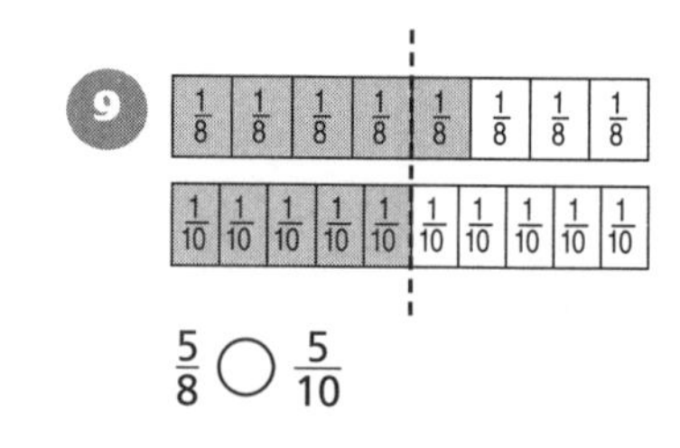

$\frac{5}{8} \bigcirc \frac{5}{10}$

Check

Compare fractions. Use <, >, or =.

10

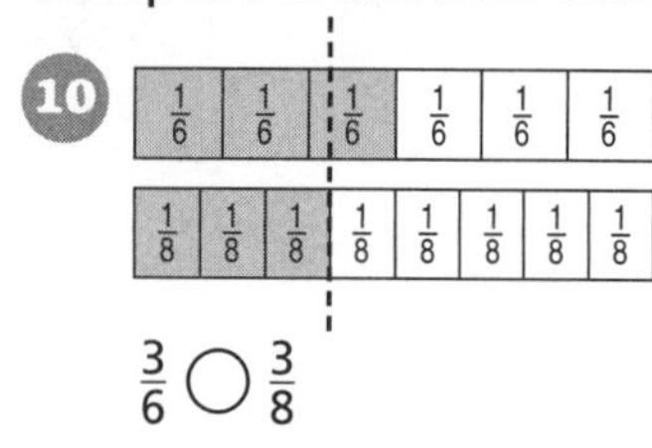

$\frac{3}{6} \bigcirc \frac{3}{8}$

11

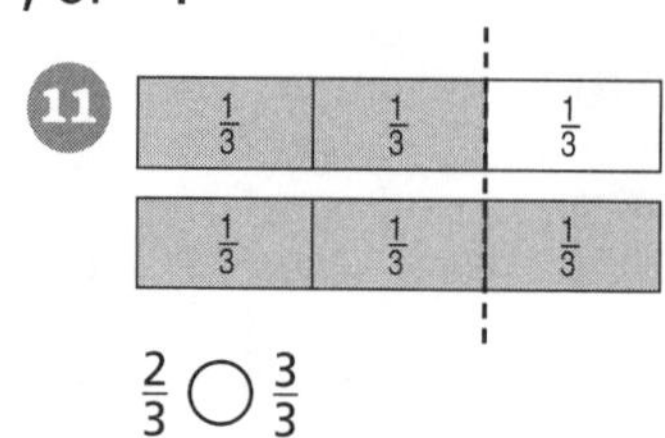

$\frac{2}{3} \bigcirc \frac{3}{3}$

12

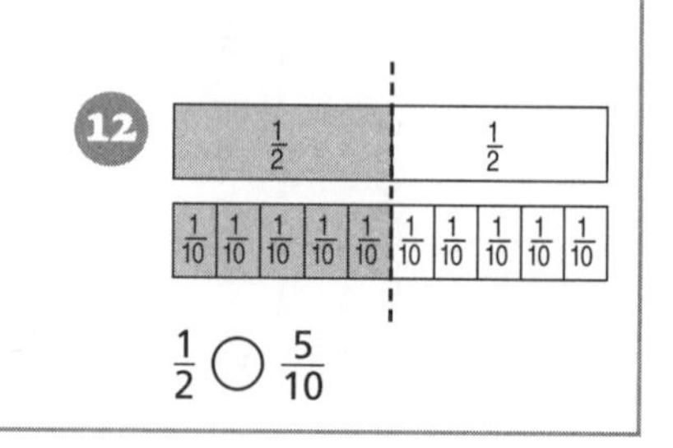

$\frac{1}{2} \bigcirc \frac{5}{10}$

Fractions on a Ruler (Nearest Eighth of an Inch)

15 Minutes

Using Skill 33

OBJECTIVE Write fractions or mixed numbers for length

Point to the rulers on the left of the page. Discuss the fact that these are inch rulers. Be sure to point out that the inches are divided into parts, or fractions, to show parts of an inch.

Ask: **What units of measure do these rulers show?** ($\frac{1}{4}$ inch units) **How many units are in 1 inch?** (4)

Have students trace the length measured to count 3 fourths.

Ask: **What is the length?** ($\frac{3}{4}$ inches)

Point to the rulers on the right of the page.

Ask: **What units of measure do these rulers show?** ($\frac{1}{8}$ inch units) **How many units are in 1 inch?** (8)

Have students trace the length measured to count 2 inches and then $\frac{3}{8}$ inches.

Ask: **What is the length?** ($2\frac{3}{8}$ inches) **What kind of number is $2\frac{3}{8}$?** (mixed number) **What does the 2 represent?** (2 whole inches) **What does the $\frac{3}{8}$ represent?** (3 eighths of an inch, or a fractional part of an inch)

TRY THESE Exercises 1–3 give students practice in writing fractions and mixed numbers for length.

- **Exercises 1–2** Measure fourths.
- **Exercise 3** Measure whole inches and eighths.

PRACTICE ON YOUR OWN Focus on the example at the top of the page. Have a student draw a fraction model that shows $\frac{2}{4} = \frac{1}{2}$ and $\frac{4}{8} = \frac{1}{2}$.

CHECK Make sure students know how to write and simplify fractions. Success is indicated by 3 out of 3 correct responses.

Students who successfully complete **Practice on Your Own** and **Check** are ready to move to the next skill.

COMMON ERRORS

- Students may forget to write the whole number part of the measurement.
- Students may forget to simplify fractions or may simplify fractions incorrectly.

Students who made more than 2 errors in the **Practice On Your Own**, or who were not successful in the **Check** section, may benefit from the **Alternative Teaching Strategy** on the next page.

Optional

Alternative Teaching Strategy
Model Fractions on a Ruler

20 Minutes

OBJECTIVE Use a fraction or a mixed number to name parts of an inch

MATERIALS paper strips, copies of rulers as shown below

Ask: **What does a fraction name?** (part of a whole) **What does the numerator tell?** (how many parts are being considered) **What does the denominator tell?** (how many equal parts are in the whole)

Distribute paper strips. Have students fold a strip in half and then lay it flat. Have them draw a line down the fold mark.

Ask: **Are the parts equal?** (yes) **How many parts are there?** (2) **What fraction names each part?** ($\frac{1}{2}$)

Have students shade $\frac{1}{2}$ of the paper strip. Have them fold the strip in half, then in half again, then lay it flat.

Have them draw lines at the fold marks.

Ask: **Are the parts equal?** (yes) **How many parts are there?** (4) **What fraction names each part?** ($\frac{1}{4}$) **How many fourths are shaded?** (2) **What is the fraction name for the shaded part?** ($\frac{2}{4}$) **What is another fraction name for the shaded part of the paper strip?** ($\frac{1}{2}$)

Have students fold the paper strip in fourths, then in half again, and then lay it flat. Have them draw lines down the fold marks.

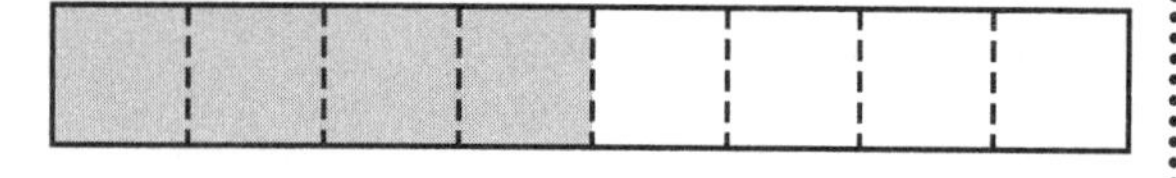

Ask: **Are the parts equal?** (yes) **How many parts are there?** (8) **What fraction names each part?** ($\frac{1}{8}$) **How many eighths are shaded?** (4) **What is the fraction name for the shaded part?** ($\frac{4}{8}$) **What are other fraction names for the shaded part of the paper strip?** ($\frac{1}{2}$ and $\frac{2}{4}$)

Finally, have students shade 2 more eighths of the paper strip.

Ask: **How many eighths are shaded?** (6) **What are two fraction names for the shaded part?** ($\frac{6}{8}$ and $\frac{3}{4}$)

Distribute copies of rulers. Have students write the fractions that name points A, B, and C on the ruler. ($\frac{1}{2}$, $\frac{2}{4}$, $\frac{4}{8}$) Then have them write the fractions or mixed numbers that name points D, E, and F. ($\frac{2}{8}$ or $\frac{1}{4}$, $1\frac{4}{8}$ or $1\frac{1}{2}$, $2\frac{6}{8}$ or $2\frac{3}{4}$)

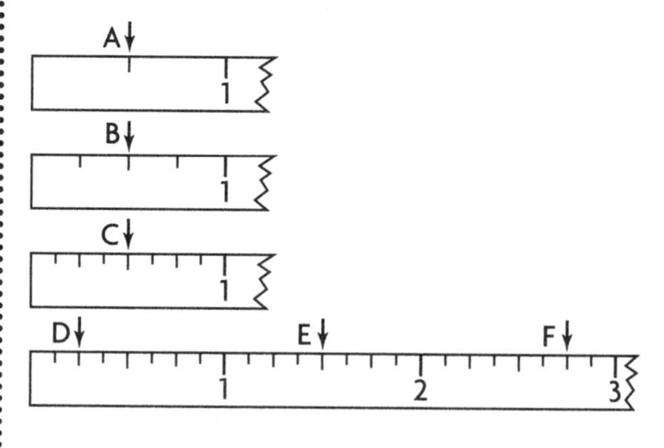

Name ______________ Skill ______________

Grade 5 Skill 33

Fractions on a Ruler (Nearest Eighth of an Inch)

You can write a fraction or a mixed number to name points on a ruler.

Fractional units, such as $\frac{1}{2}$ inch, $\frac{1}{4}$ inch and $\frac{1}{8}$ inch, are used to measure lengths that are between two whole units.

This ruler shows $\frac{1}{4}$ inch units.
There are four fourths in an inch.

Count the fourths to find the length.

Think:
Three fourths is $\frac{3}{4}$ inches.

The length is $\frac{3}{4}$ inches.

This ruler shows $\frac{1}{8}$ inch units.
There are eight eighths in an inch.

Count the whole inches, then count the eighths to find the length.

Think:
2 inches + 3 eighths = $2\frac{3}{8}$ inches

The length is $2\frac{3}{8}$ inches.

Try These

Complete. Write the fraction for the length.

1. 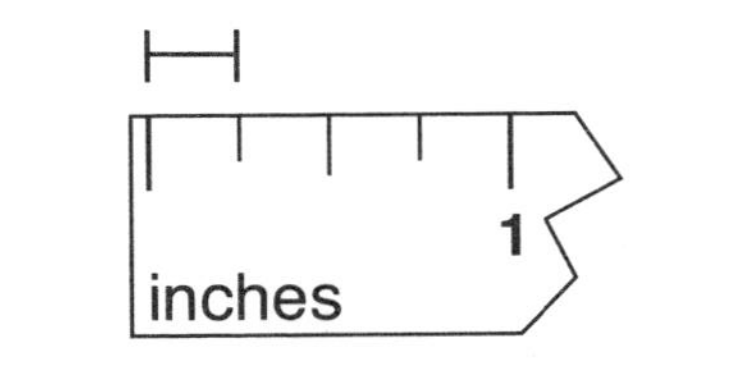

Number of fourths: ______
The length is ______ inches.

2. 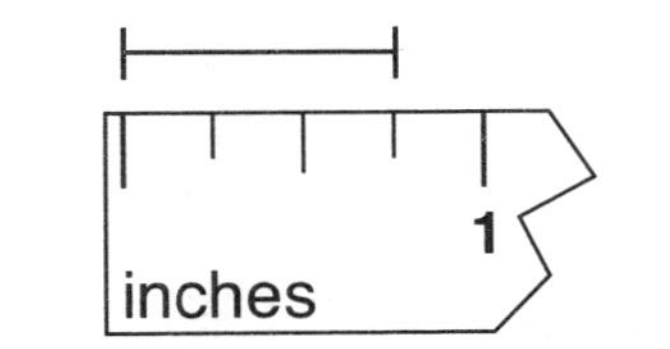

Number of fourths: ______
The length is ______ inches.

3. 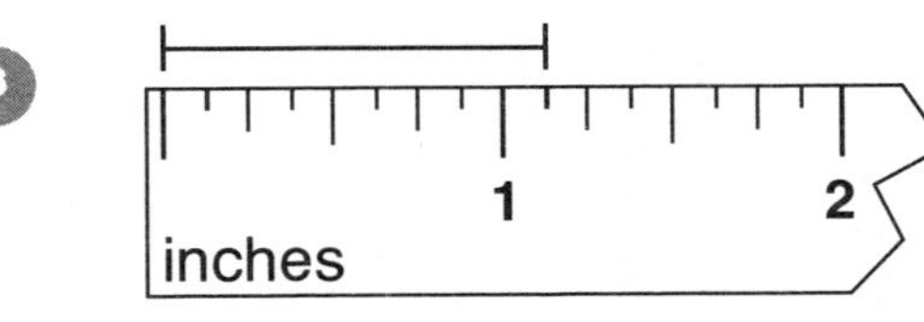

Number of whole inches: ______
Number of eighths: ______
The length is ______ inches.

Go to the next side.

Name ______________________ Skill ______________________

Practice on Your Own

Skill 33

Write a fraction or a mixed number to name each point shown on the ruler.

Think:
You can simplify the fraction to show half inches.

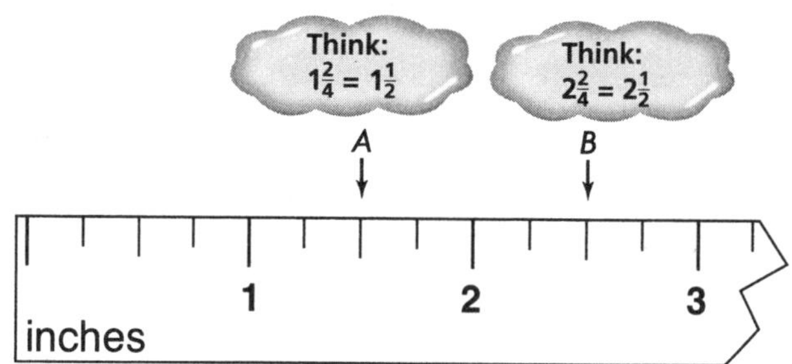

Point A is at $1\frac{1}{2}$ inches. Point B is at $2\frac{1}{2}$ inches.

Complete.

1. 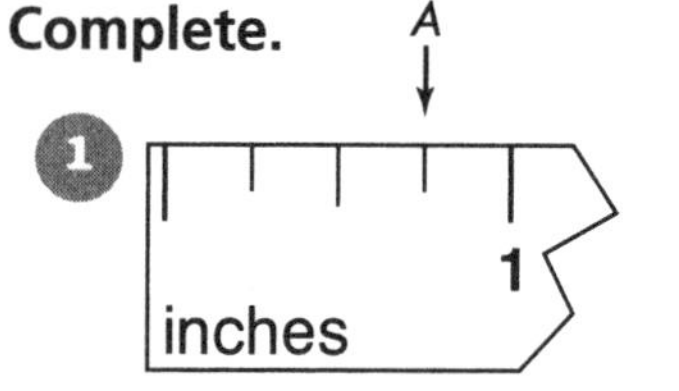
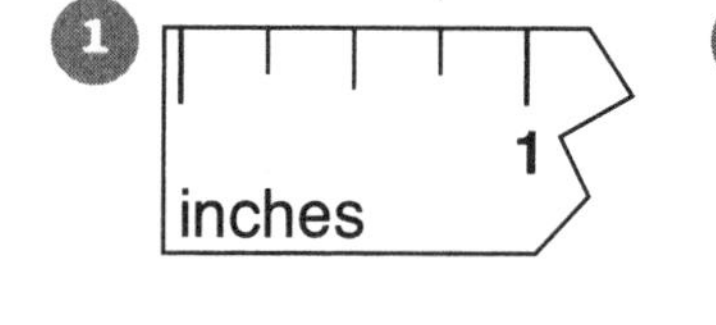

Number of fourths: ____

Point A is at ____ inch.

2. 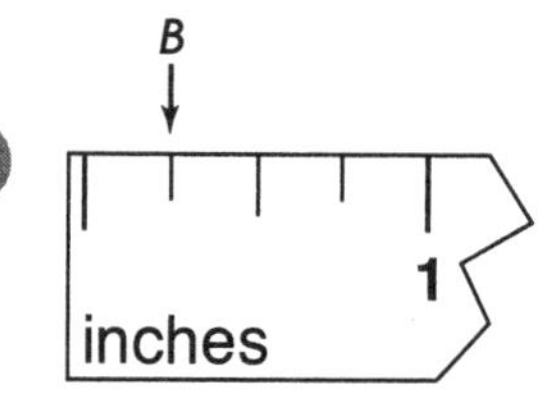
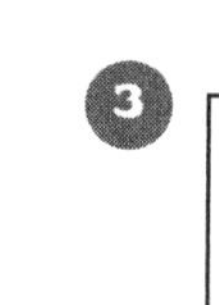

Number of whole inches: ____

Number of fourths: ____

Point B is at ____ inches.

3. 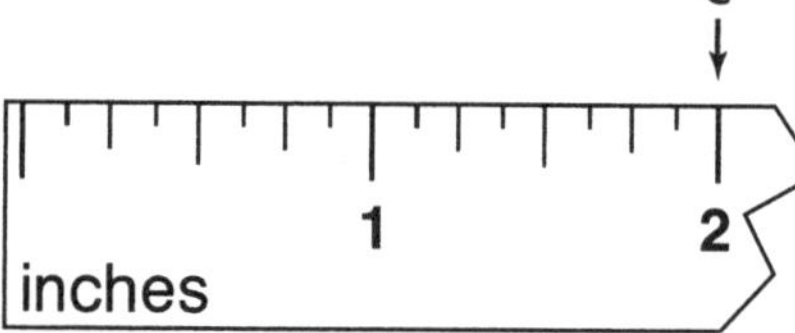

Number of whole inches: ____

Number of eighths: ____

Point C is at ____ inches.

Write a fraction or a mixed number to name each point on the ruler.

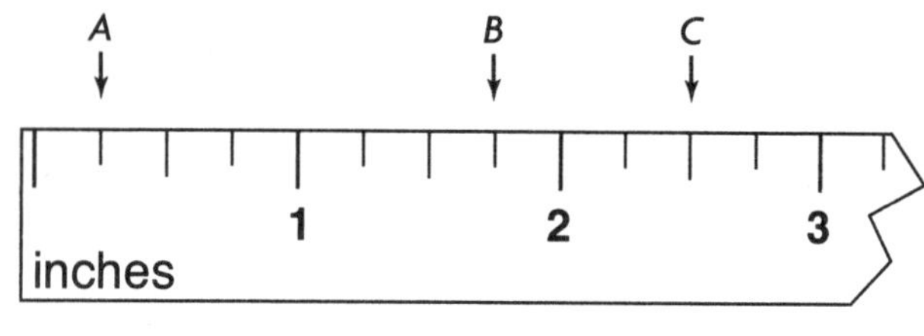

4. Point A is at ___ inches.
5. Point B is at ___ inches.
6. Point C is at ___ inches.

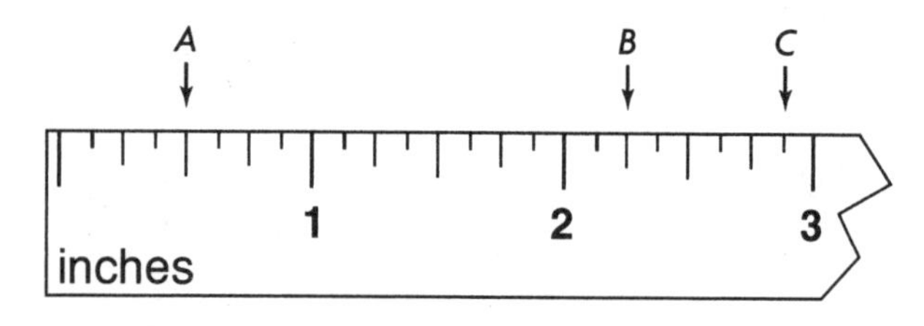

7. A. __________
8. B. __________
9. C. __________

Check

Write a fraction or a mixed number to name each point on the ruler.

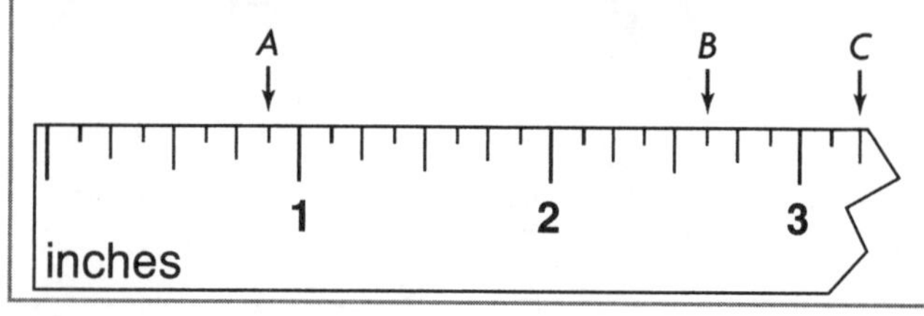

10. A. __________
11. B. __________
12. C. __________

Skill 34 Grade 5

Understand Mixed Numbers

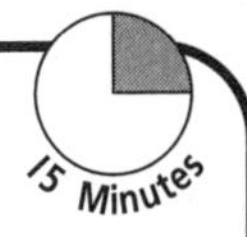

Using Skill 34

OBJECTIVE Rename fractions greater than 1 as mixed numbers; rename mixed numbers as fractions greater than 1

MATERIALS fraction models

Focus on the example for the mixed number $1\frac{3}{8}$. Discuss the fact that $1\frac{3}{8}$ describes 1 whole circle, and $\frac{3}{8}$ of another circle; together, $1\frac{3}{8}$ circles are shaded. Students should notice that another name for 1 whole is $\frac{8}{8}$. Help students use fraction models to represent other mixed numbers.

Direct students' attention to the example for renaming $\frac{5}{4}$ as a mixed number. Discuss each step together.

Ask: **What is the numerator?** (5) **What is the denominator?** (4) **What is another name for 1 whole?** ($\frac{4}{4}$) **Do you have more than 1 whole?** (yes)

Lead students to understand that when the numerator is greater than the denominator, the fraction is greater than 1 and can be renamed as a mixed number.

Work through the steps for renaming $2\frac{5}{6}$ as a fraction greater than 1. Focus on renaming 2 as 1 + 1 and then as $\frac{6}{6} + \frac{6}{6}$.

TRY THESE In Exercises 1–3 students rename fractions as mixed numbers and mixed numbers as fractions.

- **Exercises 1–2** Rename a fraction greater than 1 as a mixed number.
- **Exercises 3–4** Rename a mixed number as a fraction greater than 1.

PRACTICE ON YOUR OWN Work through the steps in the examples at the top of the page. Some students may need to use models when completing Exercises 10–12.

CHECK Determine if students can show the steps for renaming a fraction or a mixed number.

Success is indicated by 3 out of 4 correct responses.

Students who successfully complete the **Practice on Your Own** and **Check** are ready to move to the next skill.

COMMON ERRORS

- When renaming a mixed number such as $2\frac{1}{2}$ as a fraction, students may forget to write the whole number as a fraction before adding the numerator.

Students who made more than 3 errors in the **Practice on Your Own**, or who were not successful in the **Check** section, may benefit from the **Alternative Teaching Strategy** on the next page.

Optional

Alternative Teaching Strategy

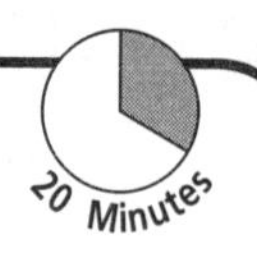

Model Mixed Numbers

OBJECTIVE Model mixed numbers

MATERIALS paper circles and fraction parts for circles, including fourths, thirds, halves, fifths, eighths

Distribute paper circles and fraction parts. Have the students cover as many circles as they can using 7 fourths.

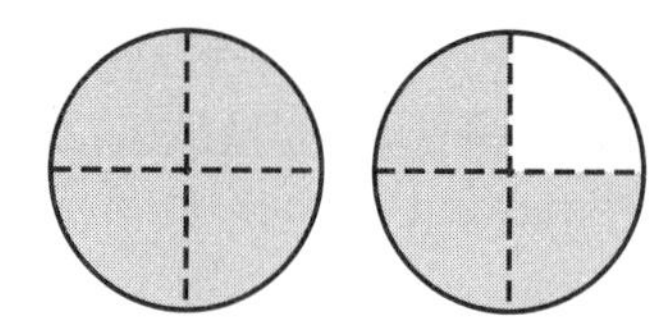

Ask: **How many circles can you make?** (1 whole circle and $\frac{3}{4}$ of another circle)

Show students how to write the result: $1\frac{3}{4}$.

Show students this model. Label the model $\frac{3}{3} + \frac{1}{3}$.

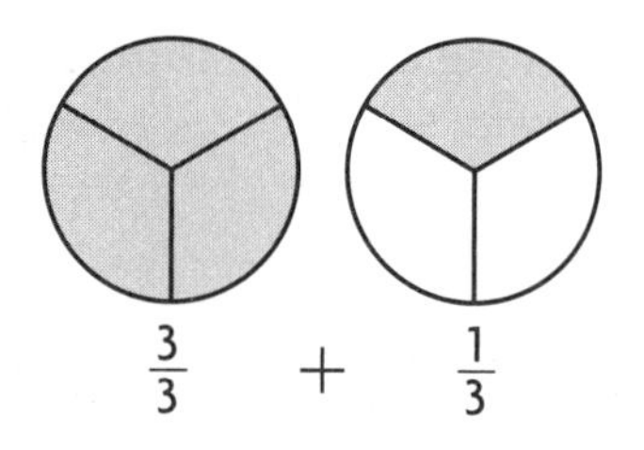

Ask: **How many thirds do you have?** (4)

Show students how to write the fraction for the four thirds: $\frac{4}{3}$

Ask: **When the numerator of a fraction is greater than the denominator—as in $\frac{4}{3}$—is the fraction greater than 1?** (yes)

Refer to the model again, but this time change the label to $1 + \frac{1}{3}$.

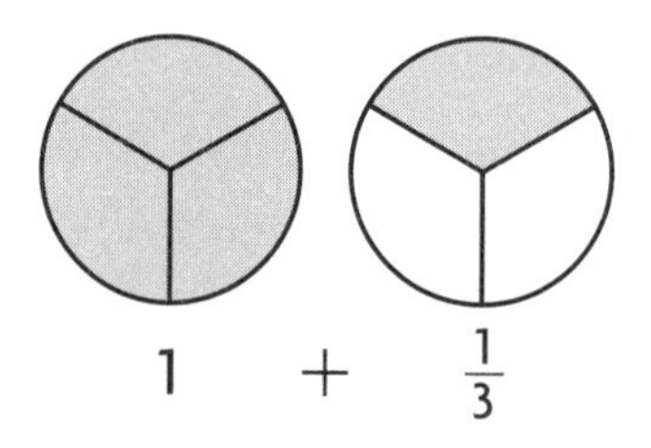

Ask: **Does $\frac{3}{3}$ equal 1 whole?** (yes) **Then how do you write $\frac{4}{3}$ as a mixed number?** ($1\frac{1}{3}$)

Have students practice making circles from fraction parts to show mixed numbers. Have them write the results as both mixed numbers and fractions greater than 1.

Name ______________________ Skill ______________________

Grade 5 **Skill 34**

Understand Mixed Numbers

A mixed number is made up of a whole number and a fraction.

Mixed Numbers

The model shows the mixed number $1\frac{3}{8}$.

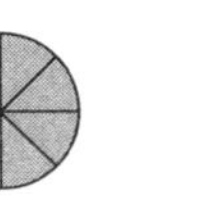

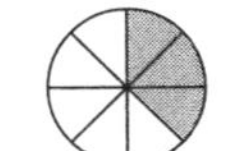

1 whole	3 of 8 parts shaded
eight eighths	three eighths
$\frac{8}{8}$	$\frac{3}{8}$
1	

Renaming Fractions Greater than 1

A fraction greater than 1 can be renamed as a mixed number. Rename $\frac{5}{4}$ as a mixed number.

$$\frac{5}{4} = \frac{4}{4} + \frac{1}{4}$$
$$= 1 + \frac{1}{4}$$
$$= 1\frac{1}{4}$$

So, $\frac{5}{4}$ written as a mixed number is $1\frac{1}{4}$.

Renaming Mixed Numbers

A mixed number can be renamed as a fraction greater than 1. Rename $2\frac{5}{6}$ as a fraction.

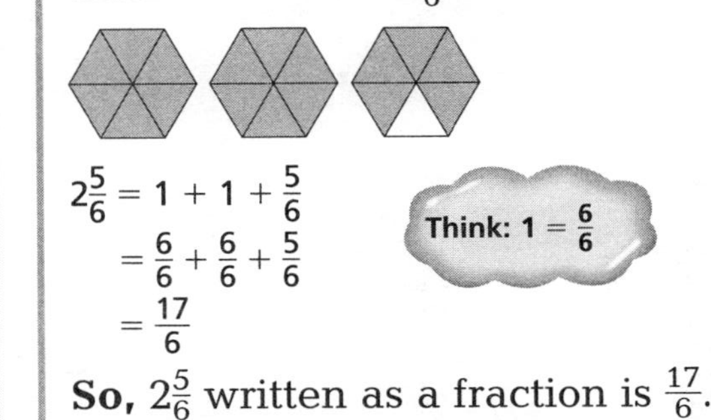

$$2\frac{5}{6} = 1 + 1 + \frac{5}{6}$$
$$= \frac{6}{6} + \frac{6}{6} + \frac{5}{6}$$
$$= \frac{17}{6}$$

So, $2\frac{5}{6}$ written as a fraction is $\frac{17}{6}$.

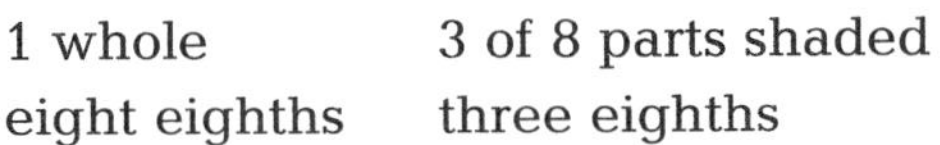

Try These

Complete.

1. Rename $\frac{9}{4}$ as a mixed number.

$\frac{9}{4} = \frac{\square}{4} + \frac{\square}{4} + \frac{\square}{4}$

$= \_\_\_ + \_\_\_ + \frac{\square}{4}$

$= \_\_\_ \frac{\square}{4}$

2. Rename $\frac{17}{10}$ as a mixed number.

$\frac{17}{10} = \frac{\square}{10} + \frac{\square}{10}$

$= \_\_\_ + \frac{\square}{10}$

$= \_\_\_ \frac{\square}{10}$

3. Rename $3\frac{1}{4}$ as a fraction.

$3\frac{1}{4} = \_\_\_ + \_\_\_ + \_\_\_ + \frac{\square}{4}$

$= \frac{\square}{4} + \frac{\square}{4} + \frac{\square}{4} + \frac{\square}{4}$

$= \frac{\square}{4}$

4. Rename $2\frac{3}{5}$ as a fraction.

$2\frac{3}{5} = \_\_\_ + \_\_\_ + \frac{\square}{5}$

$= \frac{\square}{5} + \frac{\square}{5} + \frac{\square}{5}$

$= \frac{\square}{5}$

Go to the next side.

Name ____________________ Skill ____________________

Practice on Your Own

Skill 34

Rename $\frac{8}{3}$ as a mixed number.

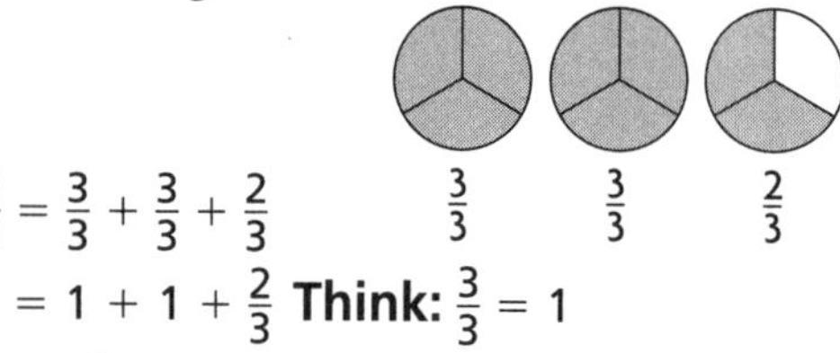

$\frac{3}{3}$ $\frac{3}{3}$ $\frac{2}{3}$

$\frac{8}{3} = \frac{3}{3} + \frac{3}{3} + \frac{2}{3}$

$= 1 + 1 + \frac{2}{3}$ **Think:** $\frac{3}{3} = 1$

$= 2\frac{2}{3}$

So, $\frac{8}{3} = 2\frac{2}{3}$.

Rename $2\frac{1}{6}$ as a fraction.

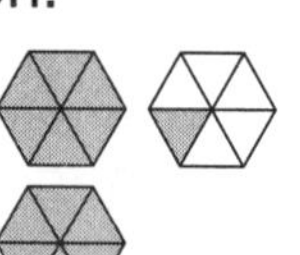

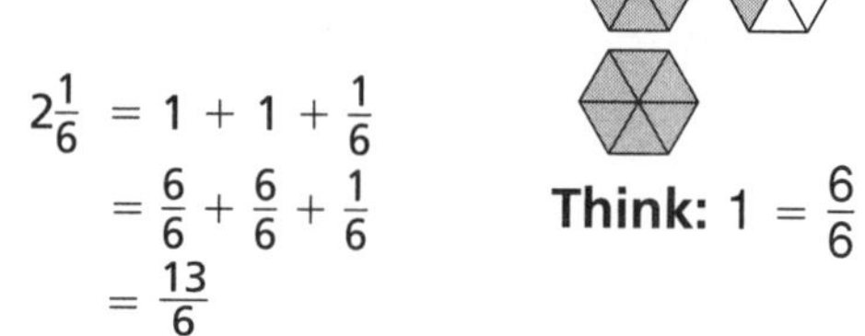

$2\frac{1}{6} = 1 + 1 + \frac{1}{6}$

$= \frac{6}{6} + \frac{6}{6} + \frac{1}{6}$

$= \frac{13}{6}$

Think: $1 = \frac{6}{6}$

So, $2\frac{1}{6} = \frac{13}{6}$.

Rename the fraction as a mixed number.

1 $\frac{23}{10}$

$\frac{23}{10} = \frac{\square}{10} + \frac{\square}{10} + \frac{\square}{10}$

$= \_\_\_ + \_\_\_ + \frac{\square}{10}$

$= \_\_\_\frac{\square}{10}$

2 $\frac{17}{5}$

$\frac{17}{5} = \frac{\square}{5} + \frac{\square}{5} + \frac{\square}{5} + \frac{\square}{5}$

$= \_\_\_ + \_\_\_ + \_\_\_ + \frac{\square}{5}$

$= \_\_\_\frac{\square}{5}$

3 $\frac{7}{6}$

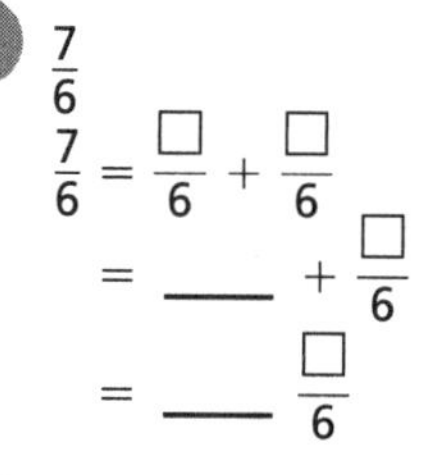

$\frac{7}{6} = \frac{\square}{6} + \frac{\square}{6}$

$= \_\_\_ + \frac{\square}{6}$

$= \_\_\_\frac{\square}{6}$

4 $\frac{11}{8}$

Mixed number: ________

5 $\frac{7}{4}$

Mixed number: ________

6 $\frac{9}{5}$

Mixed number: ________

Rename the mixed number as a fraction.

7 $1\frac{3}{5}$

$1\frac{3}{5} = \_\_\_ + \frac{\square}{5}$

$= \frac{\square}{5} + \frac{\square}{5}$

$= \frac{\square}{5}$

8 $2\frac{1}{2}$

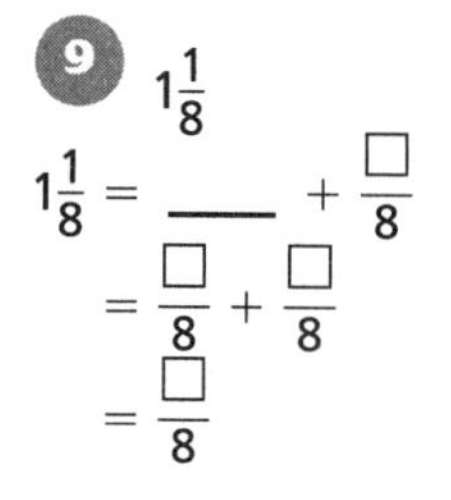

$2\frac{1}{2} = \_\_\_ + \_\_\_ + \frac{\square}{2}$

$= \frac{\square}{2} + \frac{\square}{2} + \frac{\square}{2}$

$= \frac{\square}{2}$

9 $1\frac{1}{8}$

$1\frac{1}{8} = \_\_\_ + \frac{\square}{8}$

$= \frac{\square}{8} + \frac{\square}{8}$

$= \frac{\square}{8}$

10 $4\frac{1}{6}$

Fraction: ________

11 $3\frac{1}{2}$

Fraction: ________

12 $2\frac{1}{3}$

Fraction: ________

Check

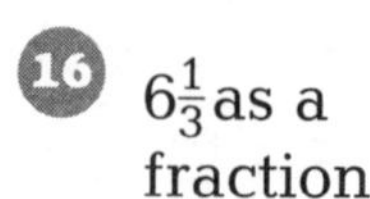

Rename.

13 $2\frac{3}{5}$ as a fraction.

____

14 $\frac{13}{8}$ as a mixed number.

____

15 $2\frac{5}{8}$ as a fraction.

____

16 $6\frac{1}{3}$ as a fraction.

____

Using Skill 35

OBJECTIVE Add fractions that have the same denominators and fractions that have different denominators

Review the terms *numerator* and *denominator.* To help students distinguish the parts of a fraction, share the mnemonic device, *d as in down,* for denominator.

Review the example for *Add Like Fractions.* Remind students that when they add fractions they add only the numerators.

In the example for *Add Unlike Fractions,* have students explain how they make equivalent fractions.

Ask: **What are some equivalent fractions for $\frac{1}{4}$?** ($\frac{2}{8}$, $\frac{3}{12}$, $\frac{4}{16}$, $\frac{5}{20}$, ...) **What are some equivalent fractions for $\frac{3}{5}$?** ($\frac{6}{10}$, $\frac{9}{15}$, $\frac{12}{20}$, $\frac{15}{25}$, ...)

Stress that 20 is the common denominator for $\frac{1}{4}$ and $\frac{3}{5}$ because 20 is the *least common multiple* of the two denominators, 4 and 5. Show students how to list multiples of 4 and 5 to find the common denominator.

4, 8, 16, **20**, 24
5, 10, 15, **20**, 25

It may help some students to consider that $\frac{4}{4}$ and $\frac{5}{5}$ are both equal to 1 and to recall using the Property of One in Multiplication.

$$\frac{1}{4} \times \frac{5}{5} = \frac{5}{20}$$
$$\frac{3}{5} \times \frac{4}{4} = \frac{12}{20}$$

TRY THESE Exercises 1–3 include verbal cues to help students add fractions and model the types of exercises found on the **Practice on Your Own** page.

- **Exercise 1** Like fractions.
- **Exercises 2–3** Unlike fractions.

PRACTICE ON YOUR OWN Review the example at the top of the page.

Ask: **What is the least common multiple for 4 and 2?** (4)

Point out that the fraction $\frac{5}{4}$, in its simplest form, is rewritten as a mixed number.

CHECK Determine if students can find the sum of two fractions less than 1 and can rewrite that sum in simplest form.

Success is indicated by 3 out of 3 correct responses.

Students who successfully complete the **Practice on Your Own** and **Check** are ready to move to the next skill.

COMMON ERRORS

- Some students may believe that when the numerators are the same, they add the denominators to find the sum.
- Some students may add both the numerators and the denominators.
- Students may not know how to write equivalent fractions.
- Students may not express the sum in simplest form.

Students who made more than 3 errors in the **Practice on Your Own**, or who were not successful in the **Check** section, may benefit from the **Alternative Teaching Strategy** on the next page.

Optional

Alternative Teaching Strategy

15 Minutes

Model Adding Like and Unlike Fractions

OBJECTIVE Use models to add like and unlike fractions

MATERIALS 2 assorted sets of colored fraction strips; 1 set cut into parts

If time permits, have students make the fraction strips.

Consider using strips that are 20 cm long and 4 cm wide to make the division into parts easier. Color-coordinate fraction strips to help students see the related fractions. For example, use one color for the strips showing halves, fourths, and eighths; another color to show thirds and sixths; and yet another color for fifths and tenths. Be sure to label two strips in each color as *one whole unit*.

Have students work with partners and take turns modeling each fraction. One student works with the fraction parts, while the other uses the strips representing ones to name the equivalent fraction and to help write the sum in simplest form.

Distribute the fraction strip parts.

Display $\frac{4}{10} + \frac{3}{10}$.

Ask: **Are these like fractions or unlike fractions?** (like) **How do you know?** (They have the same denominator.)

Which fraction strips are you going to use to model the fractions in this addition sentence? (Possible response: The strips/parts labeled tenths. Students may refer to the color for tenths.)

Monitor students as they use the fraction parts to find the sum.

Then have students add unlike fractions.

$\frac{1}{2} + \frac{3}{8}$

$\frac{1}{3} + \frac{1}{6}$

$\frac{4}{10} + \frac{3}{5}$

Use one example to demonstrate how to use the fraction strips to find equivalent fractions. Monitor students as they work on the other examples.

Repeat this activity until all students have demonstrated competence working with the models to add fractions.

1 whole unit

fifth

tenth

Name ______________________ Skill ______________________

Grade 5 **Skill 35**

Add Fractions

Like fractions have the same denominators. Unlike fractions have different denominators.

Add Like Fractions

Find $\frac{5}{9} + \frac{1}{9}$.

- The denominators are the same. → Add the numerators. → $\frac{5}{9} + \frac{1}{9} = \frac{6}{9}$
- Write the sum in simplest form. A fraction is in simplest form if the greatest common factor of the numerator and denominator is 1.

  Factors:
  6: 1, 2, **3**, 6
  9: 1, **3**, 9
- To find simplest form, find the greatest common factor of the numerator and the denominator.
- Divide the numerator and denominator by the greatest common factor.

$\frac{6}{9} = \frac{6 \div 3}{9 \div 3} = \frac{2}{3}$ Simplest form

So, $\frac{5}{9} + \frac{1}{9} = \frac{2}{3}$

Add Unlike Fractions

Find $\frac{1}{4} + \frac{3}{5}$.

The denominators are different. Write equivalent fractions with like denominators.

$\frac{1}{4} = \frac{1 \times 5}{4 \times 5} = \frac{5}{20}$

$\frac{3}{5} = \frac{3 \times 4}{5 \times 4} = \frac{12}{20}$

$\frac{5}{20} + \frac{12}{20} = \frac{17}{20}$ Simplest form

So, $\frac{1}{4} + \frac{3}{5} = \frac{17}{20}$

Try These

Find the sum. Write the answer in simplest form.

1. $\frac{1}{8} + \frac{3}{8}$

   Same denominators? ________

   Add numerators:

   $\frac{1}{8} + \frac{3}{8} = \frac{\square}{8}$

   Simplest form? ____________

   $\frac{\square}{8} = \frac{\square}{2}$

2. $\frac{1}{4} + \frac{2}{6}$

   Same denominators? ________

   $\frac{1}{4} + \frac{2}{6} =$

   $\frac{\square}{12} + \frac{\square}{12} = \frac{\square}{12}$

3. $\frac{2}{7} + \frac{2}{3}$

   Same denominators? ________

   $\frac{2}{7} + \frac{2}{3} =$

   $\frac{\square}{21} + \frac{\square}{21} = \frac{\square}{21}$

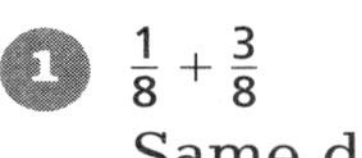

Name ______________________ Skill ______________________

Practice on Your Own

Skill 35

Find the sum. Write the answer in simplest form.

$\frac{3}{4}$
$+\frac{1}{2}$

Denominators are different.

$\frac{3}{4} = \frac{3}{4}$
$+\frac{1}{2} = \frac{2}{4}$

Write equivalent fractions.

$\frac{3}{4}$
$+\frac{2}{4}$

Add the numerators. $3 + 2 = 5$

$\frac{5}{4}$, or $1\frac{1}{4}$

So, $\frac{3}{4} + \frac{1}{2} = \frac{5}{4}$, or $1\frac{1}{4}$.

Find the sum. Write the answer in simplest form.

1. $\frac{1}{5} + \frac{3}{5}$
   Same denominators? ____
   Add numerators:
   $\frac{1}{5} + \frac{3}{5} = \frac{\square}{5}$
   Simplest form ____

2. $\frac{5}{8} + \frac{1}{4}$
   Same denominators? ____
   $\frac{5}{8} + \frac{1}{4}$
   $\frac{5}{8} + \frac{\square}{8} = \frac{\square}{8}$
   Simplest form ____

3. $\frac{7}{12} + \frac{1}{6}$
   Same denominators? ____
   $\frac{7}{12} + \frac{1}{6}$
   $\frac{7}{12} + \frac{\square}{12} = \frac{\square}{12}$
   Simplest form ____

4. $\frac{3}{4} + \frac{1}{6}$
   Same denominators? ____
   $\frac{3}{4} + \frac{1}{6}$
   $\frac{\square}{12} + \frac{\square}{12} = \frac{\square}{12}$
   Simplest form ____

5. $\frac{1}{6} + \frac{3}{6}$
   Same denominators? ____
   Add numerators: $\frac{1}{6} + \frac{3}{6} = \frac{\square}{6}$
   Simplest form ____

6. $\frac{8}{9} + \frac{5}{6}$
   Same denominators? ____
   $\frac{8}{9} + \frac{5}{6}$
   $\frac{\square}{18} + \frac{\square}{18} = \frac{\square}{18}$
   Simplest form ____

7. $\frac{3}{5} + \frac{3}{10} =$

8. $\frac{8}{9} + \frac{1}{6} =$

9. $\frac{2}{3} + \frac{1}{7} =$

Check

Find the sum. Write the answer in simplest form.

10. $\frac{3}{8} + \frac{3}{8}$

11. $\frac{7}{10} + \frac{2}{5}$

12. $\frac{3}{4} + \frac{5}{6}$

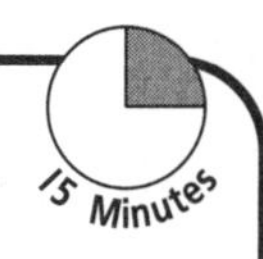

Using Skill 36

OBJECTIVE Subtract fractions with like and unlike denominators

Read together the steps for subtracting like fractions.

Ask: **Are the denominators the same?** (yes) **Then can you subtract the numerators?** (yes) **Is the fraction $\frac{4}{8}$ in simplest form?** (no) **How do you simplify the fraction?** (Divide both the numerator and denominator by the greatest common factor of the numerator and denominator.)

To review how to find the greatest common factor of two numbers. Ask: **What is the greatest common factor of two numbers?** (the greatest number that is a factor of both) **What is the greatest common factor of 4 and 8?** (4)

Read together the steps for subtracting unlike fractions.

Ask: **Are the denominators the same?** (no) **Can you subtract the numerators?** (no) Lead students through the process of rewriting the fractions as equivalent fractions with the same denominators, and then subtracting the numerators.

Focus on the denominators 5 and 3, and on the new denominator, 15. Some students will see that 15 is the least common multiple of 5 and 3.

TRY THESE Exercises 1–3 lead students through the steps for subtracting fractions.

- **Exercise 1** Subtract like fractions.
- **Exercises 2–3** Subtract unlike fractions.

PRACTICE ON YOUR OWN Focus on the examples at the top of the page. Review the process of writing equivalent fractions.

CHECK Make sure students recognize when they need to simplify. Success is indicated by 2 out of 3 correct responses.

Students who successfully complete the **Practice on Your Own** and **Check** are ready to move to the next skill.

COMMON ERRORS

- Students may subtract both numerators and denominators.
- To write the difference in simplest form, students may not find the *greatest* common factor.
- Students may not write equivalent fractions with the same denominators.

Students who made more than 3 errors in the **Practice on Your Own**, or who were not successful in the **Check** section, may benefit from the **Alternative Teaching Strategy** on the next page.

Optional

Alternative Teaching Strategy

20 Minutes

Greatest Common Factor - Least Common Multiple

OBJECTIVE Find greatest common factor and least common multiple, write fractions in simplest form, and write equivalent fractions

MATERIALS fraction strips

Have students help you list the factors of 4 in order (1, 2, 4), then the factors of 8 in order (1, 2, 4, 8). Remind students that a number is divisible by its factors. Repeat the activity several times with other pairs of numbers.

Relate greatest common factor to writing the simplest form of a fraction.

Show students the fraction $\frac{4}{8}$.

Ask: **Is $\frac{4}{8}$ in simplest form?** (no) **How do you know?** (A fraction is in simplest form when the greatest common factor of the numerator and denominator is 1. The greatest common factor of the numerator, 4, and the denominator, 8, is 4.)

Review dividing the numerator and denominator by the greatest common factor.

Use fraction bars to show that the fractions $\frac{4}{8}$ and $\frac{1}{2}$ are equivalent.

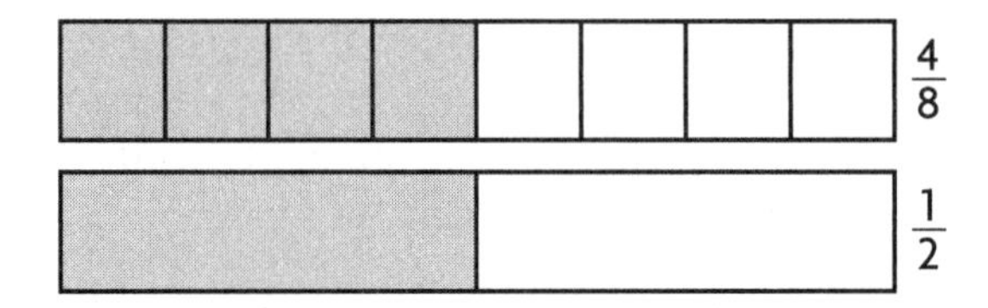

Next, practice listing some multiples of numbers, such as 3 (3, 6, 9, …) and 5 (5, 10, 15, …). Repeat several times with other numbers.

Introduce the term *least common multiple*, and show how to find the least common multiple of a pair of numbers. Use 3 and 5. Tell students to list the multiples of 5, the greater number, first. They should stop at each multiple of 5 and check to see if it is also a multiple of 3. If it is the first number that is a multiple of both 3 and 5, it is their least common multiple.

Relate least common multiple to writing equivalent fractions with the same denominators. Go over the model in Skill 36 for subtracting unlike fractions. Point out that the new denominator of the equivalent fractions is the least common multiple of the two different denominators.

Name ______________________ Skill ______________________

Subtract Fractions

Like fractions have the same denominators. Unlike fractions have different denominators.

Subtract Like Fractions

Find $\frac{5}{8} - \frac{1}{8}$.

- The denominators are the same.
- Subtract the numerators.
- Write the difference in simplest form.
- To find the simplest form, divide the numerator and denominator by their greatest common factor.

$\frac{5}{8} - \frac{1}{8} = \frac{4}{8}$ ← 5 − 1 = 4

A fraction is in simplest form if the greatest common factor of the numerator and denominator is 1.

Factors:

4: 1, 2, **4**

8: 1, 2, **4**, 8

$\frac{4}{8} = \frac{4 \div 4}{8 \div 4} = \frac{1}{2}$ simplest form

So, $\frac{5}{8} - \frac{1}{8} = \frac{1}{2}$

Subtract Unlike Fractions

Find $\frac{4}{5} - \frac{1}{3}$.

The denominators are different. Write equivalent fractions with like denominators.

$\frac{4}{5} = \frac{4 \times 3}{5 \times 3} = \frac{12}{15}$

$\frac{1}{3} = \frac{1 \times 5}{3 \times 5} = \frac{5}{15}$

Subtract the numerators.

12 − 5 = 7

$\frac{12}{15} - \frac{5}{15} = \frac{7}{15}$ simplest form

So, $\frac{4}{5} - \frac{1}{3} = \frac{7}{15}$

Try These

Find the difference. Write the answer in simplest form.

1 $\frac{7}{10} - \frac{3}{10}$

Same denominators? ________

Subtract numerators:

$\frac{7}{10} - \frac{3}{10} = \frac{\square}{10}$

Simplest form ____________

2 $\frac{3}{4} - \frac{2}{6}$

Same denominators? ________

$\frac{3}{4} - \frac{2}{6}$
↓ ↓
$\frac{\square}{12} - \frac{\square}{12} = \frac{\square}{12}$

Simplest form ____________

3 $\frac{6}{7} - \frac{2}{3}$

Same denominators? ________

$\frac{6}{7} - \frac{2}{3}$
↓ ↓
$\frac{\square}{21} - \frac{\square}{21} = \frac{\square}{21}$

Simplest form ____________

Go to the next side.

Name ______________________ Skill ______________

Practice on Your Own

Skill 36

Find the difference. Write the answer in simplest form.

$\frac{1}{4}$
$- \frac{1}{6}$

denominators are different

$\frac{1}{4} = \frac{3}{12}$
$- \frac{1}{6} = \frac{2}{12}$

Write equivalent fractions.

$\frac{3}{12}$
$- \frac{2}{12}$
$\frac{1}{12}$

Subtract the numerators.
$3 - 2 = 1$

So, $\frac{1}{4} - \frac{1}{6} = \frac{1}{12}$.

Find the difference. Write the answer in simplest form.

1 $\frac{3}{5} - \frac{1}{5}$

Same denominator? ____

Subtract numerators:

$\frac{3}{5} - \frac{1}{5} = \frac{\square}{5}$

Simplest form _____

2 $\frac{5}{8} - \frac{1}{4}$

Same denominator? ____

$\frac{5}{8} - \frac{\square}{8}$

$\frac{\square}{8} - \frac{\square}{8} = \frac{\square}{8}$

Simplest form _____

3 $\frac{8}{9} - \frac{5}{6}$

Same denominator? ____

$\frac{8}{9} - \frac{5}{6}$

$\frac{\square}{18} - \frac{\square}{18} = \frac{\square}{18}$

Simplest form _____

4 $\frac{3}{4} - \frac{1}{6}$

Same denominator? ____

$\frac{3}{4} - \frac{1}{6}$

$\frac{\square}{12} - \frac{\square}{12} = \frac{\square}{12}$

Simplest form _____

5 $\frac{5}{6} - \frac{3}{6}$

Same denominator? ____

Subtract numerators:

$\frac{5}{6} - \frac{3}{6} = \frac{\square}{6}$

Simplest form _____

6 $\frac{11}{12} - \frac{1}{6}$

Same denominator? ____

$\frac{11}{12} - \frac{1}{6}$

$\frac{11}{12} - \frac{\square}{12} = \frac{\square}{12}$

Simplest form _____

7 $\frac{3}{5} - \frac{3}{10}$

8 $\frac{7}{9} - \frac{2}{3}$

9 $\frac{3}{4} - \frac{2}{3}$

Check

Find the difference. Write the answer in simplest form.

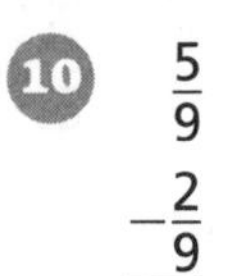

10 $\frac{5}{9} - \frac{2}{9}$

11 $\frac{7}{10} - \frac{2}{5}$

12 $\frac{1}{2} - \frac{1}{3}$

Fractions of a Whole or a Group

15 Minutes

Using Skill 37

OBJECTIVE Write fractions for part of a whole or group

MATERIALS circle showing fraction parts, counters or tiles

Use a circle showing fourths to model part of a whole. Have a volunteer shade 3 parts.

Ask: **How many equal parts are in the circle?** (4) **How many parts are shaded?** (3) **What part of the circle is shaded?** (3 fourths)

Have students write the part shaded as the fraction $\frac{3}{4}$. Review the terms *numerator* (number of equal parts shaded) and *denominator* (number of equal parts in the whole).

Use counters or tiles to model the first part of a group. Use two colors; for example, put out 2 red counters and 3 blue ones.

Ask: **How many parts are in the group?** (5) **How many are red?** (2) **What part of the group is red?** ($\frac{2}{5}$)

In the third section, each part of the group contains more than one item. Put out 3 separate groups of 2 counters each. Use two colors; for example, put out 1 group of 2 yellow counters and 2 groups of 2 green counters.

Use yarn, string, or any other method to separate each part of the group of counters.

Ask: **How many parts in the group?** (3) **How many are yellow?** (1) **What part of the group is yellow?** ($\frac{1}{3}$)

Some students may notice that without the yarn around the groups of two, $\frac{2}{6}$ of the counters are yellow, and $\frac{2}{6}$ is equivalent to $\frac{1}{3}$.

TRY THESE Exercises 1–3 take students through the process of writing fractions of a whole or group.

- **Exercise 1** Write a fraction for part of a whole.
- **Exercises 2–3** Write a fraction for part of a group.

PRACTICE ON YOUR OWN Read through the examples at the top of the page. In Exercise 2, there are 5 parts in the group. In Exercise 3, there are 3 parts in the group, even though there are 6 items; $\frac{2}{3}$ is shaded.

CHECK Determine if students recognize the role of the numerator and the denominator in a fraction. Success is indicated by 2 out of 3 correct answers.

Students who successfully complete the **Practice on Your Own** and **Check** are ready to move on to the next skill.

COMMON ERRORS

- Students may confuse the numerator and denominator.
- Students may write the numerator correctly, but write the denominator as the unshaded part.

Students who made more than 2 errors in the **Practice on Your Own**, or who were not successful in the **Check** section, may benefit from the **Alternative Teaching Strategy** on the next page.

Optional

Alternative Teaching Strategy

20 Minutes

Fractions of a Whole or Group

OBJECTIVE Model equal parts of a whole and parts of a group

MATERIALS paper circles and squares, multicolored counters

Display a circle divided into 4 unequal parts and a square divided into 4 equal parts.

Ask: **Which figure can you use to show fractions?** (square) Explain. (A fraction names equal parts of a whole or parts of a group. The circle is not divided into equal parts.)

Shade in one part of the square. Have students use a fraction to name the shaded part. ($\frac{1}{4}$) Reinforce the fact you need equal parts to show fractions. Have students name the numerator (1) and the denominator (4). They should be able to tell you that the numerator tells how many parts are shaded; the denominator tells how many parts there are in the whole.

Distribute paper circles and squares. Have students fold their own to model equal parts of a whole. Stress equal parts. Have them shade in one or more parts and write the fraction represented. Be sure that the numerator tells the number of shaded parts, and that the denominator tells the number of parts in the whole.

Distribute counters. Tell students to use 2 colors to display a group of 5 counters.

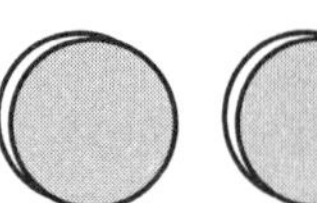
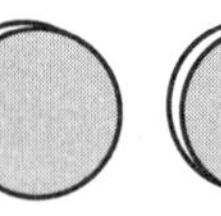
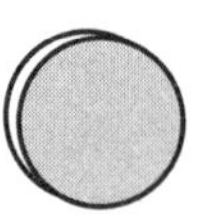

3 out of 5 counters are red

$\frac{3}{5}$ red

Ask: **How many counters are there?** (5) **How many are red?** (3) **How do you write the number of red counters as a fraction?** ($\frac{3}{5}$) **How many counters are blue?** (2) **How do you write the number of blue counters as a fraction?** ($\frac{2}{5}$)

Have students repeat the activity to display and write other fractions. Check that the numerator tells the number of counters of one color, and that the denominator tells the number of counters in the group.

Name ______________________ Skill ______________________

Grade 5 Skill 37

Fractions of a Whole or a Group

A fraction is a number that names part of a whole or part of a group.

Part of a Whole
The **whole** is divided into **4** equal parts.

3 of 4 parts are shaded.

equal parts shaded → 3
number of equal parts in the whole → 4

So, $\frac{3}{4}$ of the whole is shaded.

Part of a Group
There are **5** parts in the **group.**

2 of 5 parts are shaded.

parts shaded → 2
number of parts in the group → 5

So, $\frac{2}{5}$ of the group is shaded.

Part of a Group
There are **3** parts in the **group.**

1 of 3 parts is shaded.

parts shaded → 1
number of parts in the group → 3

So, $\frac{1}{3}$ of the group is shaded.

Try These

Complete.

1.

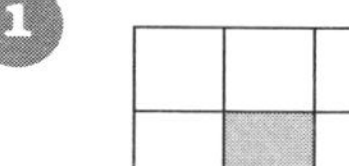

___ of ___ parts is shaded.

equal parts shaded→ ☐
number of equal parts in the whole→ ☐

2.

___ of ___ parts is shaded.

parts shaded → ☐
number of parts in the group → ☐

3.

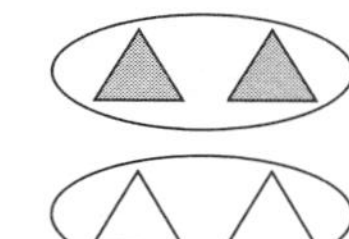

___ of ___ parts is shaded.

parts shaded→ ☐
number of parts in the group→ ☐

Go to the next side.

Name ______________________ Skill ______________________

Practice on Your Own

Skill 37

There are 7 parts in the **whole**.

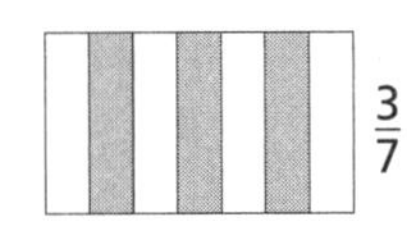

$\frac{3}{7}$ ← parts shaded
← number of equal parts in the **whole**

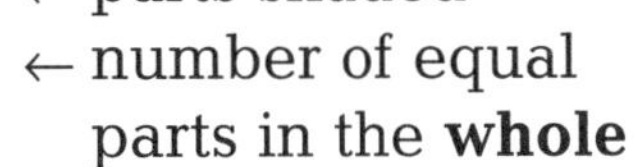

There are 3 parts in the **group**.
1 out of 3 parts is shaded.

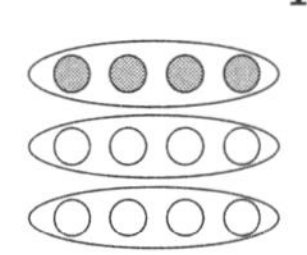

$\frac{1}{3}$ ← parts shaded
← number of parts in the **group**

Complete.

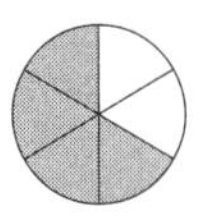

___ out of ___ parts shaded

☐ ← parts shaded
☐ ← number of equal parts in the whole

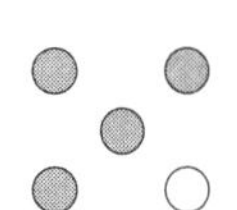

___ out of ___ parts shaded

☐ ← parts shaded
☐ ← number of parts in the group

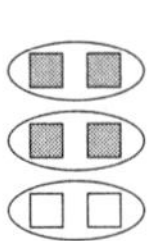

___ out of ___ parts shaded

☐ ← parts shaded
☐ ← number of parts in the group

4 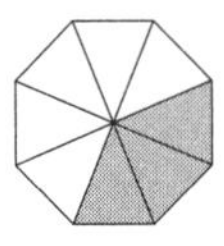

☐ ← parts shaded
☐ ← number of equal parts in the whole

5 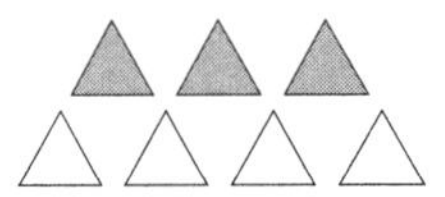

☐ ← parts shaded
☐ ← number of parts in the group

6 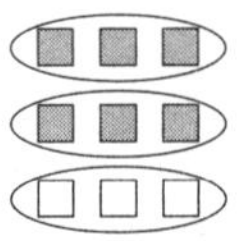

☐ ← parts shaded
☐ ← number of parts in the group

7 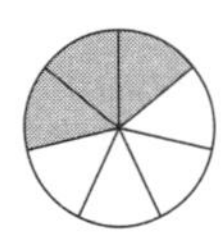 ☐/☐

8  ☐/☐

9  ☐/☐

Check

Complete.

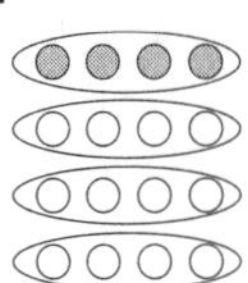 ☐/☐

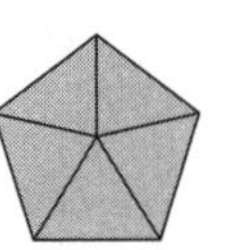 ☐/☐

12 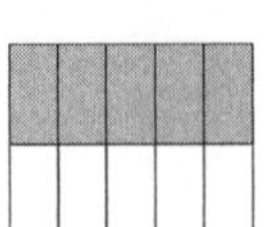 ☐/☐

Skill 38 Find the Greatest Common Factor

Grade 5

Using Skill 38

OBJECTIVE Find the common factors and the greatest common factor (GCF) for two numbers

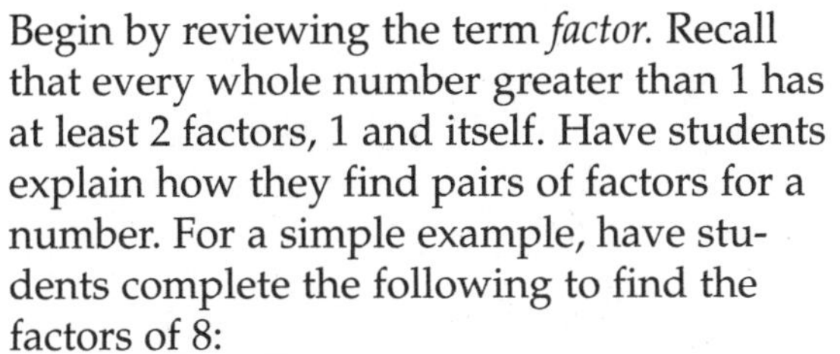

Begin by reviewing the term *factor.* Recall that every whole number greater than 1 has at least 2 factors, 1 and itself. Have students explain how they find pairs of factors for a number. For a simple example, have students complete the following to find the factors of 8:

$\square \times \square = 8$ (1 × 8)

$\square \times \square = 8$ (2 × 4)

Say: **Ask yourself, "What two numbers can I multiply to get 8?"**

List the factors for 8 from least to greatest: 1, 2, 4, 8.

Discuss division as another way to find a factor. Recall that a factor always divides the product without a remainder.

Ask: **Is 4 a factor of 8?** (yes) **How do you know?** (When you divide 8 by 4 there is no remainder.) **Is 3 a factor of 13?** (no) **How do you know?** (If you divide 13 by 3, there is a remainder.)

Refer to Step 1. Point out that the factors are listed from least to greatest.

In Step 2, point out the bold numbers in the list of factors. Then ask: **How many factors do 12 and 16 share?** (3 common factors) **What are their common factors?** (1, 2, 4)

In Step 3, point out that 1, 2, and 4 are the common factors for 12 and 16. Then ask: **Why is the number 4 circled?** (It is the greatest common factor.)

TRY THESE In Exercises 1–3 students are given common factors for each number. They then find the GCF.

- **Exercises 1–2** 2 common factors.
- **Exercise 3** 3 common factors.

PRACTICE ON YOUR OWN Review the example at the top of the page. Have students explain how they find the factors. Point out that students know the first factor for any number, 1. Recall that any number with a 5 in the ones place is divisible by 5.

Point out this short cut for finding the GCF of 15 and 45. Because 15 is a factor of 45, then 15 is the greatest possible common factor. Whenever one of the numbers can be divided by the other number without leaving a remainder, then the lesser number is the greatest common factor.

CHECK Determine if students can identify the greatest common factor of two numbers by finding the factors for each number and listing the common factors from least to greatest. Success is indicated by 2 out of 3 correct responses.

Students who successfully complete the **Practice on Your Own** and **Check** are ready to move to the next skill.

COMMON ERRORS

- Some students may lack proficiency in basic multiplication facts.
- Students may list multiples instead of factors.

Students who made more than 2 errors in the **Practice on Your Own**, or who were not successful in the **Check** section, may benefit from the **Alternative Teaching Strategy** on the next page.

Optional

Alternative Teaching Strategy

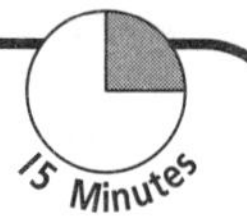

Use Division Facts to Find the GCF

OBJECTIVE Use division facts to find the common factors for 2 numbers and then identify the greatest common factor

MATERIALS number cards 1–50, triangle fact cards for multiplication and division

Begin by having students review basic facts. Demonstrate how to hold up the flash card while covering one of the factors, so that the partner can name the missing factor. Have partners take turns.

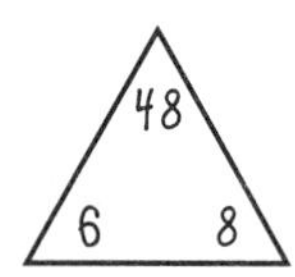

Distribute a set of number cards to each student. Assign a number (1–100) to each student, and then ask students to use their number cards to find all the factors for their number.

After students have found the factors for their numbers, have them order the factors from least to greatest.

Allow students to use the triangle fact cards to help them find all the factors.

Partners will compare their cards to find the common factors. Have them turn over all cards that are not common factors. Then have them identify their greatest common factor.

For example, have partners find the common factors for 10 and 12.

Factors for 10: 1 | 2 | 5 | 10

Factors for 12: 1 | 2 | 3 | 4 | 6 | 12

Say: **Turn over all the factors that are not common to both lists.**

Factors for 10: 1 | 2 | |

Factors for 12: 1 | 2 | | | |

Continue: **What are the common factors?** (1, 2) **What is the greatest common factor?** (2)

Repeat the activity with numbers such as 18 and 36, 20 and 30, and 12 and 48.

Name ______________________ Skill ______________________

Grade 5 Skill 38

Find the Greatest Common Factor

The greatest common factor, or GCF, is the greatest factor that two or more numbers have in common.

Find the greatest common factor of 12 and 16.

Step 1
List the factors for each number.

factors for 12: 1, 2, 3, 4, 6, 12
factors for 16: 1, 2, 4, 8, 16

Step 2
Look for common factors.

factors for 12: **1**, **2**, 3, **4**, 6, 12
factors for 16: **1**, **2**, **4**, 8, 16
The common factors for 12 and 16 are:

1, **2**, **4**

Step 3
Find the greatest common factor, GCF.

Common factors: 1, 2, (4)

4 is the greatest number that is a factor of both 12 and 16.

So, 4 is the GCF of 12 and 16.

Try These

Find the greatest common factor for each set of numbers.

1. 2 and 6
   factors for 2: 1, 2
   factors for 6: 1, 2, 3, 6
   Common factors: _____, _____
   GCF _____

2. 9 and 15
   factors for 9: 1, 3, 9
   factors for 15: 1, 3, 5, 15
   Common factors: _____, _____
   GCF _____

3. 4 and 16
   factors for 4: 1, 2, 4
   factors for 16: 1, 2, 4, 8, 16
   Common factors: ___, ___, ___
   GCF _____

Go to the next side.

Name ______________________ Skill ______________________

Practice on Your Own

Skill 38

Find the GCF of 15 and 45.
Think:
First list the factors for each number.

Factors for 15: **1**, **3**, **5**, **15**
Factors for 45: **1**, **3**, **5**, 9, **15**, 45

Then find the common factors.
Find the GCF.

Common factors: **1**, **3**, **5**, **15**
15 is the greatest common factor.
So, the GCF of 15 and 45 is 15.

Complete.

1. 3 and 7
   factors for 3: 1, 3
   factors for 7: 1, 7
   common factors: ___
   GCF ______

2. 3 and 15
   factors for 3: 1, 3
   factors for 15: 1, 3, 5, 15
   common factors: ___
   GCF ______

3. 5 and 20
   factors for 5: 1, 5
   factors for 20: 1, 2, 4, 5, 10, 20
   common factors: ___
   GCF ______

4. 4 and 12
   factors for 4: ______
   factors for 12: ______
   common factors: ___
   GCF ______

5. 6 and 9
   factors for 6: ______
   factors for 9: ______
   common factors: ___
   GCF ______

6. 7 and 21
   factors for 7: ______
   factors for 21: ______
   common factors: ___
   GCF ______

7. 12 and 21
   factors for 12: ______
   factors for 21: ______
   common factors: ___
   GCF ______

8. 18 and 25
   factors for 18: ______
   factors for 25: ______
   common factors: ___
   GCF ______

9. 20 and 36
   factors for 20: ______
   factors for 36: ______
   common factors: ___
   GCF ______

Check

Complete.

10. 16 and 40
    factors for 16: __________
    factors for 40: __________
    common factors: __________
    GCF ______

11. 28 and 42
    factors for 28: __________
    factors for 42: __________
    common factors: __________
    GCF ______

12. 20 and 50
    factors for 20: __________
    factors for 50: __________
    common factors: __________
    GCF ______

Name ______________________ Skill ______________________

Answer Card
Fractions
Grade 5

TRY THESE

1. $\frac{4}{6}$, four, 4, 6, 4, 6
2. $\frac{2}{3}$, two, 2, 3, 2, 3
3. $\frac{7}{8}$, seven, 7, 8, 7, 8

PRACTICE

1. $\frac{4}{7}$, four, 4, 7, 4, 7
2. $\frac{3}{3}$, three, 3, 3, 3, 3
3. $\frac{5}{10}$, five, 5, 10, 5, 10
4. $\frac{3}{9}$, three
5. $\frac{2}{6}$, 2, 6
6. $\frac{1}{3}$, 1, 3
7. $\frac{5}{12}$, five, 5, 12, 5, 12
8. $\frac{6}{9}$, six ninths, 6, 9, 6, 9
9. $\frac{6}{11}$, six elevenths, 6, 11, 6, 11

CHECK

10. $\frac{2}{2}$, two halves, 2, 2, 2, 2
11. $\frac{2}{7}$, two sevenths, 2, 7, 2, 7
12. $\frac{3}{8}$, three eighths, 3, 8, 3, 8

TRY THESE

1. greater than, >
2. less than, <
3. greater than, >

PRACTICE

1. greater than, >
2. less than, <
3. equal to, =
4. less than, <
5. greater than, >
6. equal to, =
7. <
8. =
9. >

CHECK

10. >
11. <
12. =

TRY THESE

1. 1, $\frac{1}{4}$
2. 3, $\frac{3}{4}$
3. 1, 9, $1\frac{1}{8}$

PRACTICE

1. 3, $\frac{3}{4}$
2. 0, 1, $\frac{1}{4}$
3. 2, 16, 2
4. $\frac{1}{4}$
5. $1\frac{3}{4}$
6. $2\frac{1}{2}$
7. $\frac{1}{2}$
8. $2\frac{1}{4}$
9. $2\frac{7}{8}$

CHECK

10. $\frac{7}{8}$
11. $2\frac{5}{8}$
12. $3\frac{1}{4}$

SKILL 34

TRY THESE

1. $\frac{4}{4} + \frac{4}{4} + \frac{1}{4}$; $1 + 1 + \frac{1}{4}$; $2\frac{1}{4}$
2. $\frac{10}{10} + \frac{7}{10}$; $1 + \frac{7}{10}$; $1\frac{7}{10}$
3. $1 + 1 + 1 + \frac{1}{4}$; $\frac{4}{4} + \frac{4}{4} + \frac{4}{4} + \frac{1}{4}$; $\frac{13}{4}$
4. $1 + 1 + \frac{3}{5}$; $\frac{5}{5} + \frac{5}{5} + \frac{3}{5}$; $\frac{13}{5}$

PRACTICE

1. $\frac{10}{10} + \frac{10}{10} + \frac{3}{10}$; $1 + 1 + \frac{3}{10}$; $2\frac{3}{10}$
2. $\frac{5}{5} + \frac{5}{5} + \frac{5}{5} + \frac{2}{5}$; $1 + 1 + 1 + \frac{2}{5}$; $3\frac{2}{5}$
3. $\frac{6}{6} + \frac{1}{6}$; $1 + \frac{1}{6}$; $1\frac{1}{6}$
4. $1\frac{3}{8}$
5. $1\frac{3}{4}$
6. $1\frac{4}{5}$
7. $1 + \frac{3}{5}$; $\frac{5}{5} + \frac{3}{5}$; $\frac{8}{5}$
8. $1 + 1 + \frac{1}{2}$; $\frac{2}{2} + \frac{2}{2} + \frac{1}{2}$; $\frac{5}{2}$
9. $1 + \frac{1}{8}$; $\frac{8}{8} + \frac{1}{8}$; $\frac{9}{8}$
10. $\frac{25}{6}$
11. $\frac{7}{2}$
12. $\frac{7}{3}$

CHECK

13. $\frac{13}{5}$
14. $1\frac{5}{8}$
15. $\frac{21}{8}$
16. $\frac{19}{3}$

Name ______________________ Skill ______________________

SKILL 35

TRY THESE

1. yes; $\frac{4}{8}$; yes; $\frac{4}{8} = \frac{1}{2}$
2. no; $\frac{3}{12} + \frac{4}{12} = \frac{7}{12}$
3. no; $\frac{6}{21} + \frac{14}{21} = \frac{20}{21}$

PRACTICE

1. yes; $\frac{4}{5}$; $\frac{4}{5}$
2. no; $\frac{5}{8} + \frac{2}{8} = \frac{7}{8}$; $\frac{7}{8}$
3. no; $\frac{7}{12} + \frac{2}{12} = \frac{9}{12}$; $\frac{9}{12} = \frac{3}{4}$
4. no; $\frac{9}{12} + \frac{2}{12} = \frac{11}{12}$; $\frac{11}{12}$
5. yes; $\frac{4}{6}$; $\frac{4}{6} = \frac{2}{3}$
6. no; $\frac{16}{18} + \frac{15}{18} = \frac{31}{18}$; $\frac{31}{18} = 1\frac{13}{18}$
7. $\frac{9}{10}$
8. $1\frac{1}{18}$
9. $\frac{17}{21}$

CHECK

10. $\frac{3}{4}$
11. $1\frac{1}{10}$
12. $1\frac{7}{12}$

SKILL 36

TRY THESE

1. yes; $\frac{4}{10}$; $\frac{4}{10} = \frac{2}{5}$
2. no; $\frac{9}{12} - \frac{4}{12} = \frac{5}{12}$; $\frac{5}{12}$
3. no; $\frac{18}{21} - \frac{14}{21} = \frac{4}{21}$; $\frac{4}{21}$

PRACTICE

1. yes; 2; $\frac{2}{5}$
2. no; 2; $\frac{5}{8} - \frac{2}{8} = \frac{3}{8}$; $\frac{3}{8}$
3. no; $\frac{16}{18} - \frac{15}{18} = \frac{1}{18}$; $\frac{1}{18}$
4. no; $\frac{9}{12} - \frac{2}{12} = \frac{7}{12}$; $\frac{7}{12}$
5. yes; 2; $\frac{2}{6} = \frac{1}{3}$
6. no; $\frac{11}{12} - \frac{2}{12} = \frac{9}{12}$; $\frac{9}{12} = \frac{3}{4}$
7. $\frac{3}{10}$
8. $\frac{1}{9}$
9. $\frac{1}{12}$

CHECK

10. $\frac{1}{3}$
11. $\frac{3}{10}$
12. $\frac{1}{6}$

SKILL 37

TRY THESE

1. 1, 9; $\frac{1}{9}$
2. 1, 3; $\frac{1}{3}$
3. 1, 2; $\frac{1}{2}$

PRACTICE

1. 4, 6; $\frac{4}{6}$
2. 4, 5; $\frac{4}{5}$
3. 2, 3; $\frac{2}{3}$
4. $\frac{3}{8}$
5. $\frac{3}{7}$
6. $\frac{2}{3}$
7. $\frac{3}{7}$
8. $\frac{4}{4}$
9. $\frac{4}{7}$

CHECK

10. $\frac{1}{4}$
11. $\frac{5}{5}$
12. $\frac{5}{10}$

SKILL 38

TRY THESE

1. 1, 2; 2
2. 1, 3; 3
3. 1, 2, 4; 4

PRACTICE

1. 1; 1
2. 1, 3; 3
3. 1, 5; 5
4. 1, 2, 4; 1, 2, 3, 4, 6, 12; 1, 2, 4; 4
5. 1, 2, 3, 6; 1, 3, 9; 1, 3; 3
6. 1, 7; 1, 3, 7, 21; 1, 7; 7
7. 1, 2, 3, 4, 6, 12; 1, 3, 7, 21; 1, 3; 3
8. 1, 2, 3, 6, 9, 18; 1, 5, 25; 1; 1
9. 1, 2, 4, 5, 10, 20; 1, 2, 3, 4, 6, 9, 12, 18, 36; 1, 2, 4; 4

CHECK

10. 1, 2, 4, 8, 16; 1, 2, 4, 5, 8, 10, 20, 40; 1, 2, 4, 8; 8
11. 1, 2, 4, 7, 14, 28; 1, 2, 3, 6, 7, 14, 21, 42; 1, 2, 7, 14: 14
12. 1, 2, 4, 5, 10, 20; 1, 2, 5, 10, 25, 50; 1, 2, 5, 10; 10

Number Sense

Decimals

Skill 39 Grade 5

Read and Write Decimals

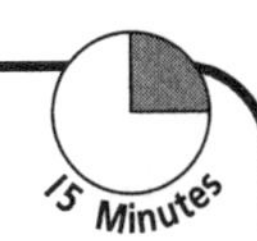

Using Skill 39

OBJECTIVE Read and write decimals in standard, expanded, and word forms

MATERIALS place-value chart

Display the following place-value chart on the board.

hundreds	tens	ones	tenths	hundredths
2	2	2 •	2	2

Ask: **What happens to the value of the digit 2 as you move one place to the left in the chart?** (The value is multiplied by 10.) **What happens to the value of the digit 2 as you move from the hundreds to the tens place?** (The value is divided by 10.) **What about when you move from the tens to the ones?** (The value is divided by 10.)

Tell students that if you move one place to the right from the ones place, the value is still divided by 10. The decimal point is placed between the ones and the tenths to show where decimal value begins.

Direct students' attention to the model of tenths. Remind them that the whole is divided into 10 equal parts and that six parts are shaded. The shaded part is six tenths. Write the decimal notation: 0.6. Discuss the part that is not shaded, 0.4.

Discuss the charts for tenths at the top of the page. Mirror this discussion for the hundredths charts.

TRY THESE In Exercises 1–3 students write decimals in expanded form and in word form.

- **Exercises 1–2** Tenths
- **Exercise 3** Hundredths

PRACTICE ON YOUR OWN Direct students' attention to the top of the page. Use the place-value chart to review the value of a digit in each place to the left of the decimal point. Then, discuss the value of a digit in each place to the right of the decimal point.

CHECK Determine if students can write the number in expanded form and word form. Success is indicated by 2 out of 2 correct responses.

Students who successfully complete the **Practice on Your Own** and **Check** are ready to move on to the next skill.

COMMON ERRORS

- Students may omit or misplace the word *and* when reading decimals.
- Students may write 0.10 and read it as 10 tenths.

Students who made more than 2 errors in the **Practice on Your Own**, or who were not successful in the **Check** section, may benefit from the **Alternative Teaching Strategy** on the next page.

Optional

15 Minutes

Alternative Teaching Strategy

Use Models to Read and Write Decimals

OBJECTIVE Use models and place-value charts to read and write decimals.

MATERIALS Transparencies showing at least 20 square regions marked in hundredths and tenths, washable markers, wet paper towels

Represent 11.11 in a place value-chart on the board. Point to the tens place.

tens	ones	tenths	hundredths
1	1 •	1	1

Ask: **What is the digit in this place?** (1) **What is its value?** (10) **What is the digit to its right?** (1) **How is the value of the 1 in the ones place different from the value of the 1 in the tens place?** (It has $\frac{1}{10}$ the value.)

Say: **In the place-value chart, the value of a digit is $\frac{1}{10}$ the value of the same digit one place to its left.**

Point to the decimal point. Discuss the role of the decimal point in separating the whole part of a number and its part less than one. Always read the decimal point as *and*.

Distribute the materials. Tell the students that you will read a number and that they will then model the number and show the number on the place-value chart.

Students work in pairs for this activity. One partner models the number by shading one of the transparencies to represent it. The other partner shows the number using the place-value chart. For each new number you read, partners switch roles.

Say: **Show eleven and three tenths.**

hundreds	tens	ones	tenths	hundredths
	1	1 •	3	

Repeat this activity for other numbers less than 20 involving tenths and hundredths: 13.07, 4.27, etc.

You may wish to vary this activity by asking the students to write the number in standard form and expanded form.

Name ______________________ Skill ______________________

Grade 5 Skill 39

Read and Write Decimals

A decimal is a number with one or more digits to the right of its decimal point.

Read and Write Tenths

The model shows tenths.

The whole is divided into 10 equal parts.
6 of 10 are shaded.
Six tenths are shaded.
0.6 is shaded.

To read the decimal 0.6, say the number and the place value of the last digit.

Ones	Tenths	Hundredths	Read
0	. 6		six tenths

Write 0.6 in standard, expanded, and word form.

Standard Form	Expanded Form	Word Form
0.6	0 + 0.6	six tenths

Read and Write Hundredths

The model shows hundredths.

The whole is divided into 100 equal parts.
32 of 100 are shaded.
Thirty-two hundredths are shaded.
0.32 is shaded.

Read 0.32.

Ones	Tenths	Hundredths	Read
0	. 3	2	thirty-two hundredths

Write 0.32 in standard, expanded, and word form.

Standard Form	Expanded Form	Word Form
0.32	0 + 0.3 + 0.02	thirty-two hundredths

Try These

Complete. Write the number in expanded form and word form.

1.

ones	tenths
0	. 3

Expanded form: ______

Word form: ______

2.

ones	tenths
1	. 7

Expanded form: ______

Word form: ______

3.

ones	tenths	hundredths
0	. 4	5

Expanded form: ______

Word form: ______

Go to the next side.

Name ______________________ Skill ______________________

Practice on Your Own

Skill 39

Think: Use place value to write and read decimals.

Thousands	Hundreds	Tens	Ones	Tenths	Hundredths
7,	4	3	8 .	0	5

Standard Form: 7,438.05
Expanded Form: 7,000 + 400 + 30 + 8 + .05
Word Form: seven thousand, four hundred thirty-eight and five hundredths

Say "and" for the decimal point.

Write the number in standard form.

1. sixteen and nine tenths
   Standard Form: ____
2. one and twenty-nine hundredths
   Standard Form: ____
3. two and five tenths
   Standard Form: ____
4. 20 + 5 + 0.6
   Standard Form: ____
5. 400 + 60 + 3 + 0.02
   Standard Form: ____
6. 3,000 + 900 + 40 + 5 + 0.8
   Standard Form: ____

Write the number in expanded form.

7. 526.18 ______________________
8. 6,092.4 ______________________

Write the number in word form.

9. 8.07 ______________________
10. 28.6 ______________________

Write the number in expanded form and word form.

11. 17.9 ______________________
12. 45.01 ______________________

Check

Write the number in expanded form and word form.

13. 586.02 ______________________
14. 8,259.6 ______________________

Skill 40 Grade 5

Round Decimals

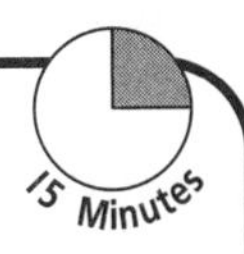

Using Skill 40

OBJECTIVE Round decimals to the nearest whole number

Begin by having students point to the first digit to the right of the decimal point. Suggest that students remember that if this digit is 5 or greater they round up. Explain that they are asked to use the number line and rounding rules to round decimals to the nearest whole number.

Direct students' attention to the number line and rounding rules.

Ask: **Between which two whole numbers is 112.7?** (112 and 113) **Which of these whole numbers is 112.7 closer to?** (113)

What whole number does 112.7 round up to? (113)

Using the *rounding rules*, what digit do you look at to round 112.7 to 113? Explain. (7, because it is the first digit to the right of the digit in the rounding place)

How did you decide to round up? (I looked at the digit to the right of 2. The digit 7 > 5, so the digit 2 increases by 1.)

Continue to ask similar questions as you work through other examples.

TRY THESE Exercises 1–3 model the type of exercises students will find on the **Practice on Your Own** page.

- **Exercise 1** Use a number line.
- **Exercises 2–3** Use rounding rules.

PRACTICE ON YOUR OWN Review the examples at the top of the page. Ask students to explain rounding decimals to the nearest whole number using the number line and the rounding rules.

CHECK Determine if students understand when and how to round up and round down. Success is indicated by 2 out of 3 correct responses.

Students who successfully complete the **Practice on Your Own** and **Check** are ready to move to the next skill.

COMMON ERRORS

- Students may use the place to which they are rounding to decide how to round.
- Students may not round up when the digit to the right is 5, thinking that the digit must be greater than 5 to apply the rule.

Students who made more than 2 errors in the **Practice on Your Own**, or who were not successful in the **Check** section, may benefit from the **Alternative Teaching Strategy** on the next page.

Optional

Alternative Teaching Strategy

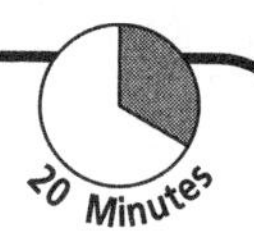

Round Decimals

OBJECTIVE Round decimals to the nearest whole number

MATERIALS blank number lines

Begin by having students use only the number lines to round "easier" decimals. Emphasize the halfway number as the benchmark, and later tie this idea to the rounding rules.

Provide a decimal to round, for example 2.4. Have students label the number line as shown below, using a bold line to indicate the halfway number.

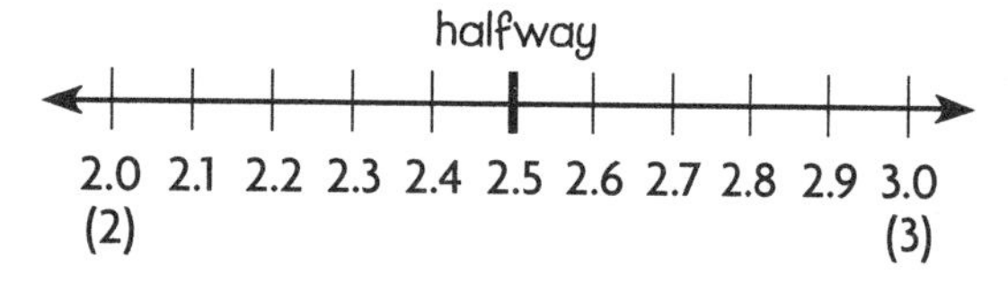

Have students locate 2.4, then count back to determine its distance from 2 (4). Have students determine its distance to 3. Help them recognize that 2.4 is closer to 2 than to 3, so they should round down to 2.

Next have students decide how to round 2.7. Repeat the same steps to determine the distance from 2 and the distance from 3. Also point out that 2.7 is to the right of the halfway number. So, 2.7 rounds up to 3.

Then have students round 2.5. They should determine that 2.5 is the same distance from 2 as it is from 3. Mention that mathematicians have agreed that halfway numbers round up.

Continue with numbers such as 14.3, 14.8, 14.5, and 286.1, 286.9, and 286.5.

When students demonstrate understanding, have them use rounding rules to round other numbers. You may wish to have them underline the digit to the right of the rounding place.

Name ______________________ Skill ______________________

Grade 5 Skill 40

Round Decimals

Use the same rules to round decimals that you use to round whole numbers.
Round 112.7 to the nearest whole number.

Number Line
Use the number line.
If a number is halfway or greater between two whole numbers, round up.

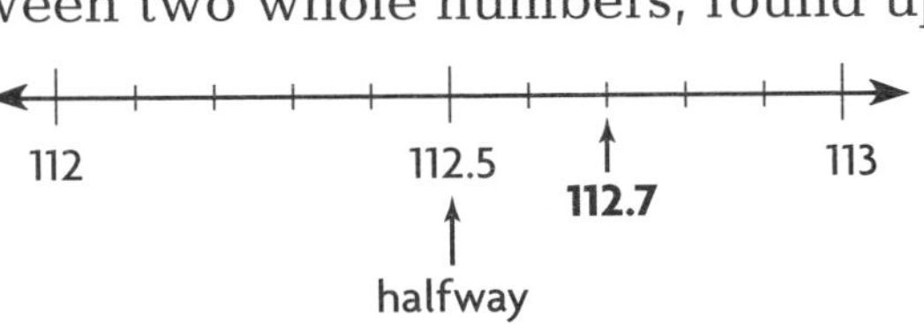

112.7 is between 112 and 113.
112.7 is closer to 113.
So, 112.7 rounds to 113.

Rounding Rules
Use the rounding rules.

Round to this place.
112.7

Look at the first digit to the right.
112.7

Rules
- Find the place you want to round.
- Look at the first digit to its right.
- If this digit is less than 5, the digit in the rounding place stays the same. This is called *rounding down*.
- If this digit is 5 or more, the digit in the rounding place increases by 1. This is called *rounding up*.

Since $7 > 5$, the digit 2 increases by 1.
So, 112.7 rounds to 113.

Try These

Complete. Round to the nearest whole number.

1 31.6

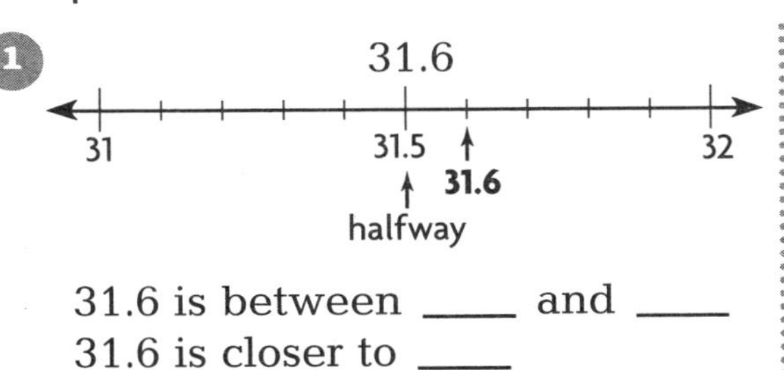

31.6 is between ____ and ____
31.6 is closer to ____
31.6 rounds to ____.

2 4.5

Digit in rounding place is ____
Digit to its right is ____
Is this digit 5 or more? ____
4.5 rounds to ____.

3 217.2

Digit in rounding place is ____
Digit to its right is ____
Is the digit 5 or more? ____
217.2 rounds to ____.

Go to the next side.

Name ______________________ Skill ______________________

Practice on Your Own

Skill 40

Round 84.3 to the nearest whole number.
Use a number line.

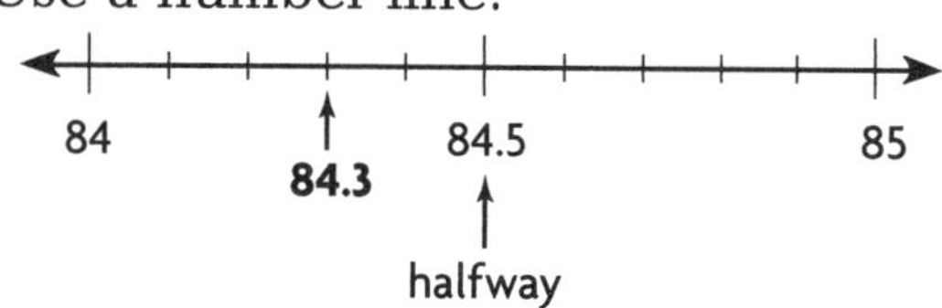

84.3 is between 84 and 85.
84.3 is closer to 84.
So, 84.3 rounds to 84.

Round 198.2 to the nearest whole number.
Use rounding rules.

Round to this place. Look at the first digit to its right.

19<u>8</u>.2

Since $2 < 5$, the digit 8 stays the same.
So, 198.2 rounds to 198.

Complete. Round to the nearest whole number.

1 5.3

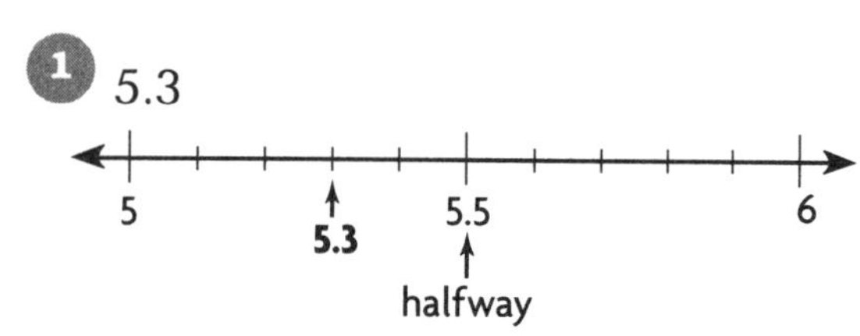

5.3 is between _____ and _____
5.3 is closer to _____
5.3 rounds to _____

2 28.7

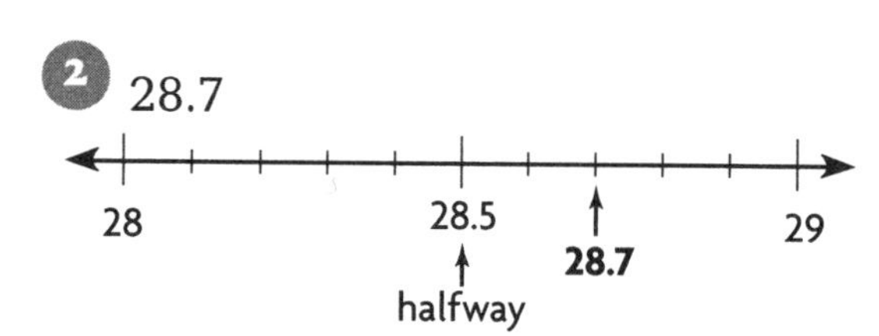

28.7 is closer to _____
28.7 rounds to _____

3 261.8

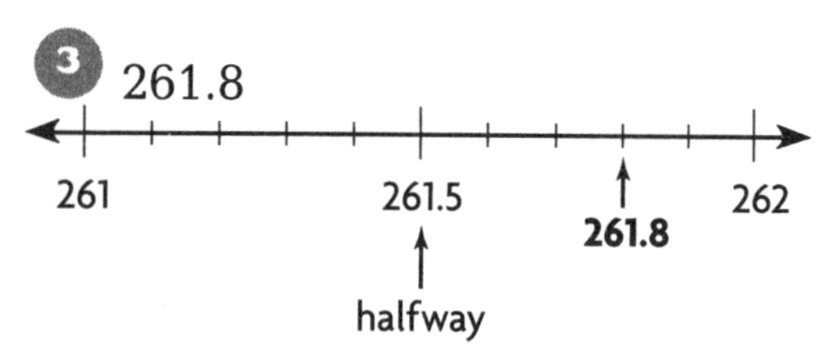

261.8 rounds to _____

4 0.5
Digit in rounding place is _____
Digit to its right is _____
Is this digit 5 or more? _____
0.5 rounds to _____

5 36.3
Digit in rounding place is _____
Digit to its right is _____
36.3 rounds to _____

6 342.6
Digit in rounding place is _____
342.6 rounds to _____

7 39.5 _____ **8** 45.2 _____ **9** 634.7 _____

Check

Round to the nearest whole number.

10 2.4 _____ **11** 24.5 _____ **12** 575.9 _____

Skill 41 Grade 5

Mental Math: Decimals

Using Skill 41

OBJECTIVE Use whole-number strategies to mentally add and subtract decimals

Begin the lesson by reviewing how to write numbers in expanded form. For example:

$$56 = 50 + 6$$

Draw students' attention to the addition example. Point out that both addends show tenths. Explain that they can add as if the addends were 2-digit whole numbers, and then, since the addends show tenths, the sum must also show tenths. Ask: **How does breaking the addend 12 into 10 + 2 help you add mentally?** (It is easier to add if you think 38 + 10 + 2.) **How do you know where to place the decimal point?** (Since the addends show tenths, the sum should show tenths.)

Be sure students understand that both addends must have the same number of decimal places before they try to add mentally using this strategy. For example, to mentally add 3.08 + 1.2, students must think about 3.08 + 1.20 and 308 + 120. In this case, the sum is a number of hundredths.

Draw attention to the subtraction example. Again, students can think of whole numbers, subtract, and then place a decimal point in the answer. Some students may find skip-counting backward to be an easier strategy.

TRY THESE In Exercises 1–4 students model the use of whole numbers to add or subtract decimals.

- **Exercises 1–2** Mental math with decimal addition.
- **Exercises 3–4** Mental math with decimal subtraction.

PRACTICE ON YOUR OWN Review the example at the top of the page. Be sure students understand that they can think about adding and subtracting decimals as they do whole numbers, and then place the decimal point.

CHECK Determine if students can use mental math strategies to add and subtract. Success is indicated by 2 out of 3 correct responses.

Students who successfully complete the **Practice on Your Own** and **Check** are ready to move to the next skill.

COMMON ERRORS

- Students might not think about a consistent label on the addends (tenths or hundredths) and its relationship to the placement of the decimal point in the sum.
- Students might not know addition or subtraction facts.
- Students might not know where to place the decimal point for whole number answers that have a zero in the ones place. For example:

1.7	think	17
0.3		+ 3
		20

Students who made more than 1 error in the **Practice on Your Own**, or who were not successful in the **Check** section, may benefit from the **Alternative Teaching Strategy** on the next page.

Optional

Alternative Teaching Strategy
Modeling Decimals With Money

15 Minutes

OBJECTIVE Use play money to model the addition and subtraction of decimals

MATERIALS play money (dollar bills, coins)

Explain to students that we can relate decimal numbers and money amounts.

Point out the similarity between tenths and dimes. Just as there are 10 tenths in a whole, there are 10 dimes in a dollar. So, $3.40, or 3 dollars and 4 dimes, can be thought of as 3.4, or 3 ones and 4 tenths.

Ask students to represent 1.6 using money. Guide students to see that the number to the left of the decimal point equals the number of dollars. The first number to the right of the decimal point equals the number of dimes.

Ask a volunteer to demonstrate how to combine, or add, 3.4 and 1.6 with money.

3.4 + 1.6

4 dollars and 10 dimes, or 5 dollars

Therefore 3.4 + 1.6 = 5.0.

Help students demonstrate subtracting 0.2 from 3.0 by suggesting the exchange of a dollar bill for 10 dimes.

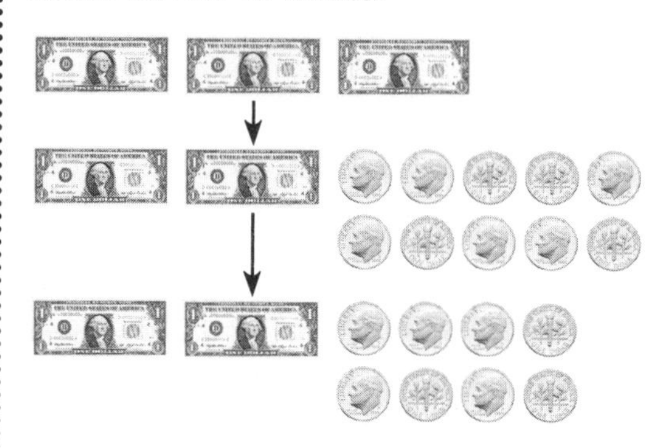

2 dollars and 8 dimes

Therefore 3.0 − 0.2 = 2.8.

Continue with similar examples by modeling the addition and subtraction of decimals with play money. Lead students toward a mental visualization of adding and subtracting dimes (tenths) and dollars (ones) with and without regrouping.

Grade 5 Skill 41

Mental Math: Decimals

Add and subtract decimals as you do whole numbers. Then place the decimal point in the answer.

Addition

- Use mental math to find $3.8 + 1.2$.

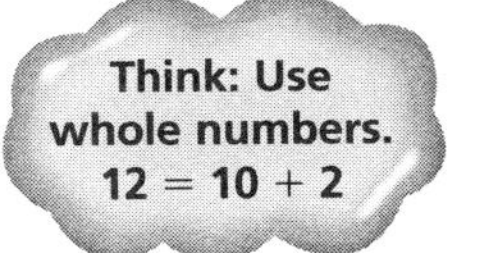

So, think $38 + 10 + 2 = 48 + 2 = 50$

Whole Numbers	Decimals
38	3.8
+12	+1.2
50	5.0

Subtraction

- Use mental math to find $1.0 - 0.3$.

Whole Numbers	Decimals
10	1.0
−3	−0.3
7	0.7

Try These

Use mental math. Add or subtract.

1. $27 + 13$ $2.7 + 1.3$
2. $49 + 11$ $4.9 + 1.1$
3. $10 - 6$ $1.0 - 0.6$
4. $20 - 5$ $2.0 - 0.5$

Go to the next side.

Name ____________________ Skill ____________________

Practice on Your Own

Skill 41

Add or subtract.

Add and subtract decimals as you do whole numbers. Then place the decimal point in the answer.

Whole Numbers	Decimals	Whole Numbers	Decimals
8 + 2 = 10	0.8 + 0.2 = 1.0	6 − 3 = 3	0.6 − 0.3 = 0.3

Use mental math. Add or subtract.

1. 7 + 3   0.7 + 0.3
2. 6 + 4   0.6 + 0.4
3. 58 + 12   5.8 + 1.2
4. 71 + 19   7.1 + 1.9
5. 9 − 4   0.9 − 0.4
6. 8 − 7   0.8 − 0.7
7. 20 − 15   2.0 − 1.5
8. 90 − 8   9.0 − 0.8
9. 1.7 + 0.3
10. 1.1 + 0.9
11. 5.0 + 0.6
12. 0.5 + 0.5 =
13. 3.2 + 1.8 =
14. 1.3 + 2.7 =
15. 0.7 − 0.3 =
16. 4.0 − 0.3 =
17. 1.0 − 0.2 =

Check

Add or subtract. Use mental math.

18. 0.1 + 0.9 =
19. 4.4 + 1.6 =
20. 5.0 − 0.8 =

Skill 42 Grade 5

Add and Subtract Decimals

15 Minutes

Using Skill 42

OBJECTIVE Add and subtract decimals

Begin by having students examine the place-value chart and notice the tenths place and the decimals. Recall that digits to the left of the decimal point name 1 or a number greater than 1; and digits to the right of the decimal point name a number less than 1.

Have students note that when adding or subtracting decimals they first line up the decimal points.

Say: **Add the tenths. How many tenths do you have?**(13 tenths)
Do you regroup? (yes) **How do you know?** (There are more than 9 tenths, so regroup 13 tenths as 1 one 3 tenths.)
Where do you place the regrouped one? (at the top of the ones column)

As students add the ones, remind them to add the regrouped one.

Ask: **What do you do after you complete the addition?** (Place the decimal point in the sum.)

Continue to ask similar questions as you work through subtracting decimals. Guide students to see that they can regroup with decimal numbers exactly as they regroup with whole numbers. Remind students that the careful crossing out and placement of regrouped digits will help them do accurate work.

TRY THESE Exercises 1–4 model the type of exercises students will find in the Practice on Your Own section:

- **Exercise 1** Add ones and tenths.
- **Exercise 2** Subtract ones and tenths.
- **Exercise 3** Add tens, ones, and tenths.
- **Exercise 4** Subtract tens, ones, and tenths.

PRACTICE ON YOUR OWN Review the examples at the top of the page. Ask students to explain the importance of lining up the decimal point in the answer.

CHECK Determine if students know when and how to regroup in the tenths place before adding or subtracting the ones and tens.

Success is indicated by 3 out of 4 correct responses.

Students who successfully complete the **Practice on Your Own** and **Check** are ready to move to the next skill.

COMMON ERRORS

- Students may regroup incorrectly.
- Students may forget to place the decimal point in the answer.

Students who made more than 4 errors in the **Practice on Your Own**, or who were not successful in the **Check** section, may benefit from the **Alternative Teaching Strategy** on the next page.

Optional

Alternative Teaching Strategy
Use Models to Add and Subtract Decimals

20 Minutes

OBJECTIVE Use decimal models to add and subtract decimals

MATERIALS decimal models (blank squares, squares showing tenths,) scissors, paper

You may wish to prepare the decimal models ahead of time, or have the students make them themselves with graph paper.

Have students work in pairs. Distribute models and discuss how the models represent ones and tenths. Then explain that they can use the squares to model addition and subtraction of decimals.

Have students model 2.3 + 0.8.

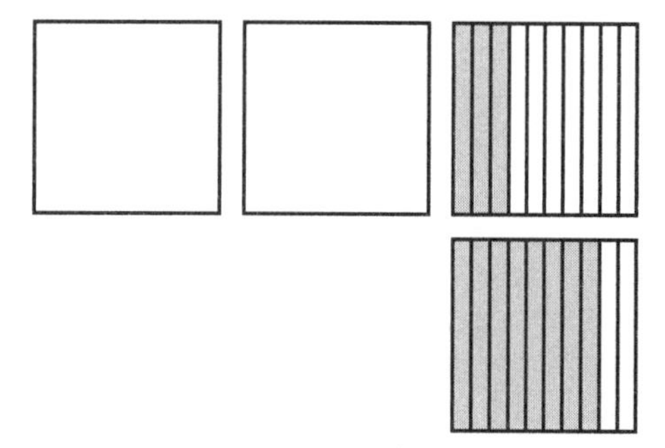

Explain that when they add the decimals they join the models. Have student work through the regrouping of 11 tenths as 1 one 1 tenth, by cutting the tenths and taping them to model a one. The models show that the sum is 3 ones 1 tenth.

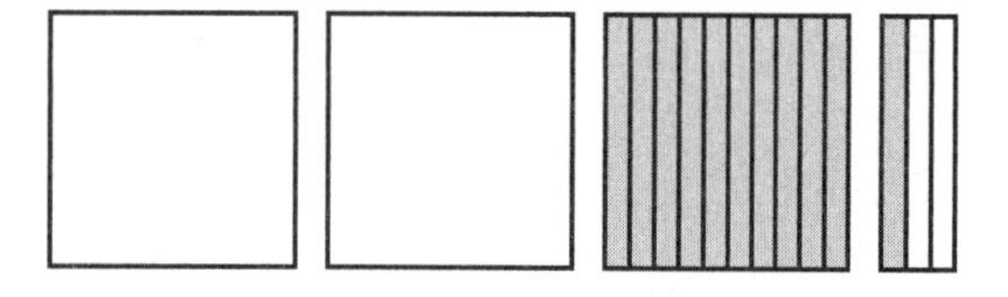

Have students record the addition on paper using a labeled place-value grid. Have them note that the decimals are aligned, and the regrouped one is placed carefully at the top of the ones column.

Remind them that for the answer to be correct they must write a decimal point in the sum.

For subtraction, have them model 1.3 − 0.9.

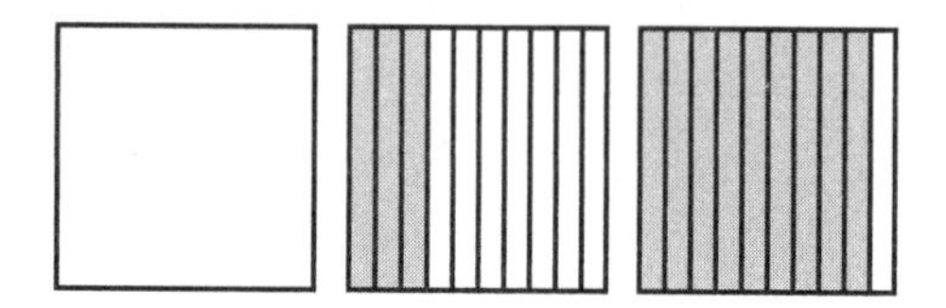

Guide students as they realize that they need to regroup the one as 10 tenths in order to subtract. Have them exchange the blank square representing the one, for a square with tenths. Have them cut away 9 tenths and find a difference of 4 tenths.

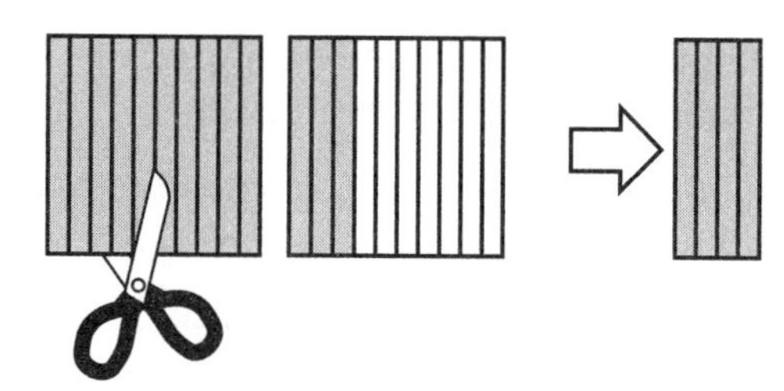

Then have them record the subtraction on paper.

Repeat the activity with other examples. When the students understand how to add and subtract decimals with regrouping, have them add and subtract without using models.

Name ____________ Skill ____________

Grade 5 Skill 42

Add and Subtract Decimals

Add and subtract decimals as you do whole numbers.

Add Decimals

Step 1 Line up the digits in each place. Add. Regroup.

	tens	ones.	tenths
		[1]	
	2	0.	4
+		5.	9
			3

9 tenths + 4 tenths = 13 tenths

Regroup 13 tenths as 1 one, 3 tenths

Step 2 Add the ones. Add the regrouped one.

	tens	ones.	tenths
		[1]	
	2	0.	4
+		5.	9
		6	3

5 ones + 1 one = 6 ones

Step 3 Write the tens.

	tens	ones.	tenths
		[1]	
	2	0.	4
+		5.	9
	2	6.	3

Write a decimal point in the answer.

So, 20.4 + 5.9 = 26.3.

Subtract Decimals

Step 1 Line up the digits. There are not enough tenths to subtract. Regroup the ones. Subtract.

	tens	ones.	tenths
		0	12
	3	~~1~~.	~~2~~
−	1	5.	8
			4

Regroup 1 one as 0 ones, 10 tenths

10 tenths + 2 tenths = 12 tenths

12 tenths − 8 tenths = 4 tenths

Step 2 There are not enough ones to subtract. Regroup the tens. Subtract the ones.

	tens	ones.	tenths
	2	10	12
	~~3~~	~~0~~ ~~1~~.	~~2~~
+	1	5.	8
		5	4

Regroup 3 tens as 2 tens, 10 ones

10 ones + 0 ones = 10 ones

10 ones − 5 ones = 5 ones

Step 3 Subtract the tens.

	tens	ones.	tenths
	2	10	12
	~~3~~	~~0~~ ~~1~~.	~~2~~
−	1	5.	8
	1	5.	4

Write a decimal point in the answer.

2 tens − 1 ten = 1 ten

So, 31.2 − 15.8 = 15.4.

Try These

Add or subtract.

1

	ones.	tenths
	2.	0
+	0.	6

2

	ones.	tenths
	3.	2
−	1.	5

3

	tens	ones.	tenths
	2	7.	5
+		2.	8

4

	tens	ones.	tenths
	1	9.	0
−	1	6.	5

Go to the next side.

Name ______________________ Skill ______________________

Practice on Your Own

Add or subtract.

Think:
Add and subtract as you do with whole numbers.

	ones.	tenths
	2. (1)	2
+	0.	9
	3.	1

Think:
Regroup 11 tenths as 1 one, 1 tenth.
Place the decimal point in the answer.

Subtract.

	tens	ones.	tenths
	~~5~~ (4)	~~0~~. (10)	5
−	4	1.	0
		9.	5

Regroup 5 tens as 4 tens, 10 ones.
0 ones + 10 ones = 10 ones
Place the decimal point in the answer.

Add.

 ones. tenths
0. 4
+ 0. 5

 ones. tenths
☐
0. 4
+ 0. 9

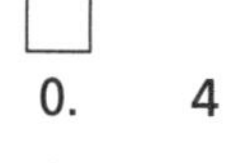

 ones. tenths
☐
4. 8
+ 3. 3

 tens ones. tenths
☐
3 0. 8
+ 8. 3

Subtract.

0. 8
− 0. 3

☐ ☐
3. 5
− 0. 6

☐ ☐
5. 8
− 4. 9

☐
☐ ☐ ☐
4 0. 0
− 8. 3

Add or subtract.

1. 5
− 0. 3

10
☐ ☐
2. 0
− 1. 3

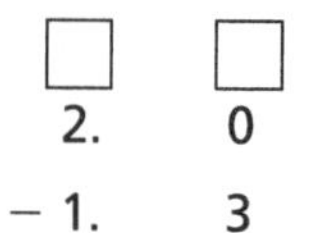

11
☐
9. 8
+ 7. 3

12
☐ ☐
5 2. 1
− 0. 3

13
9. 8
+ 8. 9

14
☐ ☐
8. 0
− 6. 1

15
2 7. 6
+ 4. 9

16
☐
☐ ☐ ☐
3 0. 0
− 2 8. 7

Add or subtract.

3.6
+ 6.9

18
7.2
− 4.8

19
10.0
− 6.3

20
28.8
+ 19.3

Skill 43 Grade 5

Repeated Addition of Decimals

Using Skill 43

OBJECTIVE Find the sum for repeated addition of decimals

Point out the models for the first example. Suggest that students use the models to help them as they add the decimals. Explain that they are asked to find the sum of repeated decimals.

Point out that in the first example the decimals are added by thinking of the place value as if it were any other label (hundredths, ones, inches, dimes, etc.). Explain that the first two decimals can be added, then the last two decimals can be added. Ask: **What is the sum of 8 tenths and 8 tenths?** (16 tenths)

How many tenths are in one whole? (10 tenths)

How many ones and tenths are in 16 tenths? (1 one and 6 tenths, 1.6)

Continue to ask similar questions as you work through the second example. Have students note that the decimal points are aligned before adding in the vertical format. After adding the numbers, the decimal is placed in the sum. This is perfectly modeled by the word form of the equation and sum. Students may benefit from practicing the exercises out loud, emphasizing the place value of the decimal parts of the numbers.

TRY THESE In Exercises 1–3 students add 2, 3, or 4 repeated decimal addends.

- **Exercise 1** 2 repeated addends.
- **Exercise 2** 4 repeated addends.
- **Exercise 3** 3 repeated addends.

PRACTICE ON YOUR OWN Review the example at the top of the page. Ask students to explain how adding decimals is like adding whole numbers and how it is different from adding whole numbers.

CHECK Determine if students know how to regroup tenths when adding repeated decimals. Success is indicated by 3 out of 3 correct responses.

Students who successfully complete the **Practice on Your Own** and **Check** are ready to move to the next skill.

COMMON ERRORS

- Students may fail to align decimals before they add.
- Students may forget to regroup the tenths as ones and tenths in the sum.

Students who made more than 3 errors in the **Practice on Your Own**, or who were not successful in the **Check** section, may benefit from the **Alternative Teaching Strategy** on the next page.

Optional

Alternative Teaching Strategy

15 Minutes

Model the Repeated Addition of Decimals

OBJECTIVE Use a number line to model the repeated addition of decimals

MATERIALS number lines from 0 to 5 showing intervals of tenths, paper

Distribute the number lines. Present this example:

$$0.5 + 0.5 + 0.5 =$$

Point out that the addends in the number sentence are the same. Explain that one way to model the addition is on a number line.

Demonstrate the addition on a number line as the students use their number lines. Remind them to start at 0. Guide them to see that they can jump an interval of 0.5 three times. The number they land on is the sum. Encourage students to skip count, emphasizing place value, as they mark the jumps on the number line (5 tenths, 10 tenths, 15 tenths).

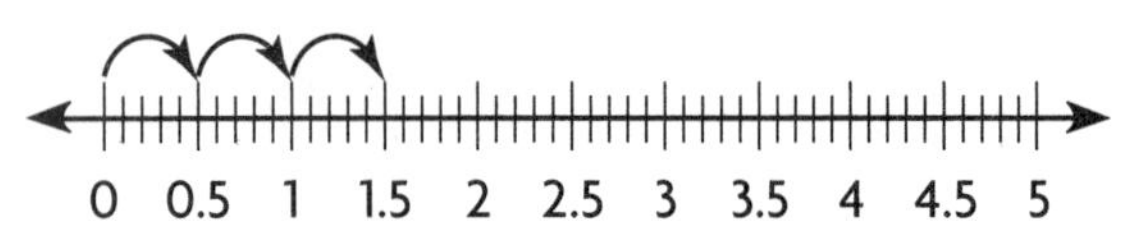

Then present the addition in vertical form.

$$\begin{array}{r} 0.5 \\ 0.5 \\ +\ 0.5 \\ \hline \end{array}$$

As students work through the addition, help them recognize that there are more than 10 tenths, so they can regroup 15 tenths as 1 one and 5 tenths. Ask: **Where did you place the decimal point?** (between the 1 one and the 5 tenths)

Can you use the number line skip-counting method when you have a vertical repeated addition exercise? (yes)

Repeat the activity for 0.3 + 0.3 + 0.3 + 0.3 and 1.5 + 1.5 + 1.5.

The goal is for the students to connect the number line model for adding repeated decimals with the addition in vertical form. When students have internalized the addition procedure, have them add repeated decimals without the number line as a guide.

Grade 5 Skill 43

Repeated Addition of Decimals

Use a model.

Find 0.4 + 0.4 + 0.4 + 0.4.

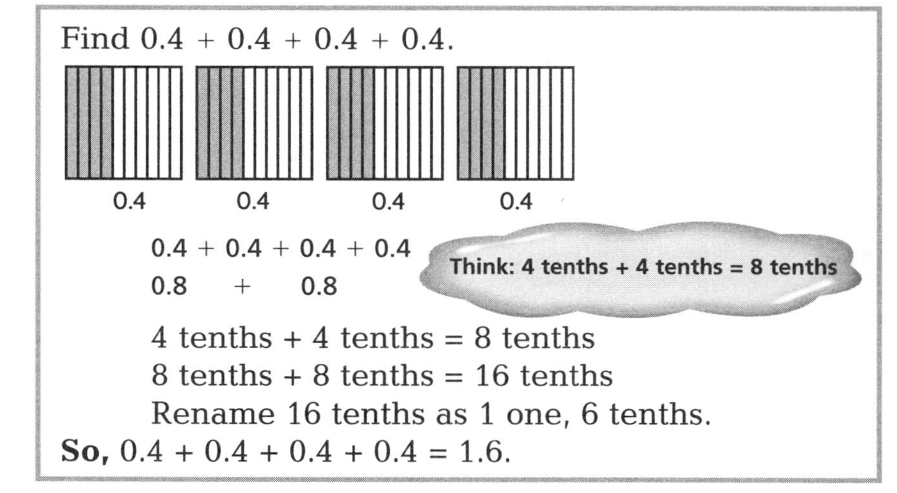

0.4 + 0.4 + 0.4 + 0.4
0.8 + 0.8

4 tenths + 4 tenths = 8 tenths
8 tenths + 8 tenths = 16 tenths
Rename 16 tenths as 1 one, 6 tenths.

So, 0.4 + 0.4 + 0.4 + 0.4 = 1.6.

Find 1.2 + 1.2 + 1.2.

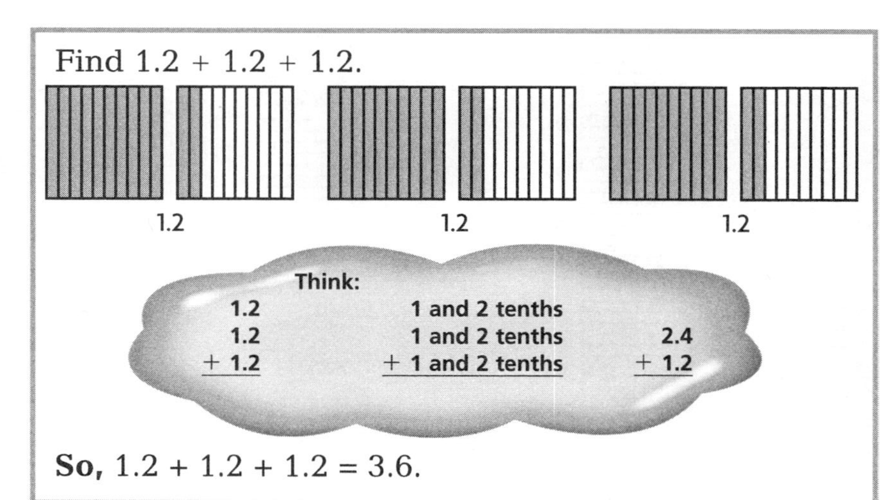

So, 1.2 + 1.2 + 1.2 = 3.6.

Try These

Add.

1. 0.2 + 0.2

0.2 + 0.2 = ______

2. 0.3 + 0.3 + 0.3 + 0.3

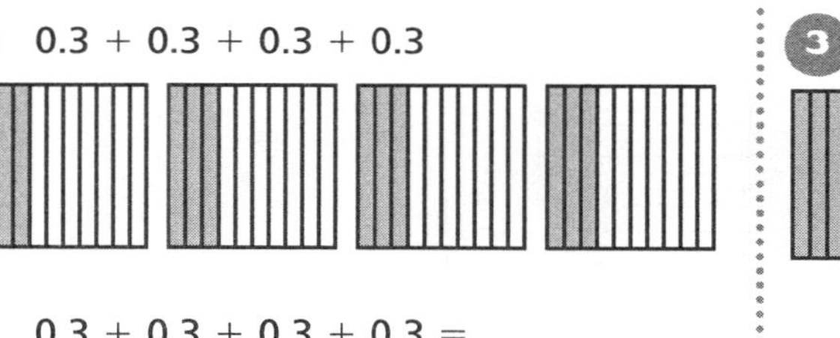

0.3 + 0.3 + 0.3 + 0.3 = ______

3. 1.4 + 1.4 + 1.4

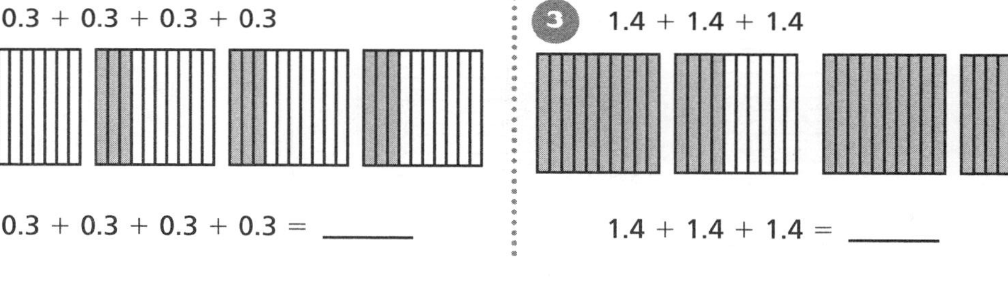

1.4 + 1.4 + 1.4 = ______

Go to the next side.

Name ______________________ Skill ______________________

Practice on Your Own

Skill 43

Think:
Add decimals like you add whole numbers. Remember to line up the decimal points.

$$\begin{array}{r} 0.5 \\ 0.5 \\ +\ 0.5 \\ \hline 1.5 \end{array}$$

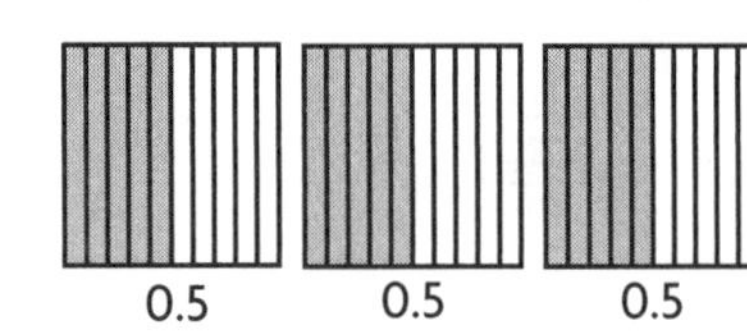

Add.

1. 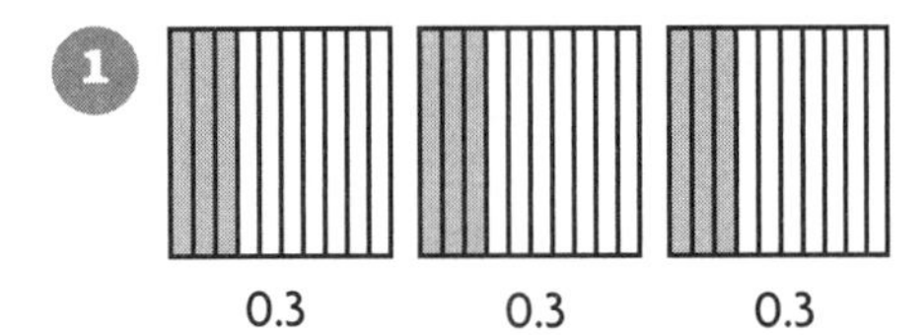

0.3 + 0.3 + 0.3 = ______

2. 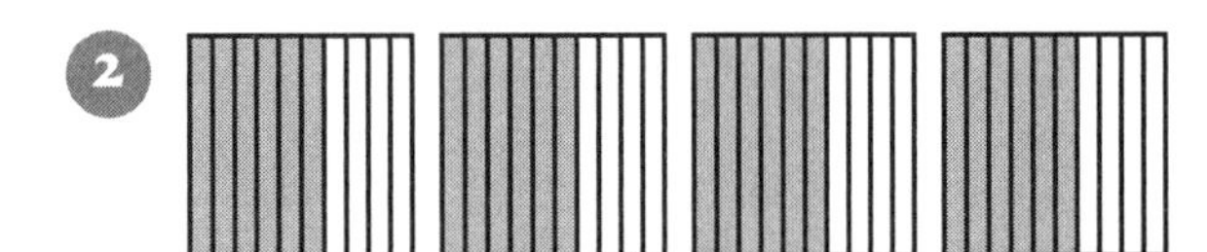

0.6 + 0.6 + 0.6 + 0.6 = ______

3. 1.1 + 1.1 = ______

4. 1.0 + 1.0 + 1.0 + 1.0 = ______

5. $\begin{array}{r} 0.5 \\ +\ 0.5 \\ \hline \end{array}$

6. $\begin{array}{r} 0.7 \\ 0.7 \\ +\ 0.7 \\ \hline \end{array}$

7. $\begin{array}{r} 0.8 \\ +\ 0.8 \\ \hline \end{array}$

8. 0.9 + 0.9 = ______

9. 1.3 + 1.3 + 1.3 = ______

10. 0.2 + 0.2 + 0.2 + 0.2 = ______

Check

Add.

11. 0.4 + 0.4 = ______

12. 1.1 + 1.1 + 1.1 = ______

13. 0.8 + 0.8 + 0.8 + 0.8 = ______

Skill 44 Grade 5

Equivalent Decimals

15 Minutes

Using Skill 44

OBJECTIVE Identify equivalent decimals as decimals that name the same amount

Review the section to the left at the top of the page. Discuss the models and note that each shows the square with the same portion shaded. These two numbers are *equivalent decimals,* or *equivalent numbers.*

In the next section, the place-value chart is used to show equivalent decimals. Stress the information in the box: zeros to the right of the final non-zero digit do not change the value of the number as long as they are *after* a decimal point.

Ask: **What happens to the value of a number when you tack on zeros to the right and there is no decimal?** (each zero multiplies the value of the whole-number part of the number by 10) **Why?** (You move each digit to the left into a place with a greater value.)

TRY THESE Exercises 1–4 provide models or place-value charts for the students to use when writing the equivalent decimal numbers.

- **Exercises 1–2** Models and word form.
- **Exercises 3–4** Place-value chart.

PRACTICE ON YOUR OWN Ask students to look at the models at the top of the page.

Ask: **What do you call decimals that name the same amount?** (equivalent decimals or equivalent numbers)

Direct students' attention to the place-value chart at the top of the page. Refer them to the place to the right of the thousandths place. This place is called the ten-thousandths place. Emphasize that the zeros to the right of the last non-zero digit to the right of a decimal point do not change the value of the number.

Ask: **What would happen to the value of the number if you put a zero right after the decimal point?** (The value of the decimal part of the number would be divided by 10.) **Why?** (You would be moving all of the non-zero digits to the right over to a place with a lesser value)

CHECK Determine if students can write two equivalent decimals for each number.

Success is indicated by 3 out of 3 correct responses.

Students who successfully complete the **Practice on Your Own** and **Check** are ready to move to the next skill.

COMMON ERRORS

- Some students may not relate tenths to hundredths correctly.
- Some students may omit the whole number parts of their decimals.

Students who made more than 2 errors in the **Practice on Your Own** or who were not successful in the **Check** section, may benefit from the **Alternative Teaching Strategy** on the next page.

Optional

Alternative Teaching Strategy

15 Minutes

Models for Equivalent Decimals: Tenths and Hundredths

OBJECTIVE Use models to show equivalent decimals for tenths and hundredths

MATERIALS 10 x 10 square regions cut from centimeter grid paper, tenths strips, hundredths squares

Students work in pairs for this activity. One partner models a decimal in tenths while the other models the equivalent decimal in hundredths.

Distribute the materials.

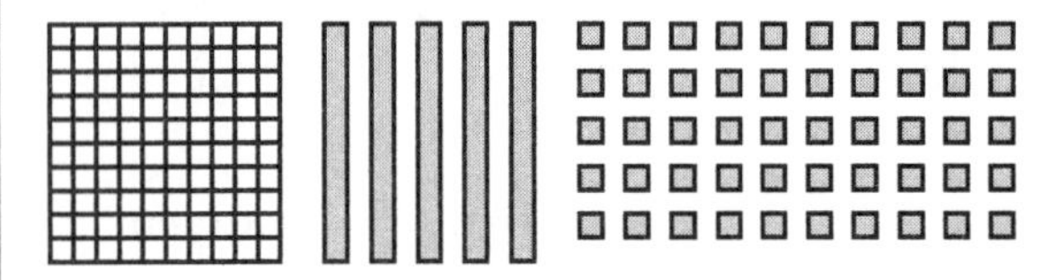

Say: **Cover two tenths of one of the large squares using the tenths strips.**

Ask: **How many hundredths squares would you use to cover the same part of this region?** (20)

The students should then verify this response by covering 20 hundredths with hundredths squares and comparing the two models.

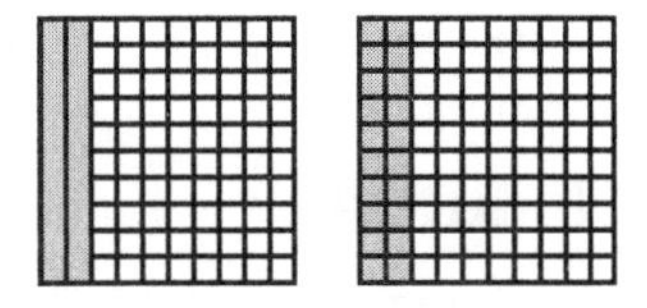

Ask: **What do you call decimals that name the same amount?** (equivalent decimals)

Ask students to write the decimal for each model. Show them how to write the equivalent fractions.

$0.2 = 0.20$

$\frac{2}{10} = \frac{20}{100}$

Repeat this activity to show $0.5 = 0.50$ and $0.3 = 0.30$

Ask students if zeros to the right of the tenths digit change the value of 0.5 or 0.3. Have them explain their thinking.

Name ______________________ Skill ______________

Grade 5 **Skill 44**

Equivalent Decimals

Equivalent decimals are decimals that name the same amount.

Compare the part shaded in the two models.

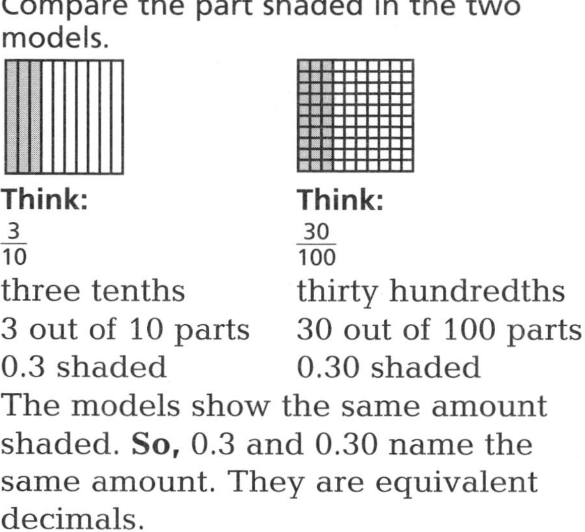

Think:	**Think:**
$\frac{3}{10}$	$\frac{30}{100}$
three tenths	thirty hundredths
3 out of 10 parts	30 out of 100 parts
0.3 shaded	0.30 shaded

The models show the same amount shaded. **So,** 0.3 and 0.30 name the same amount. They are equivalent decimals.

0.3 = 0.30

Use a place-value chart to show equivalent decimals.

ONES		TENTHS	HUNDREDTHS	THOUSANDTHS
0	.	3		
0	.	3	**0**	
0	.	3	**0**	**0**

The zeros to the right of the tenths digit do not change the value of the decimal.

Try These

Write the equivalent decimals.

1. 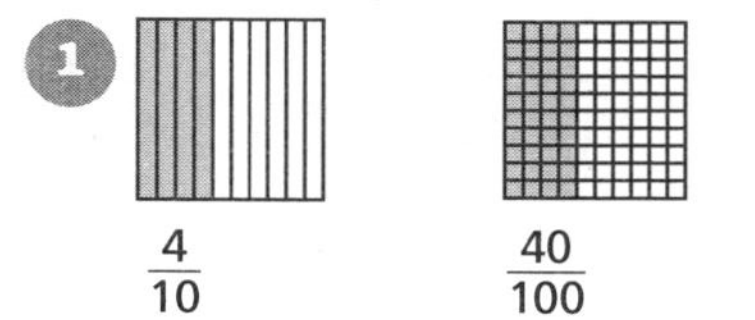

$\frac{4}{10}$ four tenths — $\frac{40}{100}$ forty hundredths

____ = ____

2. 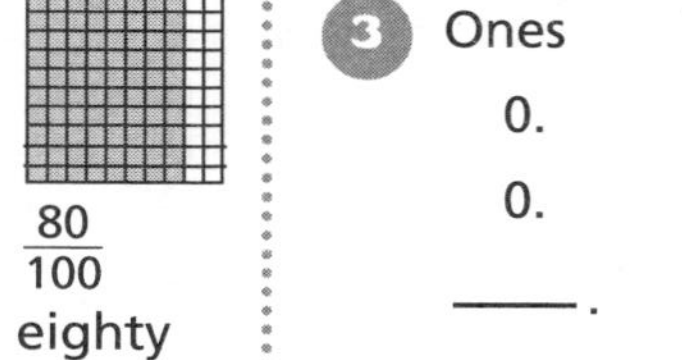

$\frac{8}{10}$ eight tenths — $\frac{80}{100}$ eighty hundredths

____ = ____

3. 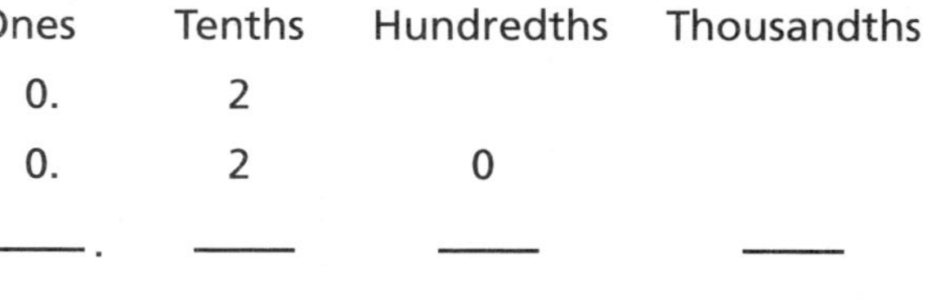

Ones	Tenths	Hundredths	Thousandths
0.	2		
0.	2	0	
____ .	____	____	____

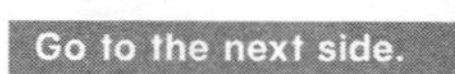

Name ______________________________ Skill ______________________

Practice on Your Own

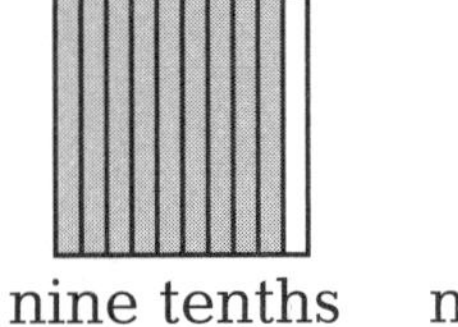
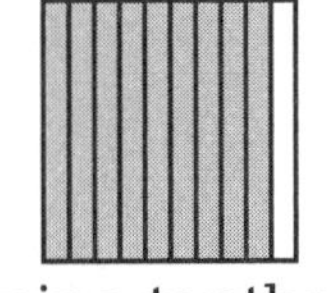

nine tenths
0.9

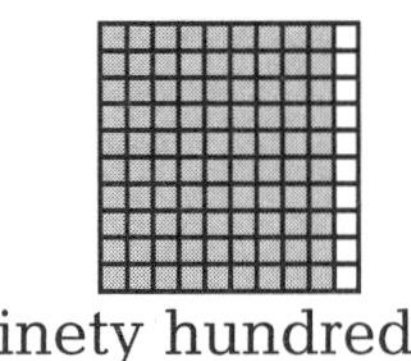

ninety hundredths
0.90

0.9 and 0.90 name the same amount.
0.9 = 0.90

ONES	TENTHS	HUNDREDTHS	THOUSANDTHS	TEN-THOUSANDTHS
0.	4	5		
0.	4	5	0	
0.	4	5	0	0

The zeros to the right of the hundredths digit do not change the value of the decimal.

Write two equivalent decimals.

1

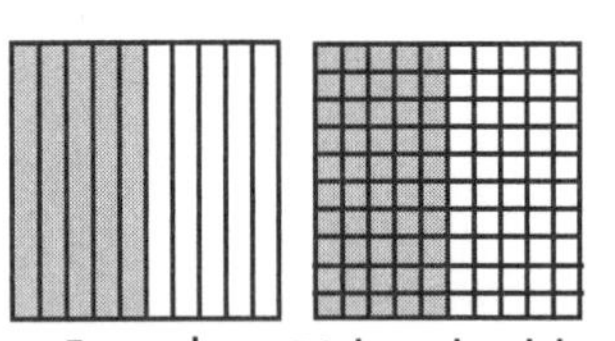

5 tenths 50 hundredths

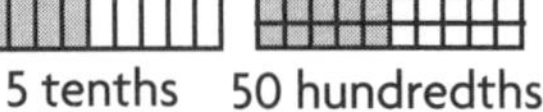

_____ = _____

2

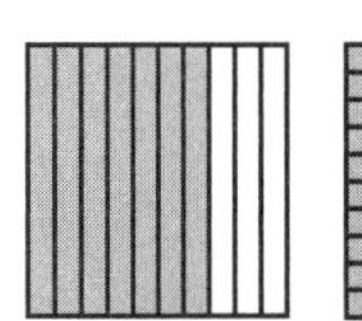

7 tenths 70 hundredths

_____ = _____

3

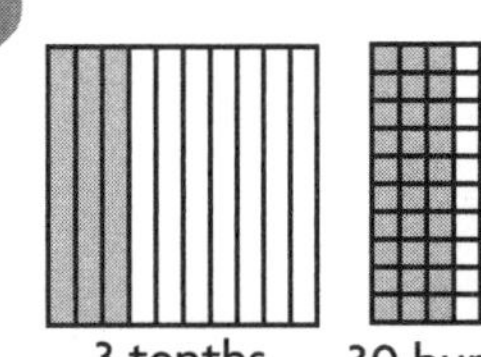

3 tenths 30 hundredths

_____ = _____

4 O T H Th TTh

0 . 5

____

____

5 O T H Th TTh

0 . 1 2

____

____

6 O T H Th TTh

1 . 0 1

____

____

7 6.5

____

____

8 0.04

____

____

9 1.4

____

____

10 4.9

____

____

11 0.90

____

____

12 2.7

____

____

Check

13 0.08 ______________

____

14 4.01 ______________

____

15 6.100 ______________

____

Divide Decimals by Whole Numbers

Using Skill 45

OBJECTIVE Divide decimals by whole numbers

Guide students to read about dividing decimals at the top of the page. Tell students that, just as they estimated quotients when they divided whole numbers, they can estimate quotients when dividing with decimals.

Ask: **In Step 1, what compatible numbers can you use?** (144 and 12)

Focus on Step 2. Be sure students understand that the steps for dividing decimals are the same as those for dividing whole numbers, except that with decimals, a decimal point is placed in the quotient.

Ask: **What do you do before you divide?** (Place the decimal point in the quotient.) Stress care in placing the decimal point in the quotient directly above the decimal point in the dividend.

Lead students through dividing tens and ones in Steps 2 and 3. Point out the arrows used to bring down digits. Emphasize the need to compare remainders with divisors to be sure that a remainder is less than a divisor.

Point out the importance of the check in Step 4. Finally, have students compare the quotient, 12.3, to the estimate, 12, to verify that the quotient is reasonable.

TRY THESE In these exercises, students divide decimals by whole numbers. They also estimate and check the quotient.

- **Exercise 1** Divide by a 3-digit number.
- **Exercise 2** Divide by a 2-digit number.

PRACTICE ON YOUR OWN Review the example at the top of the page. Point out to students that when a quotient is less than 1, they place a zero in the ones column of the quotient. Encourage students to multiply to check all their answers.

CHECK Determine if students know when and where to place zeros when dividing decimals by whole numbers. Success is indicated by 3 out of 4 correct responses.

Students who successfully complete the **Practice on Your Own** and **Check** are ready to move to the next skill.

COMMON ERRORS

- Students may not place the decimal point correctly or may leave it out.
- Students may forget to bring down digits, or may bring down an incorrect digit.

Students who made more than 2 errors in the **Practice on Your Own**, or who were not successful in the **Check** section, may benefit from the **Alternative Teaching Strategy** on the next page.

Optional

Alternative Teaching Strategy

20 Minutes

Use Money to Model Dividing Decimals

OBJECTIVE Model dividing a decimal by a whole number

MATERIALS play money

Display the example $5.25 ÷ 3.

Ask: **What is the divisor?** (3) **What is the dividend?** ($5.25)

Tell students to assume that $5.25 must be shared equally by 3 people.

Use play money to model the division. Put out 5 one-dollar bills, 2 dimes, and 5 pennies. Divide the one-dollar bills into 3 groups.

Ask: **How many ones are in each group?** (1) **How many ones are left?** (2) **Can you divide 2 ones into 3 groups?** (no)

Show students that you can exchange the 2 ones for 20 dimes and add them to the 2 dimes you already have. Divide the 22 dimes among the 3 groups.

Ask: **How many dimes are in each group?** (7) **How many dimes are left?** (1) **Can you divide 1 dime into 3 groups?** (no)

Show students that you can exchange the 1 dime for 10 pennies and add them to the 5 pennies you already have. Distribute the 15 pennies among the 3 groups.

Ask: **Do you have any pennies left?** (no) **Are the groups equal?** (yes) **How much money does each person get?** ($1.75)

Have students count all the money to be sure that there is $5.25 in all.

Finally, work through the division algorithm for $5.25 ÷ 3, relating it to the activity.

```
   $1.75
3 )$5.25
  - 3
    2 2
  - 2 1
      1 5
    - 1 5
        0
```

Name ______________________ Skill ______________________

Grade 5 Skill 45

Divide Decimals by Whole Numbers

You divide decimals as you would divide whole numbers.
Find 147.6 ÷ 12.

Step 1
Estimate the quotient to decide where to put the first digit in the quotient. Use compatible numbers.

12)147.6

12)144 = 12

Think: 12 × 12 = 144

The quotient is about 12. The first digit of the quotient will be in the tens place.

Step 2
Before you divide, write the decimal point in the quotient above the decimal point in the dividend. Divide the tens.

Think: 12 × □ = 12

```
     1 .
12)147.6
  − 12
     2
```

Multiply. 12 × 1
Subtract. 14 − 12
Compare. 2 < 12

Step 3
Divide the ones.

Think: 12 × □ = 24

```
     12.
12)147.6
  − 12↓
     27
   − 24
      3
```

Multiply. 12 × 2
Subtract. 27 − 24
Compare. 3 < 12

Step 4
Divide the tenths. Write a 3 in the tenths place.

Think: 12 × □ = 36

```
     12.3
12)147.6
  − 12
     27
   − 24
      36
    − 36
       0
```

Multiply. 12 × 3
Subtract. 36 − 36
Check

```
  12.3
 × 12
   246
  1230
 147.6
```

So, 147.6 ÷ 12 = 12.3.

Try These

Find the quotient. Remember to write the decimal point in the quotient.

1.

```
       □.□
287 )7 1 7. 5
   − □□□
     □□□□
   − □□□□
          □
```

Estimated quotient: _____
Check: □.□ × 287

2.

```
   □.□

− □□
   □□
 − □□
    □
```

Estimated quotient: _____
Check: □.□ × 11

Go to the next side.

Name ______________________ Skill ______________________

Practice on Your Own

Skill 45

Find 2.08 ÷ 4.

Think:
Divide decimals as you would divide whole numbers. Remember to write the decimal point in the quotient.

```
   0.52
4)2.08
  20↓
    8
  - 8
    0
```

When the quotient is less than 1, put a zero in the ones place.

Check

```
  0.52
×    4
  2.08
```

Think:
2.0 ÷ 4 = 0.5
The estimated quotient is less than one.

Find the quotient. Write the decimal in the quotient.

1. 8)57.6 — □.□; −□□; □□; −□□; □
   Estimated quotient: _____
   Quotient: _____
   Check: □.□ × 8

2. 13)72.8 — □.□; −□□; □□; − □□; □
   Estimated quotient: _____
   Quotient: _____
   Check: □.□ × 13

3. 6)72.6
   Estimated quotient: _____
   Quotient: _____
   Check: □□.□ × 6

4. 15)205.5
   Estimated quotient: _____
   Quotient: _____
   Check: □□.□ × 15

5. 5)32.5

6. 9)4.59

7. 7)43.54

8. 21)107.1

Check

9. 3)28.2

10. 8)5.84

11. 6)38.52

12. 15)112.5

Skill 46 Grade 5 Model Decimal Multiplication

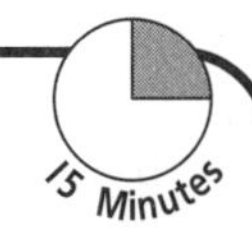

Using Skill 46

OBJECTIVE Use a model to multiply decimals

MATERIALS graph paper

Begin by explaining that multiplying with decimals can be shown with models. Point out that finding 0.4×0.7 is the same as finding 0.4 of 0.7.

In Step 1, have students note that there are ten parts in the model. Ask: **If 1 column equals 1 tenth, how many tenths are shaded?** (7 tenths) So the model represents the factor 0.7.

Explain that in Step 2 the tenths model is divided into 10 equal rows. Ask: **If 1 row equals 1 tenth, how many tenths are crossed?** (4 tenths) The crossed part of the model represents the factor 0.4.

Students may notice that the ten columns and ten rows result in 100 equal parts, or hundredths.

In Step 3, explain that the section with both shaded parts and crossed parts represents 0.4 of 0.7 or 0.4×0.7. The product is 28 hundredths, or 0.28.

Have students look at the model. Remind them that when you multiply two decimals less than 1, the product is less than either of the two factors.

TRY THESE Exercises 1–3 provide practice using the hundredths model to represent multiplication of decimals. Verbal prompts guide the students to write each multiplication sentence using decimals.

- **Exercises 1–3** Use a hundredths model to multiply tenths by tenths.

PRACTICE ON YOUR OWN Review the example at the top of the page. Have students identify the shaded and crossed sections of the hundredths model. Ask students to identify the number sentence.

CHECK Determine if students can multiply tenths and write the multiplication number sentence illustrated by the hundredths model.

Success is indicated by 2 out of 3 correct responses.

Students who successfully complete the **Practice on Your Own** and **Check** are ready to move to the next skill.

COMMON ERRORS

- Students might not recall multiplication facts.
- Students might not realize that one factor is represented by the shaded columns, and the other factor is represented by the crossed rows.
- Students might fail to count the exact number of overlapping sections in the hundredths model.
- Students might not position the decimal point to reflect the correct value of the product.

Students who made more than 2 errors in the **Practice on Your Own**, or who were not successful in the **Check** section, may benefit from the **Alternative Teaching Strategy** on the next page.

Optional

Alternative Teaching Strategy

Model Decimal Multiplication

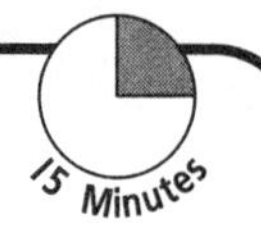

OBJECTIVE Use a model to multiply tenths

MATERIALS graph paper (inch square), flip chart

Cut a 10 × 10 square from 1-inch graph paper. On the flip chart, make a copy of the grid and label as shown:

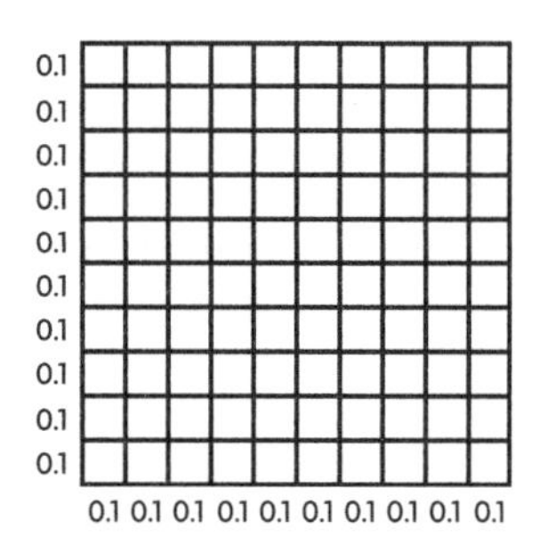

Distribute the graph paper to each student. Guide students to see how the grid is separated into equal parts. Explain that:

- Each column or row represents one tenth of the grid, or 0.1.
- Each square represents one hundredth of the grid, or 0.01.

Tell students the grid can be used as a model for multiplying decimals. For example, 0.5 × 0.7.

Direct students to count 7 tenths from the left and fold the grid so that only 7 tenths or 7 columns show. Point out that the grid shows the factor 0.7.

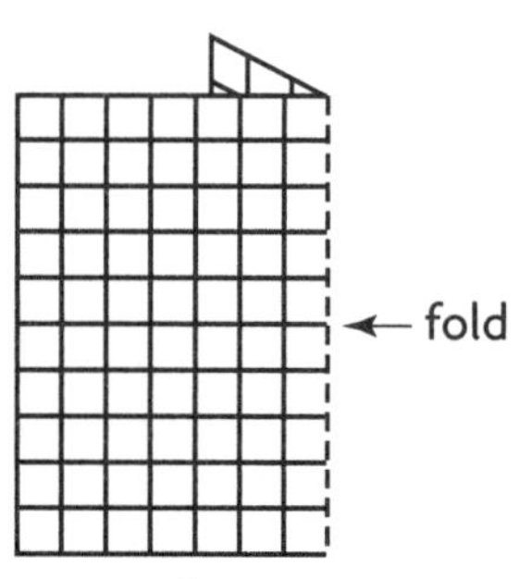

Next, direct students to count 5 tenths or rows from the top and fold them back so that only 5 tenths remain. Point out that the remaining part of the grid shows 0.5 of 0.7 or 0.5 × 0.7. Have them count the squares and find the product 35 hundredths or 0.35. So, 0.5 × 0.7 = 0.35.

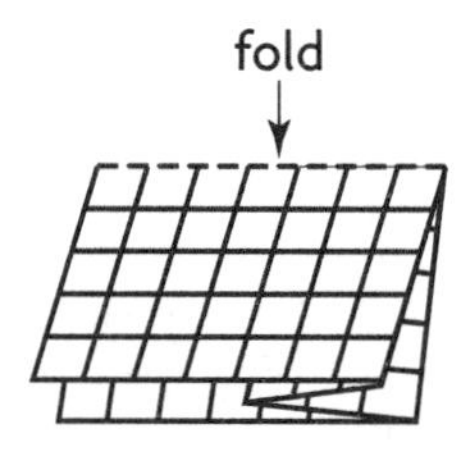

Repeat the activity with other examples until students understand how to model the multiplication of decimals. Then compare the factors and the products. Have students note that when they multiply decimals the product is less than either of the two factors.

Name ______________________ Skill ______________________

Grade 5 Skill 46

Model Decimal Multiplication

Make a model to represent 0.4 of 0.7. The word "of" means to multiply.
Model 0.4 × 0.7 to find the product.

Step 1
Use a tenths model. Shade 7 columns to show the second factor, 0.7.

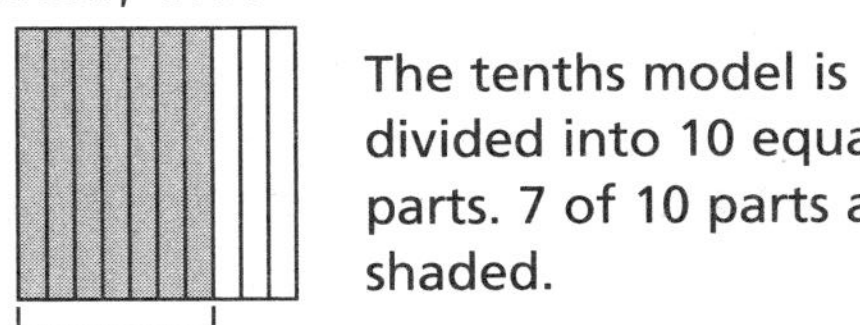

The tenths model is divided into 10 equal parts. 7 of 10 parts are shaded.

Shaded parts = *seven tenths*, or 0.7

Step 2
Divide the tenths model into 10 equal rows. Cross 4 rows to represent 0.4.

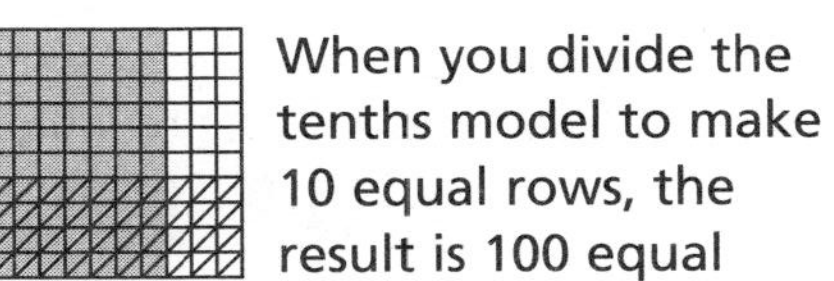

When you divide the tenths model to make 10 equal rows, the result is 100 equal parts.

Crossed parts = *four tenths* or 0.4

Step 3
The section with both shaded parts and crossed parts represents the product.

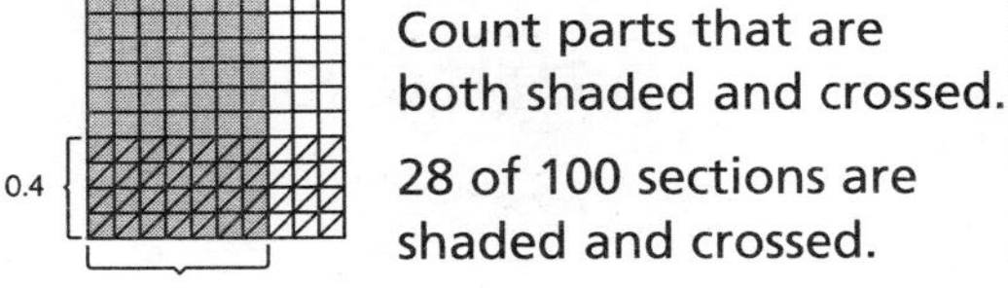

Count parts that are both shaded and crossed.

28 of 100 sections are shaded and crossed.

Shaded and crossed parts = *twenty-eight hundredths*, or 0.28.

So, you write the number sentence as 0.4 × 0.7 = 0.28.

Try These

Write the number sentence represented by each model.

1. shaded part: 0.3
crossed part: _____
shaded and crossed parts: _____
Number sentence: _____ × _____ = 0.18

2. shaded part: _____
crossed part: _____
shaded and crossed parts: _____
Number sentence: _____ × _____ = _____

3. shaded part: _____
crossed part: _____
shaded and crossed parts: _____
Number sentence: _____ × _____ = _____

Go to the next side.

Name ______________________________ Skill ______________________

Practice on Your Own

Skill 46

Think:
The section that is both shaded and crossed represents the product of the two decimals.

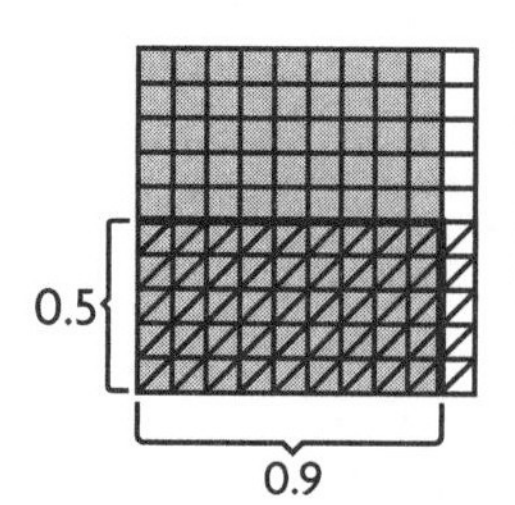

The model represents 0.5 of 0.9.

Number sentence:
$0.5 \times 0.9 = 0.45$

Complete. Write the number sentence represented by each model.

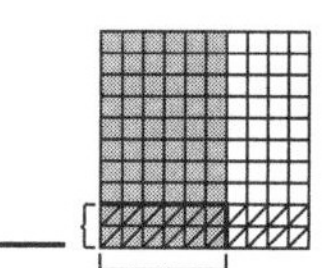

Number sentence:

____ × ____ = ____

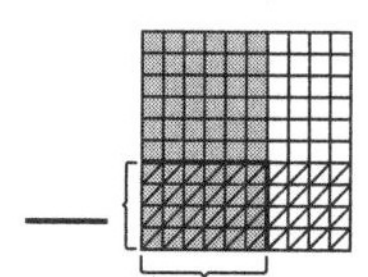

Number sentence:

____ × ____ = ____

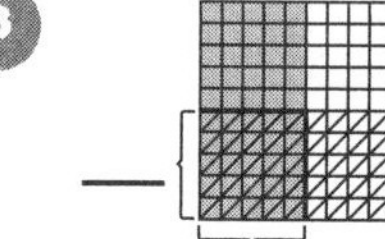

Number sentence:

____ × ____ = ____

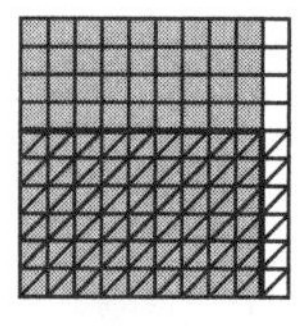

____ × ____ = ____

5
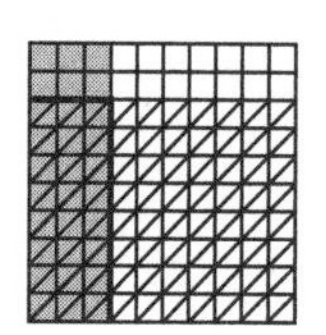

____ × ____ = ____

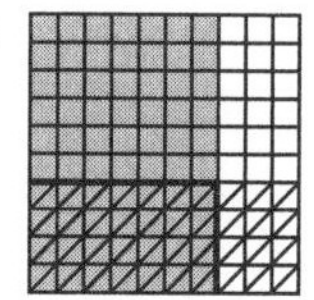

____ × ____ = ____

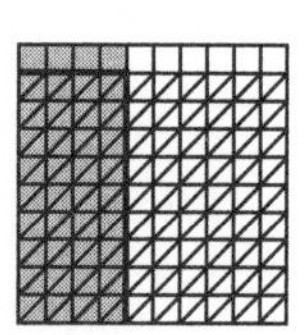

____

8
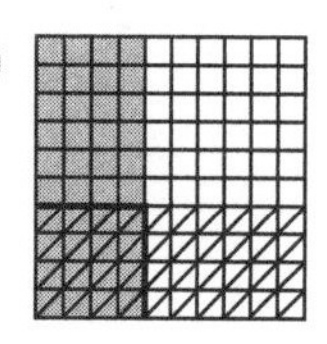

____

9
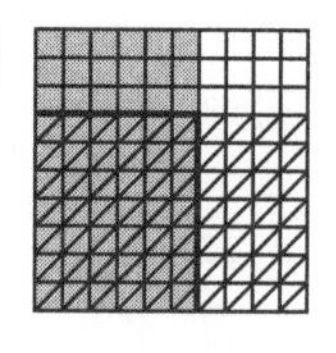

____

Check

Write the number sentence modeled.

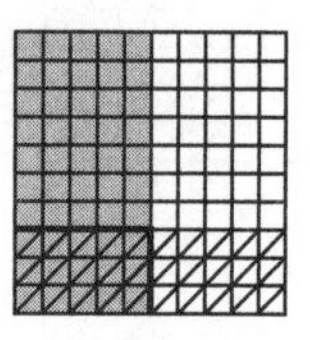

____

11
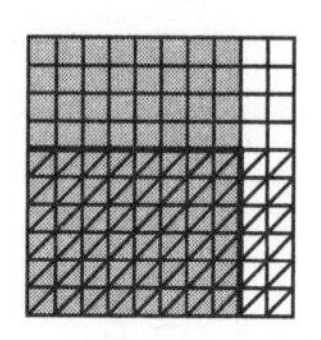

____

12
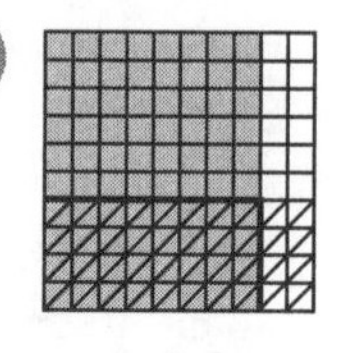

____

Understand Hundredths (for Percents)

15 Minutes

Using Skill Card 47

OBJECTIVE Write and read a fraction and a decimal to represent hundredths

MATERIALS hundredths decimal model, tenths decimal model

You may wish to review tenths with students. Recall with students that one tenth is represented by the decimal, 0.1, and the fraction, $\frac{1}{10}$. Show students the tenths decimal model. Point out that the model is divided into 10 equal parts.

Ask: **What are 10 tenths equal to?** (1 whole)

Tell students that one whole can also be divided into 100 equal parts. Show the hundredths decimal model. Tell students that each small square represents one hundredth of the whole.

Direct students' attention to the first example. Have students look at the hundredths decimal model. Point out that there are 100 equal parts in the whole and that one of the equal parts is shaded. Help students recognize that this represents one hundredth of the whole. Point out that you can write a fraction or a decimal to represent one hundredth. Ask: **How do you write the fraction to represent one hundredth?** ($\frac{1}{100}$) **How do you write the decimal to represent one hundredth?** (0.01)

Direct students' attention to the next example. Have them look at the decimal model. Ask: **How many equal parts are there in all?** (100) **How many equal parts are shaded?** (34)

Tell students that the shaded part represents thirty-four hundredths. Point out to students that thirty-four hundredths can also be written as a fraction or a decimal and that both the fraction and decimal are read as thirty-four hundredths.

TRY THESE Exercises 1–2 use models to provide practice writing and reading fractions and decimals for hundredths.

- **Exercise 1** Seven hundredths.
- **Exercise 2** Sixty-five hundredths.

PRACTICE ON YOUR OWN Review the example at the top of the page. Have students read the numbers shown in the model. In Exercises 1–4, students use models to write and read fractions and decimals. In Exercises 5–7, students are given the word name and must write the fraction and decimal.

CHECK Determine if students can write a fraction and decimal for a given word name.

Success is indicated by 3 out of 3 correct responses.

Students who successfully complete the **Practice on Your Own** and **Check** are ready to move to the next skill.

COMMON ERRORS

- Students read decimals incorrectly.
- Students write decimals incorrectly; for example, they write three hundredths as 0.30 rather than as 0.03.

Students who made more than 1 error in the **Practice on Your Own**, or who were not successful in the **Check** section, may benefit from the **Alternative Teaching Strategy** on the next page.

Optional

Alternative Teaching Strategy

20 Minutes

Model Reading and Writing Fractions and Decimals

OBJECTIVE Read and write a fraction and a decimal for a model showing hundredths

MATERIALS cards with decimal models, fractions, decimals, and word names for hundredths

Before beginning this activity, review how to write the fraction, decimal, and word name for a decimal model for hundredths.

Have students work in pairs for this activity. Distribute the fraction, decimal, and word name cards.

Ask students to match the fraction, decimal, and word name cards to the model. Show the following model.

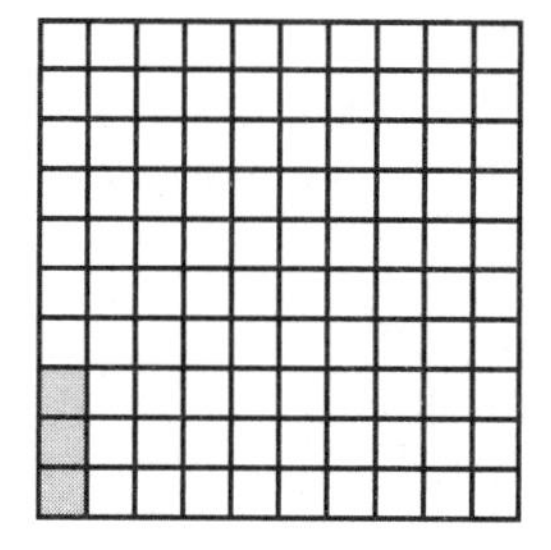

Ask: **How many of 100 equal parts are shaded?** (3)

Have one partner show the fraction card that matches the model. Have the other partner show the decimal card that matches the model. Partners take turns showing the word name card that matches the model.

Next, show the following model.

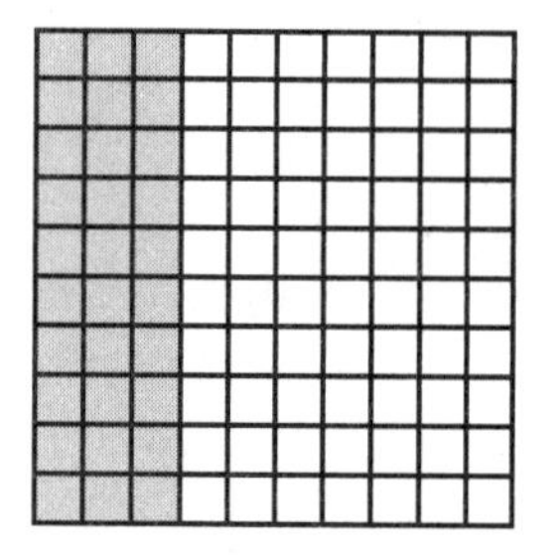

Ask: **How many of 100 equal parts are shaded?** (30)

Have students show the fraction, decimal, and word name cards that match the model. Guide students to recognize the difference between 0.03 and 0.30.

Repeat this activity for additional models.

You may wish to adapt this activity by showing the word name for a fraction, and then having students show the fraction and decimal cards that match the word name.

Name ______________________ Skill ______________________

Grade 5 **Skill** 47

Understand Hundredths (for Percents)

Write a fraction or decimal to represent hundredths.

This square represents a whole.
It is divided into 100 equal parts.
1 of 100 equal parts is shaded.
The shaded part is one hundredth of the whole.
You can write a fraction or a decimal to represent the shaded part:

Fraction: $\frac{1}{100}$ **Decimal:** 0.01

Read: one hundredth

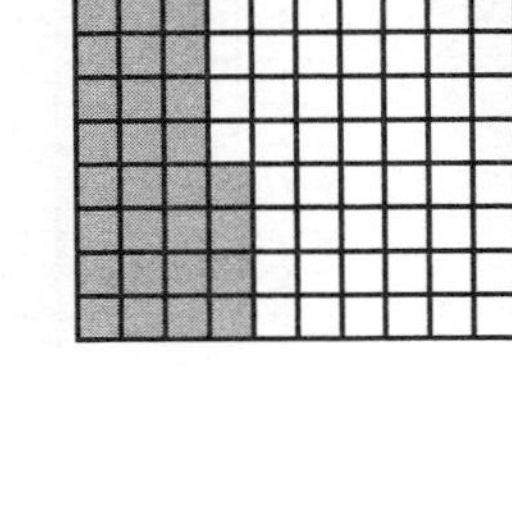

In this model, 34 of 100 equal parts are shaded.
You can write a fraction or a decimal to represent thirty-four hundredths.

Fraction: $\frac{34}{100}$ **Decimal:** 0.34

Read: thirty-four hundredths

Try These

Complete.

1. _____ of _____ parts are shaded.

   Fraction: $\frac{\square}{\square}$

   Decimal: _____

   Read: ______________________

2. _____ of _____ parts are shaded.

   Fraction: $\frac{\square}{\square}$

   Decimal: _____

   Read: ______________________

Go to the next side.

Name ______________________ Skill ______________________

Practice on Your Own

Skill 47

Describe the shaded part. Write as a fraction, as a decimal, and in word form.

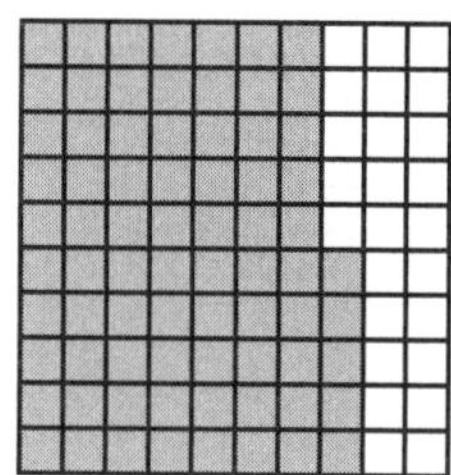

Think: 75 of 100 parts are shaded.

Fraction: $\frac{75}{100}$

Decimal: 0.75

Read: seventy-five hundredths

Describe the shaded part. Write as a fraction, as a decimal and in word form.

1. ____ of ____ parts are shaded.

   Fraction: $\frac{\square}{\square}$

   Decimal: ____

   Read: ______________

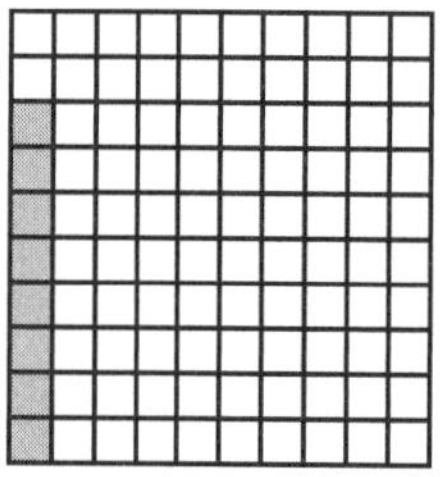

2. ____ of ____ parts are shaded.

   Fraction: $\frac{\square}{\square}$

   Decimal: ____

   Read: ______________

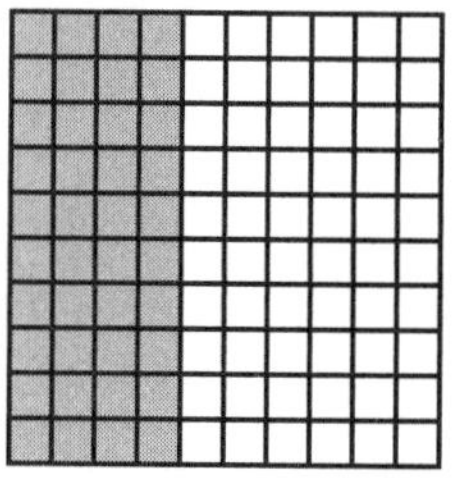

3. Fraction: $\frac{\square}{\square}$

   Decimal: ____

   Read: ______________

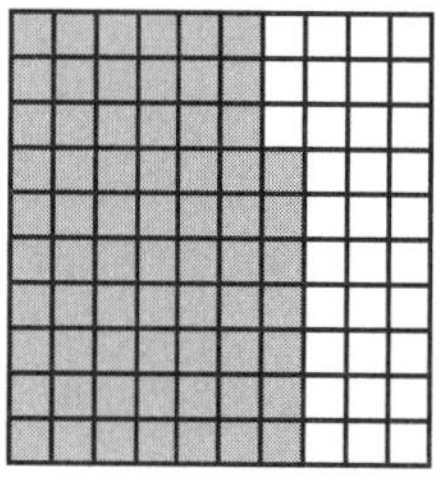

4. Fraction: $\frac{\square}{\square}$

   Decimal: ____

   Read: ______________

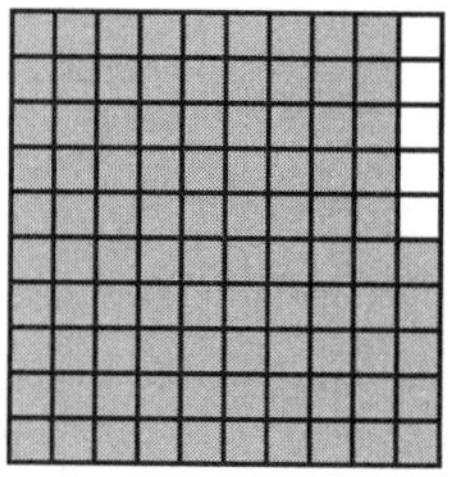

Write a fraction and a decimal.

5. forty-nine hundredths

   Fraction: ____

   Decimal: ____

6. three hundredths

   Fraction: ____

   Decimal: ____

7. eighty hundredths

   Fraction: ____

   Decimal: ____

Check

Write a fraction and a decimal.

8. five hundredths

   Fraction: ____

   Decimal: ____

9. thirty-one hundredths

   Fraction: ____

   Decimal: ____

10. sixty-eight hundredths

    Fraction: ____

    Decimal: ____

Skill 48 Grade 5

Relate Fractions and Decimals

Using Skill 48

OBJECTIVE Relate fractions and decimals

MATERIALS place-value chart

Ask students to read the information on writing a fraction or decimal to represent tenths. Then, have them look at the model and find the number of equal parts and the number of shaded parts.

Direct students' attention to the fraction $\frac{7}{10}$ and the decimal 0.7. Write 0.7 in a place-value chart and help students understand the relationship between the two numbers. Ask: **What is the denominator of the fraction?** (10) **Where is the digit 7 in the place-value chart?** (in the tenths place) So, both numbers represent 7 tenths.

Use similar questions for writing a fraction or decimal to represent hundredths.

Go through the steps for writing the equivalent fraction for $\frac{3}{4}$. Ask: **How do you know that $\frac{3}{4}$ is equivalent to $\frac{75}{100}$?** (Possible answer: multiplying $\frac{3}{4}$ and $\frac{25}{25}$ is equivalent to multiplying $\frac{3}{4}$ and 1.)

Help students understand the reason for writing an equivalent fraction with a denominator of 100. Use a place-value chart to show students that since the denominator is now 100, they can write the fraction as a decimal.

Have students study the example showing $1\frac{34}{100}$. Discuss the fact that a mixed number represents a value greater than one.

TRY THESE In Exercises 1–3 students write fractions, decimals, and mixed numbers for models of tenths and hundredths.

- **Exercise 1** Write a fraction and a decimal to represent tenths.
- **Exercise 2** Write a fraction and a decimal to represent hundredths.
- **Exercise 3** Write a mixed number as a decimal to hundredths.

PRACTICE ON YOUR OWN Review the example at the top of the page. Use a place-value chart to help students review writing other decimals.

CHECK Determine if students can relate fractions, decimals, and mixed numbers by writing equivalent forms of each number. Success is indicated by 3 out of 3 correct responses.

Students who successfully complete the **Practice on Your Own** and **Check** are ready to move to the next skill.

COMMON ERRORS

- Given a fraction in tenths or hundredths, students may write the decimal form in incorrect place-value position.
- Students may forget that to write equivalent fractions they must multiply by a fraction equal to 1.
- When writing a mixed number as a decimal, students may forget to write the whole number in the mixed decimal.

Students who made more than 2 errors in the **Practice on Your Own**, or who were not successful in the **Check** section, may benefit from the **Alternative Teaching Strategy** on the next page.

Optional

Alternative Teaching Strategy

20 Minutes

Model Relating Fractions and Decimals

OBJECTIVE Use models to write fractions as decimals

MATERIALS place-value charts, decimal models for tenths

Display a square showing tenths; color 3 tenths red.

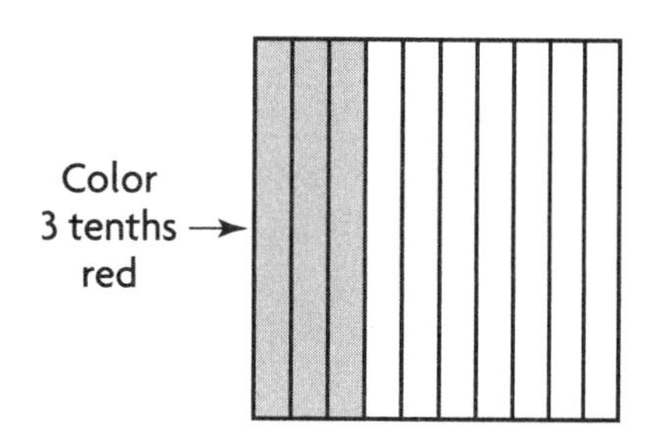

Ask: **How many equal parts are in the whole?** (10) **How many parts are red?** (3) **What fraction do you write for 3 tenths?** ($\frac{3}{10}$)

Next to the fraction, write the word form for 3 tenths. Then show students how to place the digit 3 in the tenths column of a place-value chart, writing a zero in the ones column.

Ask: **How many ones?** (0) **How many tenths?** (3) **How do you read the decimal?** (three tenths)

Pointing to the fraction, ask:

How do you read the fraction? (3 tenths)

Do the fraction and the decimal represent the same value? (yes)

How can you tell? (They both represent the red part of the decimal model.)

Add dotted lines to the square to show hundredths. Point out that the decimal square now has 100 equal parts. Shade 4 parts blue.

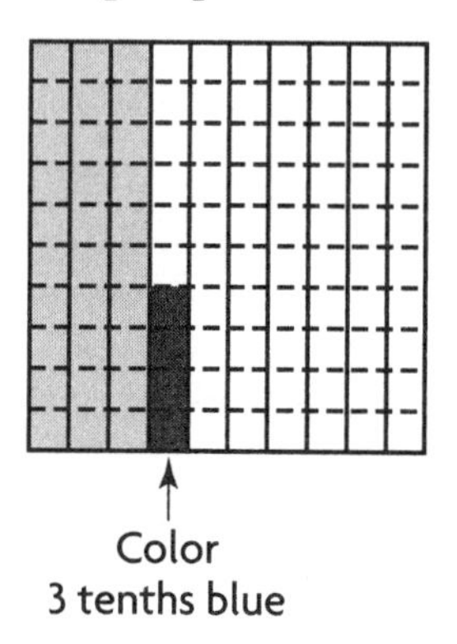

Ask students how many parts are blue and to give the fraction that represents the part. Have them tell you the word form (4 hundredths) and display the decimal in the place-value chart.

Refer to the model again.

Ask: **How many parts are blue or red?** (3 tenths and 4 hundredths, or 34 hundredths)

Help students count tenths and hundredths to verify that 3 tenths plus 4 hundredths is the same as 34 hundredths.
Write: $0.3 + 0.04 = 0.34$.

Then have a student write 0.34 in the place-value chart.

When students demonstrate understanding, have them try the exercise using only pencil and paper.

Name ______________________ Skill ______________________

Grade 5 Skill 48

Relate Fractions and Decimals

You can write an amount less than 1 as a fraction or as a decimal.

Tenths
Write a fraction or decimal to represent tenths.

The whole is divided into 10 equal parts.
7 of the 10 parts are shaded.
So, seven tenths are shaded.

Fraction: $\frac{7}{10}$ Decimal: 0.7

Hundredths
Write a fraction or decimal to represent hundredths.

The whole is divided into 100 equal parts.
9 of 100 parts are shaded.
So, nine hundredths are shaded.

Fraction: $\frac{9}{100}$ Decimal: 0.09

Write a decimal for the fraction $\frac{3}{4}$.

Fraction
$\frac{3}{4}$

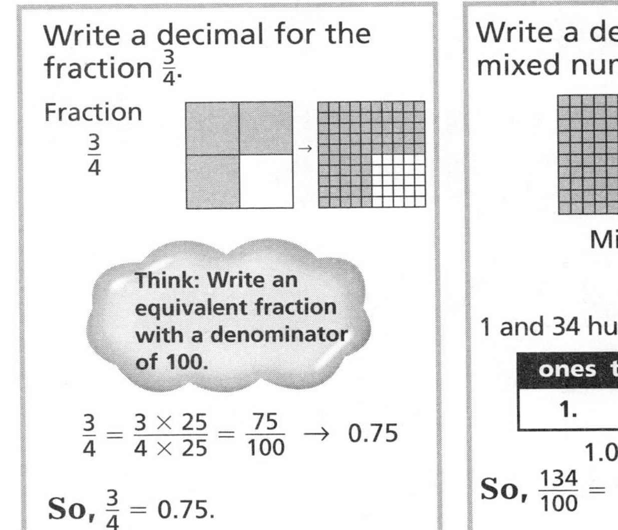

$\frac{3}{4} = \frac{3 \times 25}{4 \times 25} = \frac{75}{100} \rightarrow 0.75$

So, $\frac{3}{4} = 0.75$.

Write a decimal for the mixed number $1\frac{34}{100}$.

Mixed Number
$1\frac{34}{100}$

1 and 34 hundredths are shaded.

ones	tenths	hundredths
1.	3	4

$1.0 + 0.3 + 0.04$

So, $\frac{134}{100} = 1.34$.

Try These

Write a fraction or mixed number and a decimal to describe the shaded part.

1. Fraction: _____ Decimal: _____

2. Fraction: _____ Decimal: _____

3. Mixed Number: _____ Decimal: _____

Go to the next side.

Name ______________________ Skill ______________________

Practice on Your Own

Skill 48

Think:
Write a mixed number and a decimal to describe the shaded part.

Mixed number: $1\frac{49}{100}$

Decimal: 1.49

ones	tenths	hundredths
1.	4	9

1.0 + 0.4 + 0.09

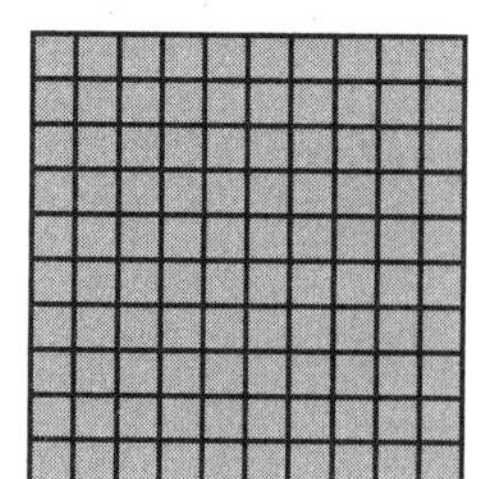

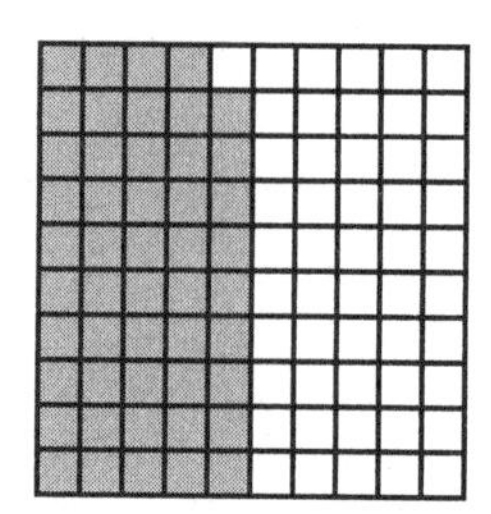

Write a fraction or mixed number and a decimal.

1. 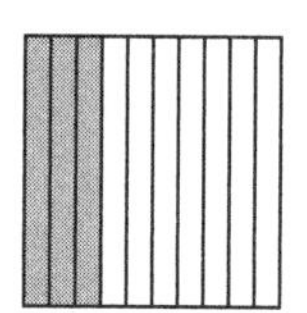

Fraction: _____

Decimal: _____

2. 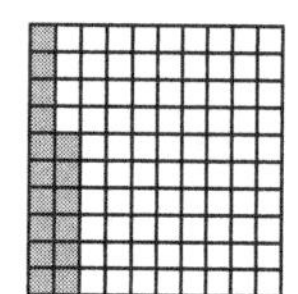

Fraction: _____

Decimal: _____

3. 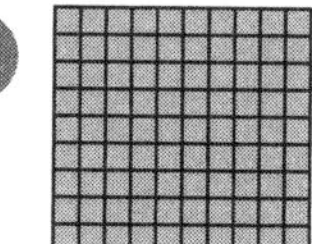

Mixed Number: _____

Decimal: _____

Write the decimal.

4. $\frac{9}{10}$

Decimal: _____

5. $\frac{48}{100}$

Decimal: _____

6. $2\frac{87}{100}$

Decimal: _____

7. $\frac{1}{4}$

Decimal: _____

8. $\frac{2}{5}$

Decimal: _____

9. $3\frac{2}{5}$

Decimal: _____

Check

Write a fraction, mixed number, or decimal.

10. 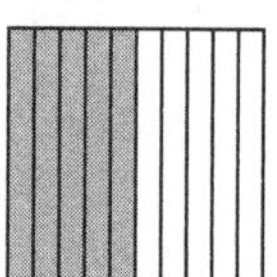

Fraction: _____

Decimal: _____

11. 

Mixed Number: _____

Decimal: _____

12. 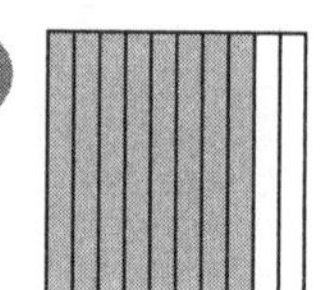

Decimal: _____

Answer Card
Decimals
Grade 5

SKILL 39

TRY THESE

1. 0 + 0.3; three tenths
2. 1 + 0.7; one and seven tenths
3. 0 + 0.4 + 0.05; forty-five hundredths

PRACTICE

1. 16.9
2. 1.29
3. 2.5
4. 25.6
5. 463.02
6. 3,945.8
7. 500 + 20 + 6 + 0.1 + 0.08
8. 6,000 + 90 + 2 + 0.4
9. eight and seven hundredths
10. twenty-eight and six tenths
11. 10 + 7 + 0.9; seventeen and nine tenths
12. 40 + 5 + 0.01; forty-five and one hundredth

SKILL 39

CHECK

13. 500 + 80 + 6 + 0.02; five hundred eighty-six and two hundredths
14. 8,000 + 200 + 50 + 9 + 0.6; eight thousand, two hundred fifty-nine and six tenths

SKILL 40

TRY THESE

1. 31, 32, 32, 32
2. 4, 5, yes, 5
3. 7, 2, no, 217

PRACTICE

1. 5, 6, 5, 5
2. 29, 29
3. 262
4. 0, 5, yes, 1
5. 6, 3, 36
6. 2, 343
7. 40
8. 45
9. 635

CHECK

10. 2
11. 25
12. 576

SKILL 41

TRY THESE

1. 40, 4.0
2. 60, 6.0
3. 4, 0.4
4. 15, 1.5

PRACTICE

1. 10, 1.0
2. 10, 1.0
3. 70, 7.0
4. 90, 9.0
5. 5, 0.5
6. 1, 0.1
7. 5, 0.5
8. 82, 8.2
9. 2.0
10. 2.0
11. 5.6
12. 1.0
13. 5.0
14. 4.0
15. 0.4
16. 3.7
17. 0.8

CHECK

18. 1.0
19. 6.0
20. 4.2

Name ______________________ Skill ______________________

Name ______________________ Skill ______________________

Answer Card
Decimals
Grade 5

SKILL 42

TRY THESE

1. 2.6
2. 1.7
3. 30.3
4. 2.5

PRACTICE

1. 0.9
2. 1.3
3. 8.1
4. 39.1
5. 0.5
6. 2.9
7. 0.9
8. 31.7
9. 1.2
10. 0.7
11. 17.1
12. 51.8
13. 18.7
14. 1.9
15. 32.5
16. 1.3

CHECK

17. 10.5
18. 2.4
19. 3.7
20. 48.1

SKILL 43

TRY THESE

1. 0.4
2. 1.2
3. 4.2

PRACTICE

1. 0.9
2. 2.4
3. 2.2
4. 4.0
5. 1.0
6. 2.1
7. 1.6
8. 1.8
9. 3.9
10. 0.8

CHECK

11. 0.8
12. 3.3
13. 3.2

SKILL 44

TRY THESE

1. 0.4 = 0.40
2. 0.8 = 0.80
3. 0.200

PRACTICE

1. 0.5, 0.50
2. 0.7, 0.70
3. 0.3, 0.30

Problems 4–15: Possible answers are given.

4. 0.50, 0.500
5. 0.120, 0.1200
6. 1.010, 1.0100
7. 6.50, 6.500
8. 0.040, 0.400
9. 1.40, 1.400
10. 4.90, 4.900
11. 0.900, 0.9000
12. 2.70, 2.700

CHECK

13. 0.080, 0.0800
14. 4.010, 4.0100
15. 6.1000, 6.10000

Name ______________________ Skill ______________________

Answer Card

Decimals

Grade 5

SKILL 45

TRY THESE

1. 2, 5, 5, 7, 4, 1, 4, 3, 5, 1, 4, 3, 5, 0; 3; 2, 5; 717.5
2. 4, 3, 4, 4, 3, 3, 3, 3, 0; 4 or 5; 4, 3; 47.3

PRACTICE

1. 7, 2, 5, 6, 1, 6, 1, 6, 0; 7; 7.2; 7, 2; 57.6
2. 5, 6, 6, 5, 7, 8, 7, 8, 0; 6; 5.6; 5, 6; 72.8
3. 12, 12.1, 1, 2, 1; 72.6
4. 13 or 14, 13.7, 1, 3, 7; 205.5
5. 6.5
6. 0.51
7. 6.22
8. 5.1

CHECK

9. 9.4
10. 0.73
11. 6.42
12. 7.5

SKILL 46

TRY THESE

1. 0.3; 0.6; 0.18; $0.6 \times 0.3 = 0.18$
2. 0.6; 0.5; 0.30; $0.5 \times 0.6 = 0.30$
3. 0.4; 0.3; 0.12; $0.3 \times 0.4 = 0.12$

PRACTICE

1. 0.2; 0.6; $0.2 \times 0.6 = 0.12$
2. 0.4; 0.6; $0.4 \times 0.6 = 0.24$
3. 0.5; 0.5; $0.5 \times 0.5 = 0.25$
4. $0.6 \times 0.9 = 0.54$
5. $0.8 \times 0.3 = 0.24$
6. $0.4 \times 0.7 = 0.28$
7. $0.9 \times 0.4 = 0.36$
8. $0.4 \times 0.4 = 0.16$
9. $0.7 \times 0.6 = 0.42$

CHECK

10. $0.3 \times 0.5 = 0.15$
11. $0.6 \times 0.8 = 0.48$
12. $0.4 \times 0.8 = 0.32$

Name ____________ Skill ____________

SKILL 47

TRY THESE

1. 7, 100; $\frac{7}{100}$; 0.07; seven hundredths
2. 65, 100; $\frac{65}{100}$; 0.65; sixty-five hundredths

PRACTICE

1. 8, 100; $\frac{8}{100}$; 0.08; eight hundredths
2. 40, 100; $\frac{40}{100}$; 0.40; forty hundredths
3. $\frac{67}{100}$; 0.67; sixty-seven hundredths
4. $\frac{95}{100}$; 0.95; ninety-five hundredths
5. $\frac{49}{100}$; 0.49
6. $\frac{3}{100}$; 0.03
7. $\frac{80}{100}$; 0.80

CHECK

8. $\frac{5}{100}$; 0.05
9. $\frac{31}{100}$; 0.31
10. $\frac{68}{100}$; 0.68

SKILL 48

TRY THESE

1. $\frac{4}{10}$ or $\frac{2}{5}$; 0.4
2. $\frac{6}{100}$ or $\frac{3}{50}$; 0.06
3. $1\frac{67}{100}$; 1.67

PRACTICE

1. $\frac{3}{10}$; 0.3
2. $\frac{16}{100}$; 0.16
3. $1\frac{30}{100}$ or $1\frac{3}{10}$; 1.30 or 1.3
4. 0.9
5. 0.48
6. 2.87
7. 0.25
8. 0.40 or 0.4
9. 3.40 or 3.4

CHECK

10. $\frac{5}{10}$ or $\frac{1}{2}$; 0.5
11. $1\frac{40}{100}$ or $1\frac{2}{5}$; 1.40 or 1.4
12. 0.8 or 0.80

Answer Card

Decimals

Grade 5

Algebra and Functions

Skill 49 Grade 5

Mental Math: Function Tables

Using Skill 49

OBJECTIVE Use mental math to complete a function table.

You may wish to review addition and subtraction facts.

For the addition rule, be sure students are able to read the function machine. Ask: **What numbers represent the input on the function machine?** (3, 5, 8, 10, 12) **What operation is used as a rule in the function machine?** (addition) **What is the rule used in the function machine?** (add 6, or + 6) **What numbers represent the output?** (9, 11, 14, 16, 18)

Draw students' attention to the function machine for the subtraction rule. Ask: **What are the input numbers?** (6, 9, 17, 20, 25) **What operation is used in the function machine?** (subtraction) **What is the rule used in the function machine?** (subtract 4, or − 4) **What numbers represent the output?** (2, 5, 13, 16, 21)

Point to the function table below each function machine. The table is a condensed version of the function machine. Both tables show a function, or a rule that is applied to the input number to result in the output number.

TRY THESE Exercises 1–2 model a function machine and function table.

- **Exercise 1** A function machine and table with an addition rule.
- **Exercise 2** A function machine and table with a subtraction rule.

PRACTICE ON YOUR OWN Review the example at the top of the page. Ask students to name the input, the rule, and the output.

CHECK Determine if students can use the given rule to arrive at the output.

Success is indicated by 2 out of 2 correct responses.

Students who successfully complete the **Practice on Your Own** and **Check** are ready to move to the next skill.

COMMON ERRORS

- Students might interchange the output and the input in a subtraction rule.
- Students might have difficulty with words that indicate an operation (increase, decrease, plus, minus, and so on).
- Students might use incorrect addition or subtraction facts.

Students who made more than 4 errors in the **Practice on Your Own**, or who were not successful in the **Check** section, may benefit from the **Alternative Teaching Strategy** on the next page.

Optional

Alternative Teaching Strategy
Working with Function Machines

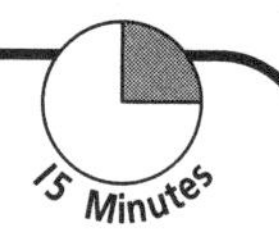

OBJECTIVE Use a function machine model

MATERIALS tag board, sticky pad paper, shoebox

On tag board, prepare an input/output diagram like the one below.

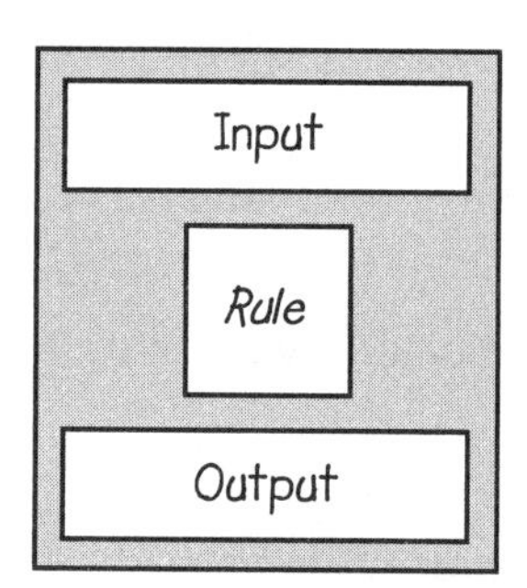

On sticky pad paper, write several addition and subtraction rules. For example:

- add 5
- add 3
- subtract 1
- subtract 4

Put the rules in the shoebox.

Ask a volunteer to write a number between 5 and 15 on sticky pad paper and place the number in the input area on the diagram.

Ask another student to pick a rule from the shoebox, read the rule aloud, and place it on the diagram. Guide students to compute as directed by the rule. Then post the result in the output area on the tag board.

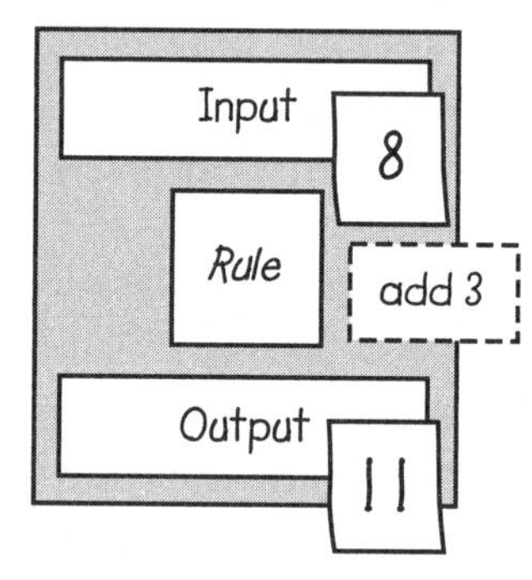

Repeat the same rule for other numbers between 5 and 15. Then select a different rule. After a few tries using the board, you might give students a function table to record their numbers and the results.

Input	8		
Output	11		

Name ______ Skill ______

Grade 5 Skill 49

Mental Math: Function Tables

Use mental math to complete the function table.

Rule: Add 6.

Function Machine

Input	3	5	8	10	12
Function Machine	+ 6	+ 6	+ 6	↓	↓
Output	9	11	14	16	18

Function Table

Input	3	5	8	10	12
Output	9	11	14	16	18

Rule: Subtract 4.

Function Machine

Input	6	9	17	20	25
Function Machine	− 4	− 4	− 4	↓	↓
Output	2	5	13	16	21

Function Table

Input	6	9	17	20	25
Output	2	5	13	16	21

Try These

1. Use mental math to complete the function table.

Rule: Add 9.

Function Machine

Input	3	6	7	10	12	15
Function Machine	+ 9	+ 9	+ 9	↓	↓	↓
Output	12	15	16	____	____	____

Function Table

Input	3	6	7	10	12	15
Output	12	15	16	____	____	____

2. Use mental math to complete the function table.

Rule: Subtract 7.

Function Machine

Input	9	14	16	20	26	30
Function Machine	− 7	− 7	− 7	↓	↓	↓
Output	2	7	9	____	____	____

Function Table

Input	9	14	16	20	26	30
Output	2	7	9	____	____	____

Go to the next side.

Name ______________________ Skill ______________________

Practice on Your Own

Think:
Use the rule to change the input number.
The output number is the result.

Use mental math to complete the function table.

Rule: Subtract 6.

Function Table

Input	9	14	16	20	26
Output	3	8	10	14	20

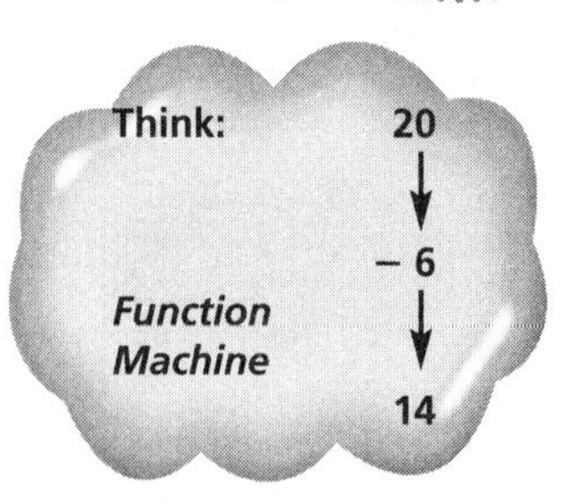

Use mental math to complete.

1 Add 8.

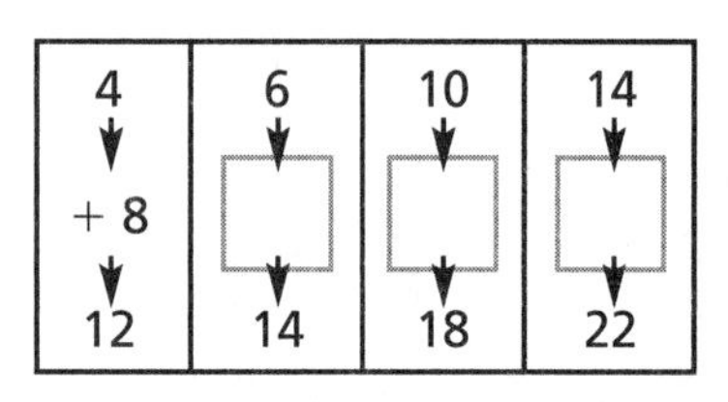

Input	4	6	10	14
Function Machine	+ 8			
Output	12	14	18	22

2 Subtract 5.

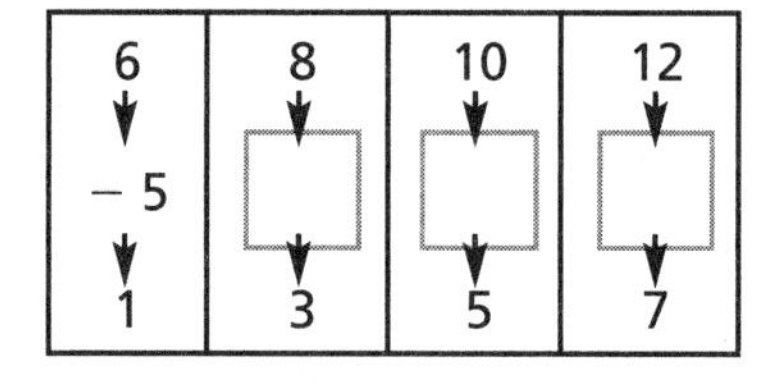

Input	6	8	10	12
Function Machine	− 5			
Output	1	3	5	7

3 Subtract 15.

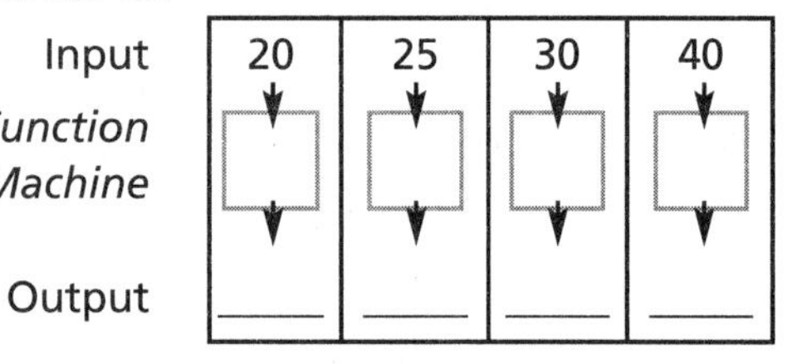

Input	20	25	30	40
Function Machine				
Output	___	___	___	___

4 Increase by 7.

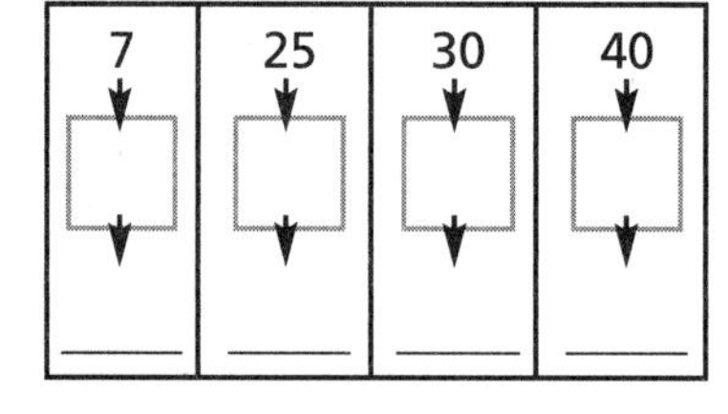

Input	7	25	30	40
Function Machine				
Output	___	___	___	___

5 Decrease by 10.

Input	20	25	30	40
Output	10	15	___	___

6 Add 4.

Input	7	10	17	20
Output	11	14	___	___

7 Subtract 9.

Input	36	27	18	9
Output	___	___	___	___

8 Add 11.

Input	11	22	33	44
Output	___	___	___	___

Check

Use mental math to complete the function table.

9 Increase by 6.

Input	24	34	44	54
Output	___	___	___	___

10 Decrease by 20.

Input	100	80	60	40
Output	___	___	___	___

Addition and Subtraction Equations

15 Minutes

Using Skill 50

OBJECTIVE Write the value for the variable in an addition or subtraction equation

You may wish to review the meaning of an equation. Remind students that it is a number sentence with an equals sign (=). The expressions on both sides of the equals sign name the same value.

Draw attention to the addition equation. Ask: **What is the equation?** ($3 + 5 = n$) **What is the sum of the numbers to the left of the equals sign?** (8) **What number must the letter *n* represent?** (8)

Direct students' attention to the subtraction equation. Ask: **What equation would make the scale balance?** ($7 - b = 2$) **What number must the right side of the equation represent?** (2) **How many squares must be removed from the left side so the pans will balance?** (5)

Help students recognize that b equals 5. The left side of the equation represents the same number of squares as the right side of the equation. A balanced scale represents an equation.

Continue to ask similar questions as you discuss the equation that has 3 addends. Help students relate the equation to the balance scale. Guide them to find a value for x such that the pans will balance.

TRY THESE Exercises 1–3 model finding the value for a variable in an equation.

- **Exercise 1** Addition equation.
- **Exercise 2** Subtraction equation.
- **Exercise 3** Addition equation with 3 addends.

PRACTICE ON YOUR OWN Review the example at the top of the page. Ask students to identify the equation and variable, and to name the number that has the same value as the expression on the left side of the equation. Then ask students to name the value of the variable.

CHECK Determine if students can find the value of a variable in an addition or subtraction equation.

Success is indicated by 3 out of 3 correct responses.

Students who successfully complete the **Practice on Your Own** and **Check** are ready to move to the next skill.

COMMON ERRORS

- Students might not understand that both sides of the equals sign in an equation must name the same value.
- Students might not recall addition or subtraction facts.
- Students might not understand the meaning of the variable in the equation.

Student who made more than 4 errors in the **Practice on Your Own**, or who were not successful in the **Check** section, may benefit from the **Alternative Teaching Strategy** on the next page.

Optional

Alternative Teaching Strategy
Modeling Addition Equations

15 Minutes

OBJECTIVE Use models to represent addition equations

MATERIALS linking cubes (2 colors)

Prepare two groups of linking cubes. In one group show 5 red cubes and 10 blue cubes. In the other group show 15 cubes. Ask a volunteer to find the number of cubes in each group.

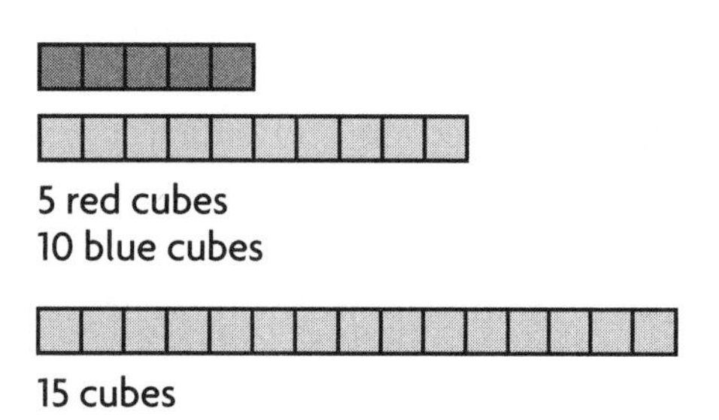

Help students recognize that the number of cubes is the same in both groups. Because of this relationship, students can use an equal sign to relate the two groups.

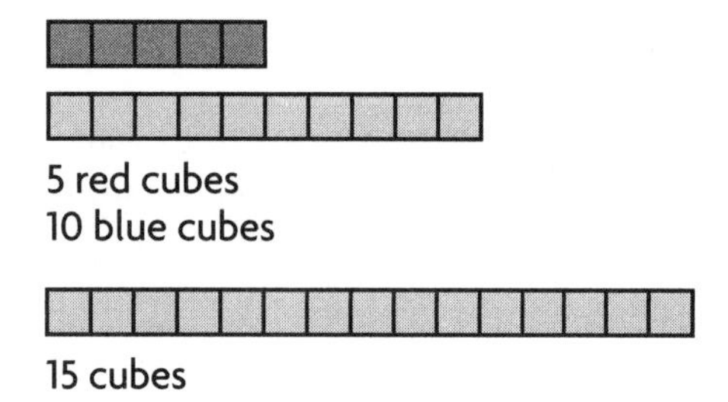

Next, guide students to see that an equation can be used to describe this relationship.

$$5 + 10 = 15$$

Tell students that letters are sometimes used to represent numbers in an equation. For example, if you couldn't see the red cubes, the letter r could be used to represent the number of red cubes.

$$r + 10 = 15$$

In this case, the letter r is equal to 5. Ask students for other ways to write the equation, for example:

$$5 + b = 15$$

The letter b represents the number of blue cubes. In this case, the letter b is equal to 10.

Give students simple addition equations with variables. Ask them to name the value of the variable. Allow students to use their cubes to model the equations until they are comfortable, and accurate enough to solve simple equations without models. For example, try these.

$$m + 7 = 12$$

$$9 + p = 14$$

You can use a similar strategy to help students model subtraction equations.

Name ______________________ Skill ______________________

Grade 5 **Skill 50**

Addition and Subtraction Equations

An equation is a numerical or algebraic sentence that states that two quantities are equal.

Addition Equation
$3 + 5 = n$
Find the value of n.

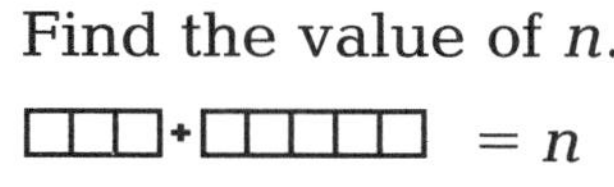

$8 = n$

So, the value of n is 8, or $n = 8$.

Subtraction Equation
$7 - b = 2$
Find the value of b.
Use a balance scale to picture the equation:

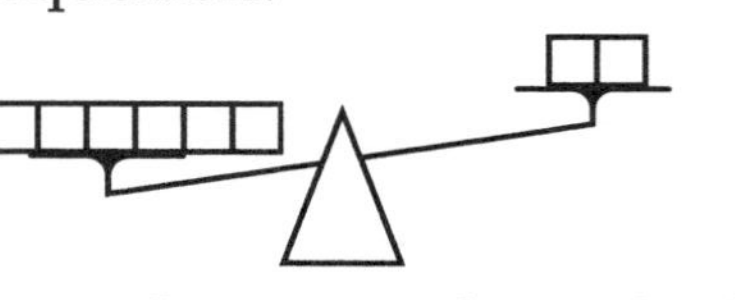

Remove 5 squares from the left side to balance the scale.
So, the value of b is 5, or $b = 5$.

Equation with 3 Addends
$2 + x + 3 = 9$
Find the value of x.
Use a balance scale to picture the equation.

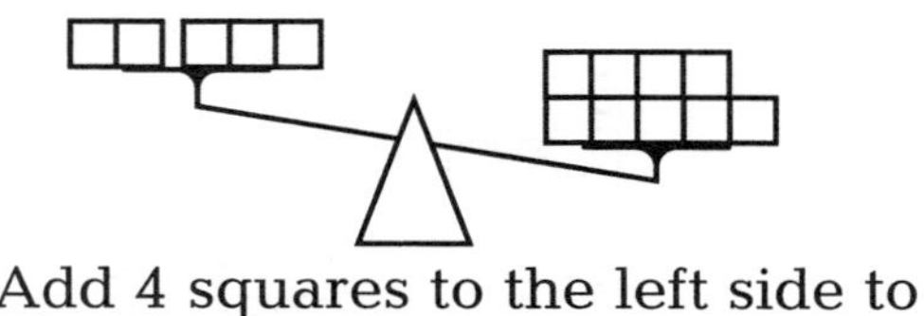

Add 4 squares to the left side to balance the scale.
So, the value of x is 4, or x = 4.

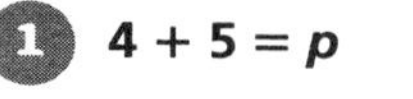

Try These

Find the value of each variable.

1. $4 + 5 = p$

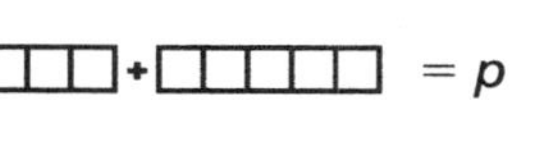

The value of p is _____, or p = _____.

2. $11 - z = 4$

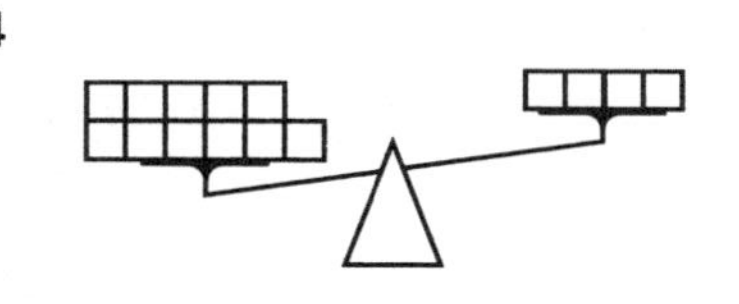

Remove _____ squares from the left side to balance the scale.

The value of z is _____, or z = _____.

3. $5 + n + 4 = 12$

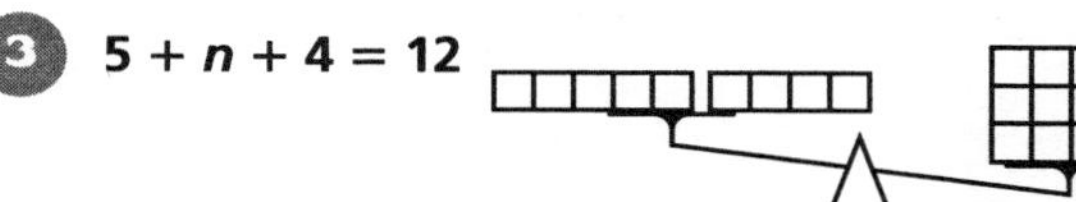

Add _____ squares to the left side to balance the scale.

The value of n is _____, or n = _____.

Go to the next side.

Name ______________________ Skill ______________________

Practice on Your Own

Skill 50

Picture a scale to help you find the value of the variable.

You can add or subtract to make the equation true.

Find the value of the variable.

$2 + 5 + m = 13$

The value of m is 6, or $m = 6$.

Find the value of each variable.

1 $6 + n = 9$

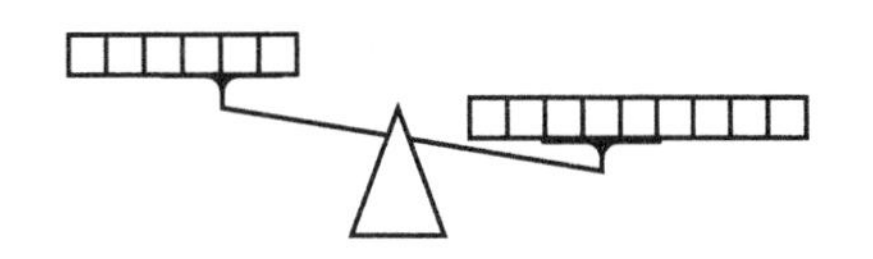

Add ________ to balance the scale.

The value of n is ________, or n = ________.

2 $8 - m = 5$

Subtract ________ to balance the scale.

The value of m is ________, or m = ________.

3 $4 + n + 3 = 11$

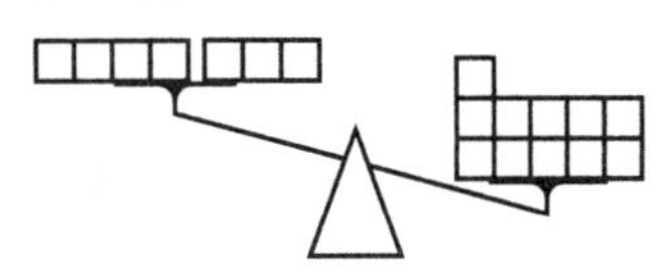

The value of n is ________,

or ________ = ________.

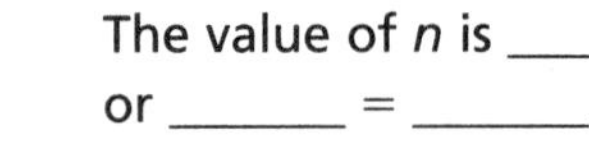

4 $12 - t = 8$

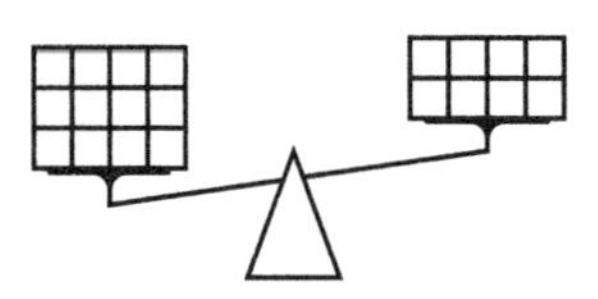

The value of t is ________,

or ________ = ________.

5 $9 + 4 + n = 16$

n = ________

6 $7 + 4 = d$

d = ________

7 $15 - t = 11$

t = ________

8 $p + 7 = 9$

____

9 $14 - q = 8$

____

10 $3 + y + 8 = 16$

____

Check

Find the value of each variable.

11 $12 - r = 9$

____

12 $4 + b + 3 = 15$

____

13 $17 + 4 = m$

____

Using Skill 51

OBJECTIVE Find the value of an expression containing parentheses

Read about parentheses at the top of the page. Work through the steps of Examples A and B.

Ask: **Are the numbers in the two expressions the same?** (yes) **What is different?** (the placement of the parentheses) **What do you do first in Example A?** (Multiply 2 and 4.) **In Example B?** (Add 4 and 6.) **What is the next step after you do the work inside the parentheses?** (Perform the other operation.)

Practice with a few more expressions; be sure to include some with division and subtraction. Finally, have a few volunteers place parentheses so that expressions have a given value. For example, have them place parentheses to show a value of 6 in the expression $2 \times 9 - 6$. ($2 \times (9 - 6)$)

TRY THESE Exercises 1 and 2 provide practice in finding the value of an expression by performing the operation in parentheses first.

- **Exercise 1** Find the expression that has a value of 0.
- **Exercise 2** Find the expression that has a value of 40.

PRACTICE ON YOUR OWN Focus on the example at the top of the page. Students should realize that the difference between the values of the two expressions is determined by the placement of the parentheses

CHECK Determine that students are performing the operation in the parentheses first. Success is indicated by 4 out of 4 correct responses.

Students who successfully complete the **Practice On Your Own** and **Check** are ready to move to the next skill.

COMMON ERRORS

- Students may perform the operation inside the parentheses first, but forget to perform the second operation.
- Students may disregard the parentheses and perform the operations from left to right.

Students who made more than 3 errors in the **Practice On Your Own**, or who were not successful in the **Check** section, may benefit from the **Alternative Teaching Strategy** on the next page.

Optional

Alternative Teaching Strategy

Use Parentheses

20 Minutes

OBJECTIVE Use flow charts to model finding the value of expressions containing parentheses

Use a simple flow chart to model finding the value of an expression containing parentheses. Draw a flow chart for 12 ÷ (4 + 2). Remind students that flow charts show the steps needed to perform a job. Review the meaning of the symbols used in the flow chart.

Ask: **What do the circles show?** (start and stop) **What do the arrows do?** (point to the next step) **What do the rectangles show?** (directions for completing the steps)

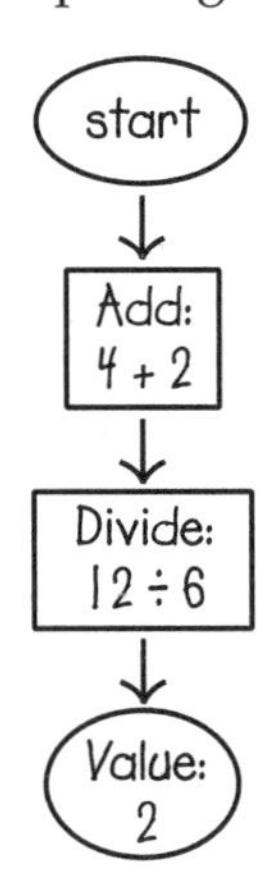

Ask: **What do you do first?** (Add 4 and 2.) **What do you do next?** (Divide 12 by 6.) **What is the value of the expression?** (2)

Have students think about how they can place parentheses to change the value of the expression. Help them write the expression (12 ÷ 4) + 2. Then have them make a flow chart.

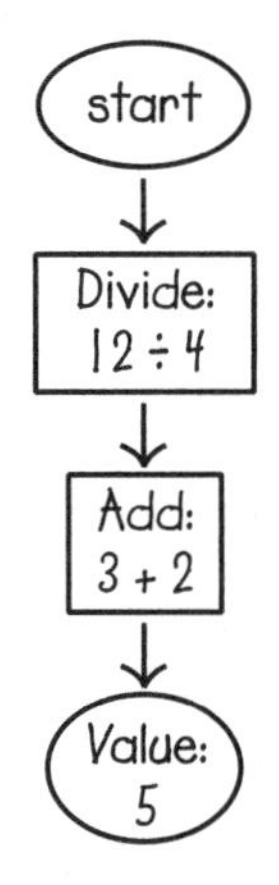

Ask: **What do you do first?** (Divide 12 by 4.) **What do you do next?** (Add 3 and 2.) **What is the value of the expression?** (5)

Grade 5 Skill 51

Use Parentheses

Parentheses in a mathematical expression tell you which operations to perform first.
Sometimes the position of parentheses can change the value of an expression.
The expression $2 \times 4 + 6$ can have different values, depending on where you place parentheses.
Compare Example A and Example B to see which gives a value of 20.

Example A

$(2 \times 4) + 6$ — **Multiply 2 and 4.**

↓ ↓

$8 + 6$ — **Add 8 and 6.**

↓

14

So, $(2 \times 4) + 6$ is 14.

Example B

$2 \times (4 + 6)$ — **Add 4 and 6.**

↓ ↓

2×10 — **Multiply 2 and 10.**

↓

20

So, $2 \times (4 + 6)$ is 20.

So, the expression $2 \times (4 + 6)$ has a value of 20.

Try These

Find the value of each expression.

1. Which expression has a value of 0?

$18 \div (6 - 3)$ → ___ ÷ ___ → ___

$(18 \div 6) - 3$ → ___ − ___ → ___

The expression ______________ has a value of 0.

2. Which expression has a value of 40?

$(10 \times 8) - 4$ → ___ − ___ → ___

$10 \times (8 - 4)$ → ___ × ___ → ___

The expression ______________ has a value of 40.

Name ______________ Skill ______________

Go to the next side.

Name ______________________ Skill ______________________

Practice on Your Own

Skill 51

Think:
Do the operation in parentheses first.

Which expression has a value of 4?

$(28 \div 4) + 3$
$\downarrow \quad \downarrow$
$7 + 3$
$\downarrow$
10

$28 \div (4 + 3)$
$\downarrow \quad \downarrow$
$28 \div 7$
$\downarrow$
4

So, the expression $28 \div (4 + 3)$ has a value of 4.

Find the value of each expression.

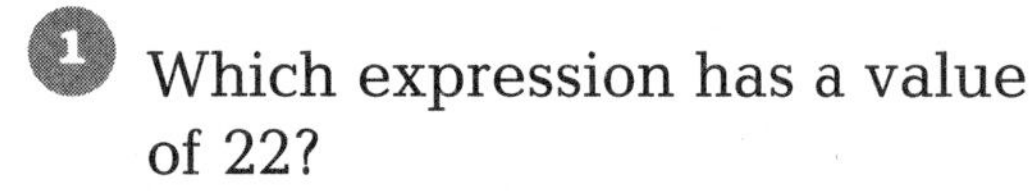

1. Which expression has a value of 22?

   $(2 \times 5) + 12$ → ___ + ___ → ___

   $2 \times (5 + 12)$ → ___ × ___ → ___

   The expression ______________ has a value of 22.

2. Which expression has a value of 1?

   $6 \times (12 \div 2)$ → ___ × ___ → ___

   $(6 \times 2) \div 12$ → ___ ÷ ___ → ___

   The expression ______________ has a value of 1.

3. $(3 \times 5) \div 5$ __________
4. $(17 - 5) \div 3$ __________
5. $13 + (12 \div 4)$ __________
6. $27 \div (3 \times 3)$ __________

Place parentheses to show the given value.

7. $4 \times 5 - 3$ is 17.
8. $10 - 3 \times 2$ is 4.
9. $36 \div 6 - 6$ is 0.
10. $12 + 8 \div 2$ is 16.

Circle the expression that shows the given value.

11. 52
    $10 + (3 \times 4)$
    $(10 + 3) \times 4$
12. 13
    $(15 - 10) \div 5$
    $15 - (10 \div 5)$
13. 60
    $(12 \times 2) + 3$
    $12 \times (2 + 3)$
14. 2
    $(28 - 14) \div 7$
    $28 - (14 \div 7)$

Check

Circle the expression that shows the given value.

15. 5
    $(2 \times 4) - 3$
    $2 \times (4 - 3)$
16. 72
    $(9 \times 10) - 2$
    $9 \times (10 - 2)$
17. 7
    $(8 + 20) \div 4$
    $8 + (20 \div 4)$
18. 4
    $(16 - 2) \times 6$
    $16 - (2 \times 6)$

Skill 52 Grade 5 Graph Ordered Pairs

15 Minutes

Using Skill 52

OBJECTIVE Locate a point on a coordinate grid for an ordered pair

MATERIALS graph paper

Have students point to the intersection of the vertical and horizontal lines. Ask: **What number is assigned to this point?** (0)

Draw student's attention to Step 1. Point out that on this grid, movement is to the right of zero and then up. Ask: **What is the name of the horizontal axis and the name of the vertical axis?** (x-axis, y-axis) **What does the number 5 in (5,3) indicate?** (number of units to move along the x-axis)

Direct students to look at Step 2. Ask: **At what point do you start to move upward?** (at 5 on the x-axis) **How many units do you move up?** (3)

Continue with Step 3. Ask: **What is the ordered pair that names point *A*?** (5,3)

Guide students to see that an ordered pair shows how to locate a point on a coordinate grid. The first number in the ordered pair indicates movement along the x-axis. The second number indicates how far up to move.

TRY THESE In Exercises 1–3 students use ordered pairs to locate points.

- **Exercises 1–3** Given an ordered pair, name the point.

PRACTICE ON YOUR OWN Review the example at the top of the page. Have students show you how they locate the point for (7,3).

CHECK Determine if students can locate a point on the coordinate grid given an ordered pair.

Success is indicated by 4 out of 6 correct responses.

Students who successfully complete the **Practice on Your Own** and **Check** are ready to move to the next skill.

COMMON ERRORS

- Students might not start at 0, thus locating the point incorrectly
- Students might confuse the direction of each number in the ordered pair.
- Students might miscount the intervals on the x-axis or the y-axis

Students who made more than 2 errors in the **Practice on Your Own,** or who were not successful in the **Check** section, may benefit from the **Alternative Teaching Strategy** on the next page.

Optional

Alternative Teaching Strategy

Ordered Pairs in One Quadrant

20 Minutes

OBJECTIVE Locate a point on a coordinate grid for an ordered pair

MATERIALS graph paper

You may wish to use an overhead projector for this activity. Prepare the grid below before class. Color code the horizontal and vertical axes and the numbers used to label the axes.

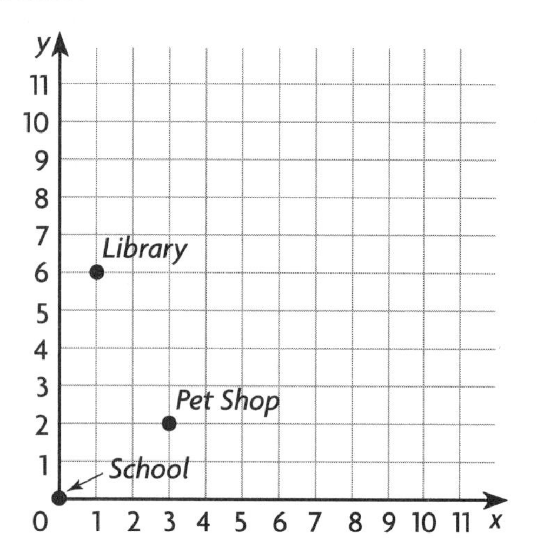

Review the parts of the grid (x-axis, y-axis) and the direction for each axis. Remind students that the numbers in an ordered pair refer to the movement on the two axes.

Starting from the location named SCHOOL, ask students to name the ordered pair that would locate the PET SHOP. (3,2) Next, name the ordered pair to locate the LIBRARY. (1,6)

It may be helpful for some students to put their fingers on the starting point and move in the direction of the x-coordinate. Stop at the x-coordinate, then move upward to locate the y-coordinate.

Continue by adding other locations on the coordinate grid. For example, SOCCER FIELD, HIGH SCHOOL, MARKET, POST OFFICE, HOME, and so on.

Students should conclude that in an ordered pair, the first number directs how far to move along the x-axis; and the second number directs how far up to move from the first point reached.

Grade 5
Skill 52

Graph Ordered Pairs

Use the coordinate grid to find the point for the ordered pair (5,3).

Step 1
Start at 0. Move 5 spaces to the right.

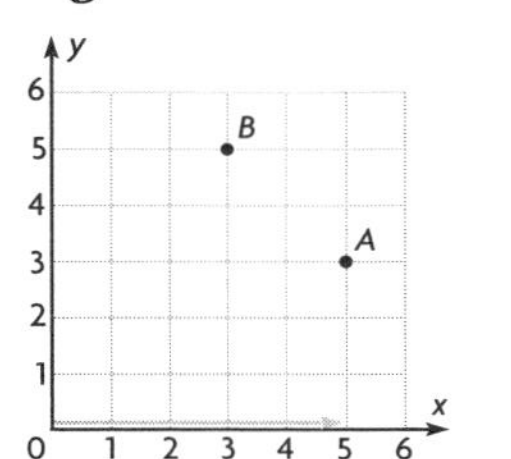

(**5**,3)

Number of spaces from zero on the x-axis.

Step 2
Start at 5 on the x-axis. Move 3 spaces up.

(5,**3**)

Number of spaces from zero on the y-axis.

Step 3
Name the point.

The letter *A* names the point for (5,3).

So, the ordered pair (5,3) names point *A*.

Try These

Use the ordered pair to name the point.

1 (6,2)

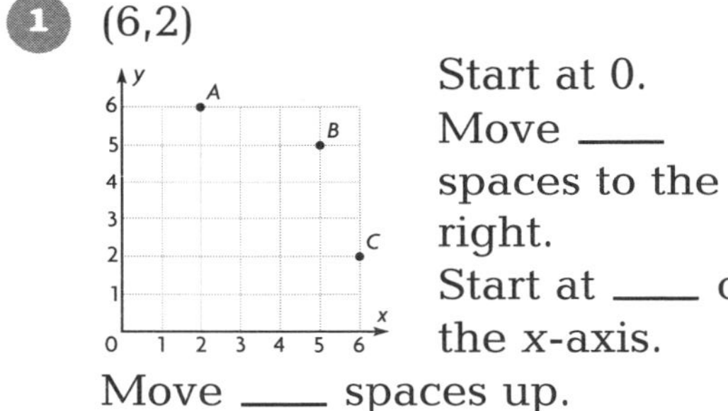

Start at 0.
Move ____ spaces to the right.
Start at ____ on the x-axis.
Move ____ spaces up.
The ordered pair (6,2) names point ____.

2 (4,7)

Start at 0.
Move ____ spaces to the right.
Start at ____ on the x-axis.
Move ____ spaces up.
The ordered pair (4,7) names point ____.

3 (8,5)

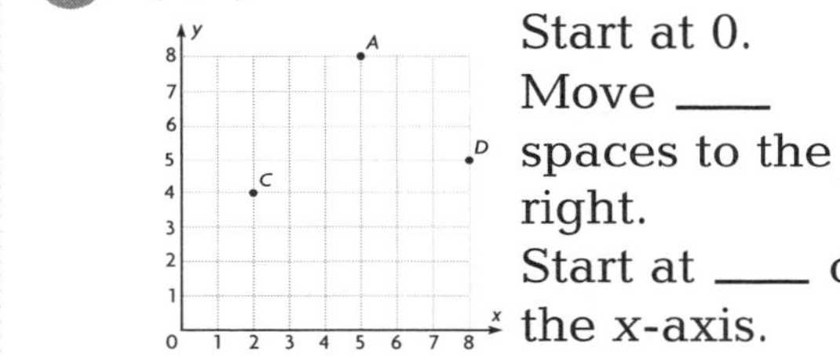

Start at 0.
Move ____ spaces to the right.
Start at ____ on the x-axis.
Move ____ spaces up.
The ordered pair (8,5) names point ____.

Go to the next side.

Practice on Your Own

Skill 52

Think:
You can use an **ordered pair** of numbers to locate a point on a coordinate grid.

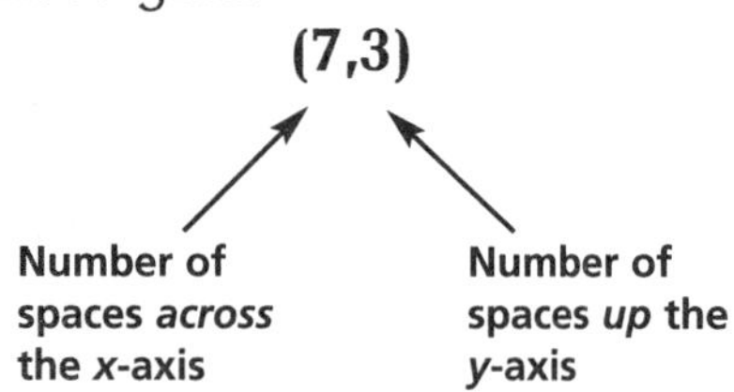

Find the point for the ordered pair (7,3) on the coordinate grid.

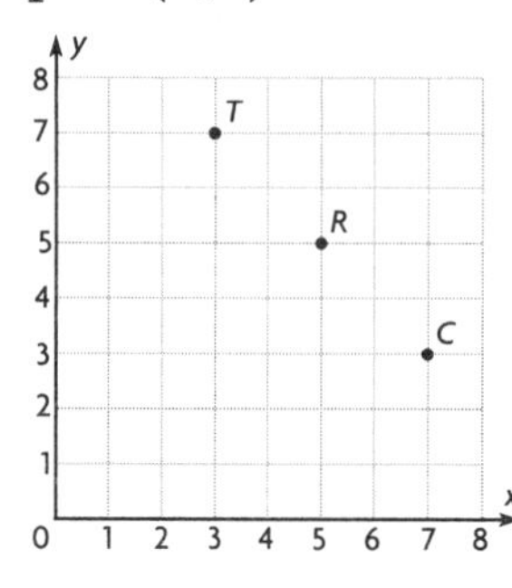

Start at 0. Move 7 spaces to the right.
Start at 7 on the x-axis. Move 3 spaces up.
The ordered pair (7,3) names point *C*.

Find the point for the ordered pair on the coordinate grid. Use Grid 1.

1 (3,4)
Start at 0. Move ____ spaces to the right.
Start at ____ on the x-axis. Move ____ spaces up.
The ordered pair (3,4) names point ____.

2 (1,5)
Start at ____. Move ____ space to the right.
Move ____ spaces up.
The ordered pair (1,5) names point ____.

3 (4,0)
Move ____ spaces to the right.
Move ____ spaces up.
point ____

Grid 1

4 (9,2) ____ **5** (5,1) ____ **6** (4,3) ____ **7** (7,7) ____ **8** (8,5) ____ **9** (2,9) ____

Check

10 (3,3) ____ **11** (6,5) ____ **12** (8,3) ____

13 (3,8) ____ **14** (0,2) ____ **15** (5,2) ____

Expressions with Exponents (2 and 3)

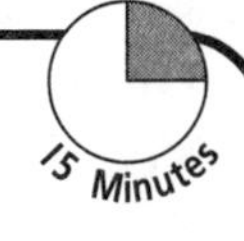

Using Skill 53

OBJECTIVE Write expressions in exponent form; find the value of expressions with exponents

Direct students' attention to the first example. Read about exponents and how to write an expression in exponent form. Be sure students understand that exponents show repeated multiplication.

Refer to 5^3. Ask: **What is the base?** (5) **How many times is it used as a factor?** (3 times) **What do you call the number 3?** (the exponent) Continue: **If the exponent were 4, how many times would 5 be a factor?** (4 times)

Next, look at the second example. Explain that they can use what they know about the exponent form to find the value of the expression 2^3.

Ask: **What is the base?** (2) **What is the exponent?** (3) **How many times is 2 used as a factor?** (3 times)

Lead students through the steps of the repeated multiplication.

Tell students that you read 2^3 as "the third power of 2" or "2 cubed."

TRY THESE Exercises 1–4 provide practice with exponents.

- **Exercises 1–2** Write an expression in exponent form.
- **Exercises 3–4** Find the value of an expression with exponents.

PRACTICE ON YOUR OWN Focus on the examples at the top of the page. Have students identify each base and how many times it is used as a factor. Have students work through the repeated multiplication.

CHECK Determine that students can write expressions in exponent form and evaluate expressions with exponents. Success is indicated by 5 out of 6 correct responses.

Students who successfully complete the **Practice on Your Own** and **Check** are ready to move to the next skill.

COMMON ERRORS

- Students may confuse the base with the exponent.
- Students may understand exponents but make mistakes multiplying, or use addition instead of multiplication.
- Students may not know multiplication facts.

Students who made more than 3 errors in the **Practice on Your Own**, or who were not successful in the **Check** section, may benefit from the **Alternative Teaching Strategy** on the next page.

Optional

20 Minutes

Alternative Teaching Strategy

Model Repeated Multiplication to Write Expressions with Exponents

OBJECTIVE Model repeated multiplication and write expressions in exponent form

MATERIALS counters

Review the meaning of *base* and *exponent* by displaying:

base → 3^2 ← exponent

Point out that in exponent form, the factor 3 is called the base, and 2 is called the exponent. Recall that the exponent tells how many times the base is used as a factor in repeated multiplication.

$3^2 \rightarrow 3 \times 3$

Display $3 \times 3 \times 3 \times 3 \times 3$. Ask: **What is the base?** (3)

How many times is 3 used as a factor? (5 times)

Write the expression in exponent form. (3^5)

Then display $2 \times 2 \times 2 \times 2$. Use counters to model the repeated multiplication.

First show 2 rows of 2 counters. Count 4 counters. Write $2 \times 2 = 4$. Then show 4 rows of 2 counters. Count 8 counters. Write $(2 \times 2) \times 2 = 4 \times 2 = 8$. Next show 8 rows of 2 counters. Count 16 counters. Write $(2 \times 2 \times 2) \times 2 = 8 \times 2 = 16$.

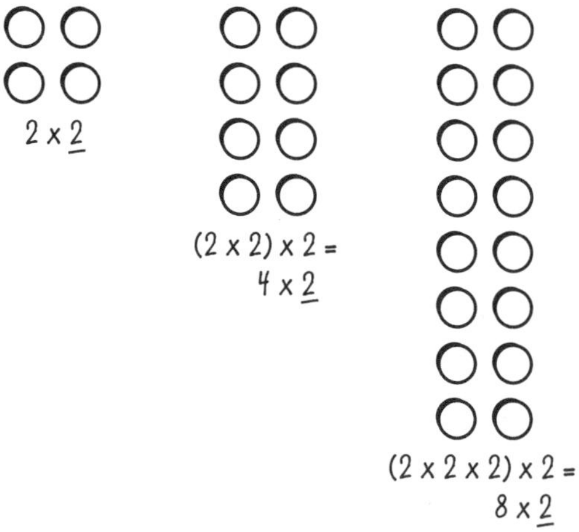

$2 \times \underline{2}$

$(2 \times 2) \times 2 =$
$4 \times \underline{2}$

$(2 \times 2 \times 2) \times 2 =$
$8 \times \underline{2}$

Ask: **How many times is 2 used as a factor in 16?** (4)

What is the exponent? (4) **How do you know?** (because 2 is used as a factor 4 times: $2 \times 2 \times 2 \times 2$)

Have students write the expression $2 \times 2 \times 2 \times 2$ in exponent form. (2^4) Then ask: **What is the value of the expression 2^4?** (16)

Repeat the examples until students are comfortable using the exponential form. Then provide expressions in exponential form, and have students find the value of the expressions without models.

Name ______ Skill ______

Grade 5 Skill 53

Expressions with Exponents (2 and 3)

Use exponents to show repeated multiplication.

An **exponent** shows how many times a number called the **base** is used as a factor.

Write the expression $5 \times 5 \times 5$ in exponent form.
In the expression above, 5 is used as a factor 3 times.
Use the exponent 3 to show this.

5^3
- The exponent tells how many times the factor is used.
- 5 is the base

So, $5 \times 5 \times 5$ written in exponent form is 5^3.

Find the value of the expression 2^3.
In the expression 2^3, the base 2 is used as a factor 3 times.
To find the value of the expression, rewrite it as a multiplication problem.

$$2^3 = 2 \times 2 \times 2$$
$$= (2 \times 2) \times 2$$
$$= 4 \times 2$$
$$= 8$$

So, the value of 2^3 is 8.

Try These

Complete. Write in exponent form.

1. 4×4
   Base: ______
   Number of times base is a factor: ______
   Exponent form: ______

2. 6×6
   Base: ______
   Number of times base is a factor: ______
   Exponent form: ______

Find the value of each expression.

3. 7^2 = ______ × ______
   = ______

   Think: Rewrite 7^2 as a multiplication problem.

4. 6^3 = ______ × ______ × ______
   = ______ × ______
   = ______

   Think: Rewrite 6^3 as a multiplication problem.

Go to the next side.

Name ______________________ Skill ______________________

Practice on Your Own

Think:
How many times is the base used as a factor?

Write in exponent form.

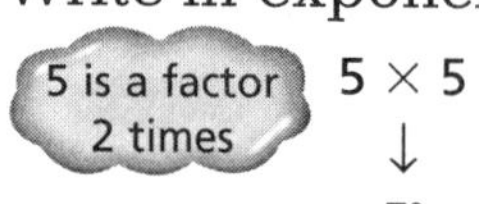

5×5
↓
5^2

Find the value.

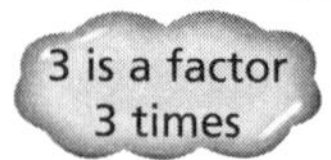

$3^3 = (3 \times 3 \times 3)$
↓
$= 27$

Complete.

1. 7×7
   Base: ____
   Exponent: ____
   Exponent form: ____

2. $4 \times 4 \times 4$
   Base: ____
   Exponent: ____
   Exponent form: ____

3. 9×9
   Base: ____
   Exponent: ____
   Exponent form: ____

4. $3^2 \times 3^2 = (\_\_\_\_ \times \_\_\_\_) \times (\_\_\_\_ \times \_\_\_\_)$

5. $2^3 \times 4^2 = (\_\_\_\_ \times \_\_\_\_ \times \_\_\_\_) \times (\_\_\_\_ \times \_\_\_\_)$

6. $5^3 \times 3^3 =$ ______________________

Find the value of each expression.

7. 4^3
8. 7^3
9. 5^3
10. 9^2
11. 6^2
12. 3^2

Check

Use exponents to write each expression.

13. 7×7 ____________
14. $9 \times 9 \times 9 \times 9$ ____________
15. $4 \times 4 \times 8 \times 8 \times 8$ ____________

Find the value of each expression.

16. 8^2 ____________
17. 5^2 ____________
18. $5^2 \times 2^3$ ____________

Skill 54 Grade 5 — Compare Whole Numbers

15 Minutes

Using Skill 54

OBJECTIVE Use place value to compare whole numbers

Begin the lesson by having a student read the 2 numbers to be compared. Tell the students that when you compare 2 whole numbers you start at the left, or the greatest, place-value position. Tell them that you compare until the digits are different. Direct students' attention to the place-value chart. Tell them that they will be using the chart to help them compare the digits in each number.

Have students look at Step 1. Stress the importance of aligning the digits in each number by place-value. Say: **Look at the first digit in each number. It is in the billions place.** Ask: **Are the digits the same?** (yes) Tell students that since the digits are the same, they now move one place to the right to compare the digits in the hundred millions place.

Direct students' attention to Step 2. Ask: **Are the digits in the hundred millions place the same?** (yes) Tell students that they continue moving one place to the right until the digits are different.

Direct students' attention to Step 3. Ask: **Are the digits in the ten millions place the same?** (no) **Is 4 less than or greater than 9?** (less than) Tell students that since 4 is less than 9, then 2,340,348,092 is less than 2,392,065,976.

TRY THESE In Exercises 1–3 students compare numbers to billions.

- **Exercise 1** Compare millions.
- **Exercise 2** Compare ten millions.
- **Exercise 3** Compare billions.

PRACTICE ON YOUR OWN Direct students' attention to the example at the top of the page. Have students identify which digits are different and in which place-value position the difference occurs. In Exercises 1–6, students compare whole numbers using prompts and alignment of digits. In Exercises 7–10, students compare whole numbers without the use of aids.

CHECK Determine if students can compare greater whole numbers by aligning digits by place-value and comparing until the digits are different.

Success is indicated by 4 out of 4 correct responses.

Students who successfully complete the **Practice on Your Own** and **Check** are ready to move to the next skill.

COMMON ERRORS

- Students may confuse the symbols for greater than and less than.
- Students may align digits incorrectly.

Students who made more than 2 errors in the **Practice on Your Own,** or who were not successful in the **Check** section, may benefit from the **Alternative Teaching Strategy** on the next page.

Optional

Alternative Teaching Strategy
Model Comparing Numbers

15 Minutes

OBJECTIVE Compare whole numbers

MATERIALS place-value pocket charts, an opaque container containing 3 sets of number cards 0–9

Begin by explaining to students that the goal of this activity is to create the greater number.

Have students work in pairs. Provide each student with a place-value pocket chart. Have one partner draw a card from the container and place it in a pocket on his or her place-value pocket chart. The other partner should not be able to see this place-value pocket chart. Now have the other partner repeat the procedure using his or her place-value pocket chart. The partners take turns drawing cards and placing them on their respective place-value pocket charts until all of the pockets have been filled.

Student 1

Thousands			Ones		
Hundreds	tens	ones	Hundreds	tens	ones
8	7	5	9	0	4

Student 2

Thousands			Ones		
Hundreds	tens	ones	Hundreds	tens	ones
8	6	5	9	1	5

Now have students show their place-value pocket charts and compare the numbers. Have them write the numbers and the appropriate comparison symbol. Remind students that the greater than or less than symbol is like an arrow that always points to the lesser number.

Name ______________________ Skill ______________________

Grade 5 Skill 54

Compare Whole Numbers

You can use place value to compare whole numbers. Begin at the greatest place-value position and compare until the digits are different.
Use the place-value chart to compare: 2,340,348,092 ○ 2,392,065,976. Use <, >, or =.

BILLIONS			MILLIONS			THOUSANDS			ONES		
Hundreds	Tens	Ones	Hundreds	Tens	Ones	Hundreds	Tens	Ones	Hundreds	Tens	Ones
		2	3	4	0	3	4	8	0	9	2
		2	3	9	2	0	6	5	9	7	6

Step 1
Begin with the greatest place.
Compare the billions.
2, 3 4 0, 3 4 8, 0 9 2
↑
↓
2, 3 9 2, 0 6 5, 9 7 6

Think: 2 = 2
The billions are the same.

Step 2
Move one place to the right.
Compare hundred millions.
2, **3** 4 0, 3 4 8, 0 9 2
↑
↓
2, **3** 9 2, 0 6 5, 9 7 6

Think: 3 = 3
The hundred millions are the same.

Step 3
Move another place to the right.
Compare ten millions.
2, 3 **4** 0, 3 4 8, 0 9 2
↑ different
↓
2, 3 **9** 2, 0 6 5, 9 7 6

Think: 4 < 9

So, 2,340,348,092 < 2,392,065,976.

Try These

Compare. Write <, >, or = for each ◯.

1. 2, 4 7 6, 8 3 5
↑
↓
2, 4 7 5, 8 6 2

Start at the left. Move to the right one place at a time.

2,476,835 ◯ 2,475,862

2. 3 2, 4 5 9, 6 0 3
↑
↓
3 2, 4 9 5, 9 2 0

Start at the left.

32,459,603 ◯ 32,495,920

3. 1, 8 3 4, 5 2 0, 9 7 0
↑
↓
1, 8 3 4, 5 2 1, 6 5 2

1,834,520,970 ◯ 1,834,521,652

Go to the next side.

Name ______________________ Skill ______________________

Practice on Your Own

Compare 4,578,392 ○ 4,578,421.
Begin with the greatest place, millions.
Move one place to the right until you find different digits.

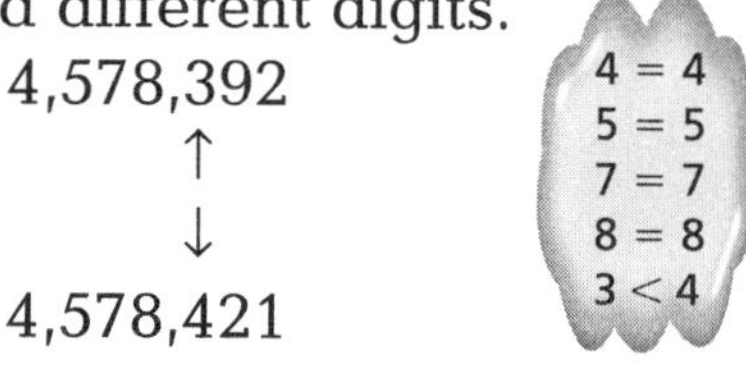

MILLIONS			THOUSANDS			ONES		
Hundreds	Tens	Ones	Hundreds	Tens	Ones	Hundreds	Tens	Ones
		4	5	7	8	3	9	2
		4	5	7	8	4	2	1

So, 4,578,392 < 4,578,421.

Compare. Write <, >, or = for each ○.

1. 4 6, 7 2 3
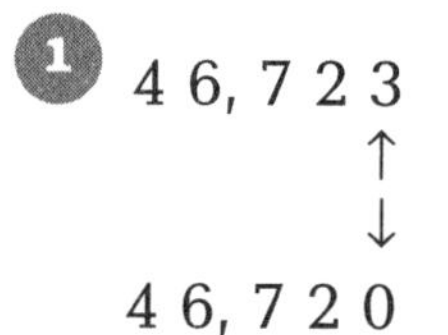
4 6, 7 2 0
46,723 ○ 46,720

2. 6, 0 9 1, 3 3 7
↑
↓
6, 0 9 1, 3 4 7
6,091,337 ○ 6,091,347

3. 8 7, 3 5 9, 6 0 1
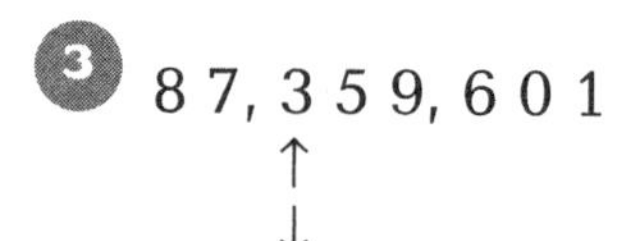
8 7, 2 5 9, 6 1 5
87,359,601 ○ 87,259,615

4. 21,537
21,637
21,537 ○ 21,637

5. 104,892
104,890
104,892 ○ 104,890

6. 5, 2 9 1, 3 3 6, 8 1 8
5, 2 9 1, 3 3 6, 8 1 8
5,291,336,818 ○ 5,291,336,818

Compare. You may wish to use the space to align the numbers by place value.

7. 235,091 ○ 235,091

8. 382,451,973 ○ 282,905,943

9. 34,567,121 ○ 34,576,212

10. 4,102,675,118 ○ 4,102,675,118

Check

Compare. You may wish to use the space to align the numbers by place value.

11. 430,742 ○ 430,742

12. 28,014,687 ○ 28,014,678

13. 5,234,901 ○ 5,243,876

14. 8,420,579,015 ○ 8,402,579,015

Skill 55 Grade 5

15 Minutes

Using Skill 55

OBJECTIVE Use a function table to find the relationship between pairs of numbers

On Skill 55, draw attention to the function tables. Be sure students understand the difference between *input* and *output*. Explain that operations are performed on the input numbers. Ask:

In which column in the table will you find the input numbers? (the x column) **The results of the operations are called the output numbers. In which column in the table will you find the output numbers?** (the y column)

As students read the first example, point out that that they can discover the rule by comparing the numbers in the x and y columns. Ask:

If the output numbers are greater than the input numbers, what does that suggest to you? (that the rule is add or multiply)

Next, have students look for a pattern in the numbers. Guide students as they discover that each number in the y column is 4 more than the number in the x column. Ask:

What is the rule? (Add 4.)

For the second function table, students will find that the pattern requires multiplication.

How did you know that addition was not the rule? (Adding the same number to the input numbers did not give the output numbers.) **What is the rule in the second function table?** (Multiply by 3.)

Ask similar questions as students work through the second example.

TRY THESE In Exercises 1–3 students examine the pattern and state the rule.

- **Exercise 1** Subtract.
- **Exercise 2** Divide.
- **Exercise 3** Multiply.

PRACTICE ON YOUR OWN Review the example at the top of the page. Have students identify the pattern and state the rule for the function table.

CHECK Determine if students can find the pattern and state the rule for each function table.

Success is indicated by 2 out of 3 correct responses.

Students who successfully complete the **Practice on Your Own** and **Check** are ready to move to the next skill.

COMMON ERRORS

- Students may not understand how to look for a pattern.
- Students may recognize that there is a pattern, but may not understand how one of the 4 operations is used to create the pattern.

Students who made more than 2 errors in the **Practice on Your Own**, or who were not successful in the **Check** section, may benefit from the **Alternative Teaching Strategy** on the next page.

Optional

Alternative Teaching Strategy
Use Function Tables

15 Minutes

OBJECTIVE Use a rule and a pattern to complete a function table

Review with students that a function table uses an input number and a rule to find output numbers. The rule tells you what to do with the input numbers. Display this function table.

Input	1	2	3	4	5
Output	6	7			

Ask: **What are the input numbers?** (1, 2, 3, 4, 5) **What is a rule?** (Add 5.)

What output numbers are there? (6, 7)

Guide students as they recognize that 1 + 5 gave an output number of 6. Then have them look at the second row and see that 2 + 5 = 7. Ask:

If you start with 3 and apply the rule, what will the output number be? (8)

Continue until the function table is complete.

Then display the following function table. Suggest that the students examine both the input numbers and the output numbers to find a rule.

Input	10	20	30	40	50
Output	5	15			

To find a rule, look for a pattern in the table. What pattern do the input numbers show? (The input numbers increase by 10.)

What pattern do you see in the output numbers? (Each output number is 5 less than the input number.)

Guide students as they discover that when an output number is less than the input number, the operation can be subtraction or division. In this case, students discover that they can subtract 5 from each input number to get the output number. Have students use a rule they found to complete the table.

When students understand how to work with function tables, present this table and have them use a rule to find the input numbers. Have students think: What number times 3 is 9?

Input					
Output	9	12	15	18	21

Rule: $y = 3x$

Name ________________ Skill ________________

Grade 5 Skill 55

Function Tables

A **function table** is an input-output table. Perform an operation on the input number to get the output number. Look for a pattern to write a rule for an input-output table.

When the number in the y-column is *greater* than the number in the x-column, think of addition or multiplication.

Each number in the *y*-column is greater than the number in the *x*-column.

input output

x	y
1	5
3	7
7	11
10	14

ADDITION PATTERN

1 + **4** = 5
3 + **4** = 7
7 + **4** = 11
10 + **4** = 14

A rule is: Add 4 to the number in column *x*.

Each number in the *y*-column is greater than the number in the *x*-column.

x	y
2	6
3	9
4	12
5	15

MULTIPLICATION PATTERN

2 × **3** = 6
3 × **3** = 9
4 × **3** = 12
5 × **3** = 15

A rule is: Multiply the number in column *x* by 3.

When the number in the y-column is *less* than the number in the x-column, think of subtraction or division.

Each number in the *y*-column is less than the number in the *x*-column.

x	y
22	20
18	16
15	13
10	8

SUBTRACTION PATTERN

22 − **2** = 20
18 − **2** = 16
15 − **2** = 13
10 − **2** = 8

A rule is: Subtract 2 from the number in column *x*.

Each number in the *y*-column is less than the number in the *x*-column.

x	y
35	7
30	6
25	5
20	4

DIVISION PATTERN

35 ÷ **5** = 7
30 ÷ **5** = 6
25 ÷ **5** = 5
20 ÷ **5** = 4

A rule is: Divide the number in column *x* by 5.

Try These

Complete. Write a rule for the function table.

1. The number in column y is less.

x	y
18	15
15	12
14	11
10	7

18 − 3 = 15
15 ☐ 3 = 12
14 ☐ 3 = 11
10 ☐ 3 = 7

Rule: ________________

2. The number in column y is ____.

x	y
24	6
20	5
16	4
12	3

24 ☐ 4 = 6
20 ☐ 4 = 5
16 ☐ 4 = 4
12 ☐ 4 = 3

Rule: ________________

3. The number in column y is ____.

x	y
2	12
3	18
4	24
5	30

2 ☐ 6 = 12
3 ☐ 6 = 18
4 ☐ 6 = 24
5 ☐ 6 = 18

Rule: ________________

Go to the next side.

Name ______________________ Skill ______________________

Practice on Your Own

Skill 55

Write a rule for the function table.

The number in column y is *greater* than the number in column x. So, the rule probably involves addition or multiplication.

MULTIPLICATION PATTERN

x	y
2	12
4	24
6	36
8	48

$2 \times \mathbf{6} = 12$
$4 \times \mathbf{6} = 24$
$6 \times \mathbf{6} = 36$
$8 \times \mathbf{6} = 48$

Rule: Multiply the number in column x by 6.

Complete. Write a rule for the function table.

1

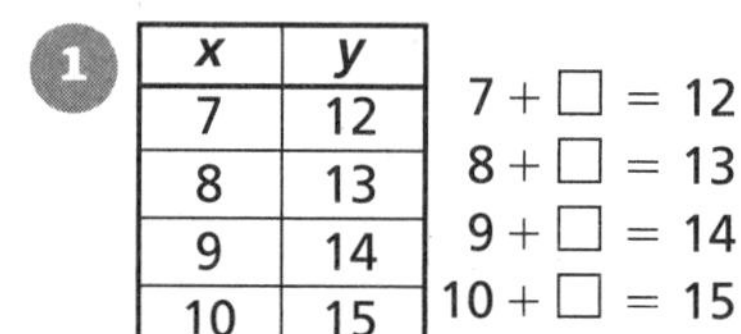

x	y
7	12
8	13
9	14
10	15

$7 + \square = 12$
$8 + \square = 13$
$9 + \square = 14$
$10 + \square = 15$

Rule: Add ____ to number in column x.

2

x	y
10	8
9	7
8	6
7	5

$10 - \square = 8$
$9 - \square = \square$
$8 - \square = \square$
$7 - \square = \square$

Rule: Subtract ____ from number in column x.

3

x	y
15	5
12	4
9	3
6	2

$15 \square 3 = 5$
$12 \square 3 = \square$
$9 \square 3 = \square$
$6 \square 3 = \square$

Rule: Divide number in column x by ____.

4

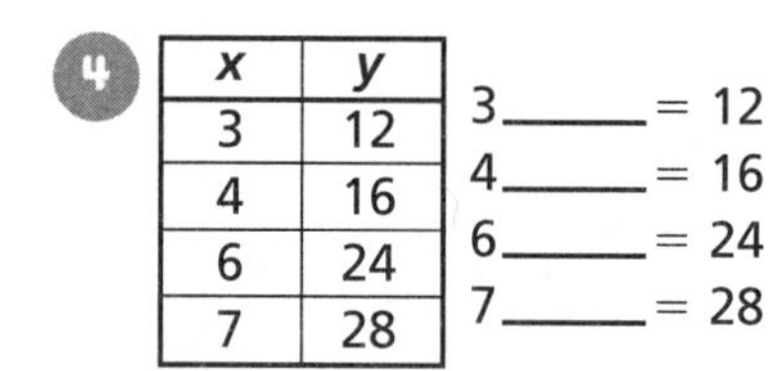

x	y
3	12
4	16
6	24
7	28

3______ = 12
4______ = 16
6______ = 24
7______ = 28

Rule: ______________

5

x	y
19	14
18	13
15	10
12	7

19______ = 14
18______ = 13
15______ = 10
12______ = 7

Rule: ______________

6

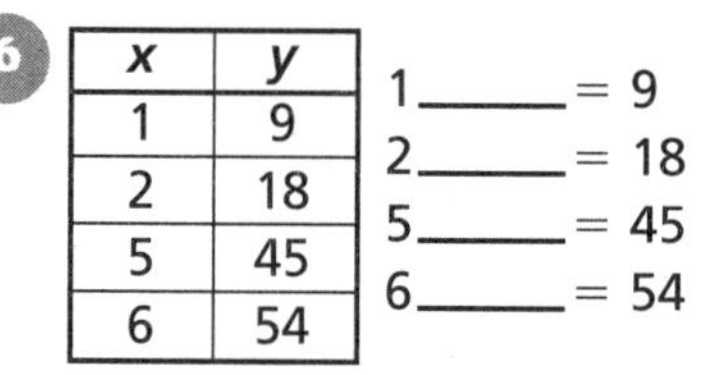

x	y
1	9
2	18
5	45
6	54

1______ = 9
2______ = 18
5______ = 45
6______ = 54

Rule: ______________

7

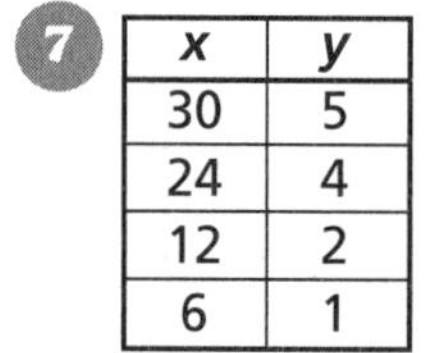

x	y
30	5
24	4
12	2
6	1

Rule: ______________

8

x	y
11	14
13	16
16	19
20	23

Rule: ______________

9

x	y
3	24
5	40
8	64
9	72

Rule: ______________

Check

Write a rule for the function table.

10

x	y
20	13
19	12
18	11
17	10

Rule: ______________

11

x	y
9	27
6	18
5	15
2	6

Rule: ______________

12

x	y
15	24
12	21
3	12
8	17

Rule: ______________

Name ______________________ Skill ______________________

Answer Card

Algebra and Functions

Grade 5

SKILL 49

TRY THESE

1. 19, 21, 24; 19, 21, 24
2. 13, 19, 23; 13, 19, 23

PRACTICE

1. + 8, + 8, + 8
2. - 5, - 5, - 5
3. - 15, - 15, - 15, - 15, 5, 10, 15, 25
4. + 7, + 7, + 7, + 7; 14, 32, 37, 47
5. 20, 30
6. 21, 24
7. 27, 18, 9, 0
8. 22, 33, 44, 55

CHECK

9. 30, 40, 50, 60
10. 80, 60, 40, 20

SKILL 50

TRY THESE

1. 9; 9
2. 7; 7; 7
3. 3; 3; 3

PRACTICE

1. 3; 3; 3
2. 3; 3; 3
3. 4; $n = 4$
4. 4; $t = 4$
5. 3
6. 11
7. 4
8. $p = 2$
9. $q = 6$
10. $y = 5$

CHECK

11. $r = 3$
12. $b = 8$
13. $m = 21$

SKILL 51

TRY THESE

1. $18 \div 3$; $3 - 3$; 6; 0; $(18 \div 6) - 3$
2. $80 - 4$; 10×4; 76; 40; $10 \times (8 - 4)$

PRACTICE

1. $10 + 12$; 2×17; 22; 34; $(2 \times 5) + 12$
2. 6×6; $12 \div 12$; 36; 1; $(6 \times 2) \div 12$
3. 3
4. 4
5. 16
6. 3
7. $(4 \times 5) - 3$
8. $10 - (3 \times 2)$
9. $(36 \div 6) - 6$
10. $12 + (8 \div 2)$
11. $(10 + 3) \times 4$
12. $15 - (10 \div 5)$
13. $12 \times (2 + 3)$
14. $(28 - 14) \div 7$

CHECK

15. $(2 \times 4) - 3$
16. $9 \times (10 - 2)$
17. $(8 + 20) \div 4$
18. $16 - (2 \times 6)$

SKILL 52

TRY THESE

1. 6; 6; 2; *C*
2. 4; 4; 7; *P*
3. 8; 8; 5; *D*

PRACTICE

1. 3; 3; 4, *H*
2. 0; 1; 5; *G*
3. 4; 0; *W*
4. *Q*
5. *V*
6. *O*
7. *K*
8. *N*
9. *I*

CHECK

10. *E*
11. *A*
12. *R*
13. *P*
14. *S*
15. *Y*

Name ______________________ Skill ______________________

Answer Card

Algebra and Functions

Grade 5

SKILL 53

TRY THESE

1. 4, 2, 4^2
2. 6, 2, 6^2
3. 7×7; 49
4. $6 \times 6 \times 6$; 36×6; 216

PRACTICE

1. 7, 2, 7^2
2. 4, 3, 4^3
3. 9, 2, 9^2
4. $(3 \times 3) \times (3 \times 3)$
5. $(2 \times 2 \times 2) \times (4 \times 4)$
6. $(5 \times 5 \times 5) \times (3 \times 3 \times 3)$
7. 64
8. 343
9. 125
10. 81
11. 36
12. 9

CHECK

13. 7^2
14. 9^4
15. $4^2 \times 8^3$
16. 64
17. 25
18. 200

SKILL 54

TRY THESE

1. $>$
2. $<$
3. $<$

PRACTICE

1. $>$
2. $<$
3. $>$
4. $<$
5. $>$
6. $=$
7. $=$
8. $>$
9. $<$
10. $=$

CHECK

11. $=$
12. $>$
13. $<$
14. $>$

SKILL 55

TRY THESE

1. $-, -, -$, subtract 3 from the number in column x
2. less, $\div, \div, \div, \div$, divide the number in column x by 4
3. greater, $\times, \times, \times, \times$, multiply the number in column x by 6

PRACTICE

1. 5, 5, 5, 5, 5
2. 2; 2, 7; 2, 6; 2, 5; 2
3. $\div$; $\div$, 4; $\div$, 3; $\div$, 2; 3
4. $\times 4, \times 4, \times 4, \times 4$; Multiply the number in column x by 4.
5. $-5, -5, -5, -5$; Subtract 5 from the number in column x.

SKILL 55

6. $\times 9, \times 9, \times 9, \times 9$; Multiply the number in column x by 9.
7. Divide the number in column x by 6.
8. Add 3 to the number in column x.
9. Multiply the number in column x by 8.

CHECK

10. Subtract 7 from the number in column x.
11. Multiply the number in column x by 3.
12. Add 9 to the number in column x.

Measurement and Geometry

Measurement

Skill 56 Grade 5 Customary Units and Tools

15 Minutes

Using Skill 56

OBJECTIVE Choose the correct tool and unit of measurement

MATERIALS measuring tools, including inch ruler; yardstick; scale; and measuring containers for tablespoon, cup, pint, quart, and gallon capacities

Read together the information about customary units and tools. Practice saying the words with students, and make note of the abbreviations for the units.

As you read, provide access to as many of the tools pictured and listed as possible. It is important for students to examine and handle the tools.

For each category of customary measures, have students give other examples of items that the units may measure. For example, for inch, objects smaller than a foot might include a piece of paper, a tile, a pencil, and so on. Encourage students to point out objects in the classroom that they might measure and to explain how they would measure them.

TRY THESE Exercises 1–4 require students to distinguish between tools and units of measure.

- **Exercise 1** Choose a small linear unit.
- **Exercise 2** Choose a larger linear unit.
- **Exercise 3** Choose a unit of capacity.
- **Exercise 4** Choose a unit of weight.

PRACTICE ON YOUR OWN Read and discuss the material at the top of the page. Ask students to give an example of when they would use a mile, a pound, and a cup.

CHECK Make sure students understand that the larger the amount or item being measured, the larger the unit of measure they should choose. Success is indicated by 3 out of 3 correct responses.

Students who successfully complete the **Practice On Your Own** and **Check** are ready to move on to the next skill.

COMMON ERRORS

- Students may have no concept of the relative size of some of the units/tools they are to use, and therefore not know which to choose to measure given amounts or items.
- Students may confuse weight and capacity.

Students who made more than 2 errors in the **Practice On Your Own**, or who were not successful in the **Check** section, may benefit from the **Alternative Teaching Strategy** on the next page.

Optional

Alternative Teaching Strategy

20 Minutes

Customary Units and Tools

OBJECTIVE Choose the appropriate unit of measurement

MATERIALS measuring tools, including inch ruler, yardstick, scale, measuring containers for tablespoon, cup, pint, quart, and gallon capacities, eyedropper

Ask students to name some classroom objects they could measure with an inch ruler or a yardstick. Responses might include desk, poster, person, bulletin board, floor, window, door, and the distance from window to door. Ask: **When do you use linear measures?** (when you want to find length, width, height, or distance) Have students discuss the markings on an inch ruler and a yardstick.

Next, discuss weight. Ask students to name some classroom objects they could weigh. Responses might include book, pen, block, and lunchbox. Ask: **Why do we weigh objects?** (to find out how heavy they are) Have students examine a scale and discuss how it is used.

Next, discuss capacity. Ask students to name some classroom objects they could find the capacity of. Responses might include glass of water, carton of juice, and aquarium.

Ask: **Why do we measure capacity?** (to find out how much a container holds)

Have students examine and discuss the measuring containers.

Tell two volunteers you want them to measure the length of the room. Give one a yardstick and the other an inch ruler. After measuring the room, have them measure the width of a piece of notebook paper.

Fill a pint container with water. Give one student a cup to measure the water and the other a tablespoon. Then have them measure the amount of water in an eyedropper.

Discuss the process and the results. Students should conclude that it's easier to measure large amounts with large units of measure and small amounts with small units.

Name ________________________ Skill ____________

Customary Units and Tools

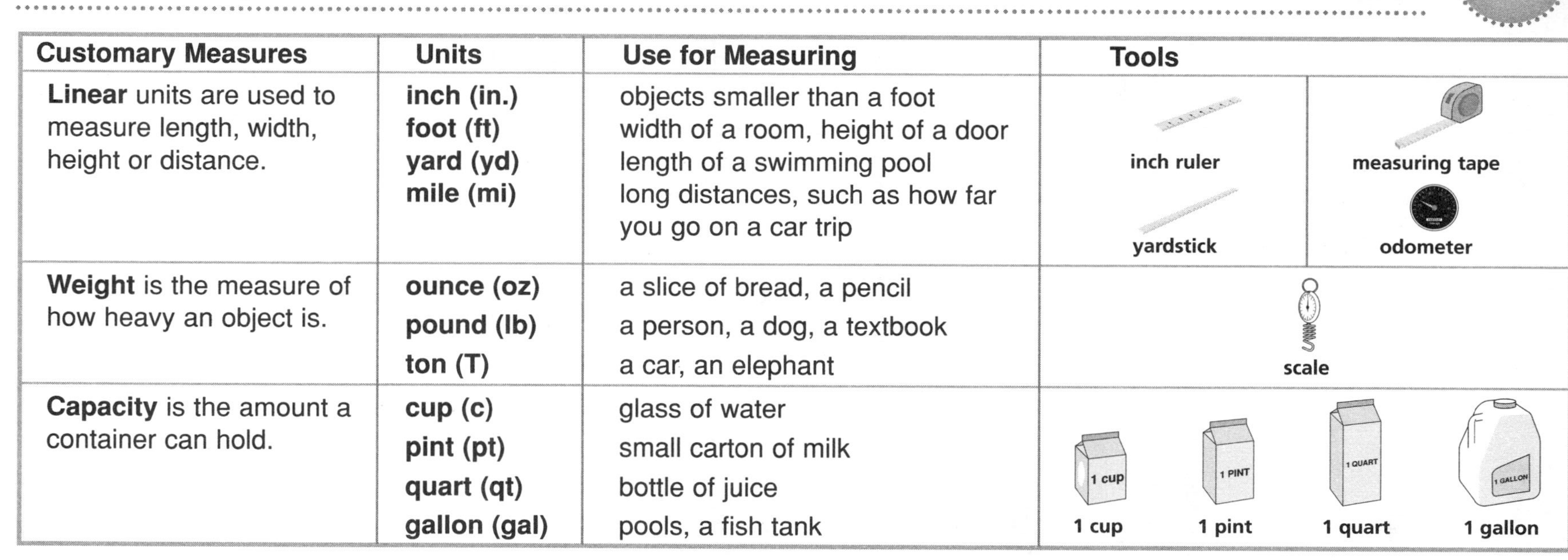

Customary Measures	Units	Use for Measuring	Tools	
Linear units are used to measure length, width, height or distance.	**inch (in.)** **foot (ft)** **yard (yd)** **mile (mi)**	objects smaller than a foot width of a room, height of a door length of a swimming pool long distances, such as how far you go on a car trip	inch ruler yardstick	measuring tape odometer
Weight is the measure of how heavy an object is.	**ounce (oz)** **pound (lb)** **ton (T)**	a slice of bread, a pencil a person, a dog, a textbook a car, an elephant	scale	
Capacity is the amount a container can hold.	**cup (c)** **pint (pt)** **quart (qt)** **gallon (gal)**	glass of water small carton of milk bottle of juice pools, a fish tank	1 cup 1 pint	1 quart 1 gallon

Try These

Complete. Use the chart above. Choose the tool and units that you would use to measure each.

1. the width of your hand

   Tool ____________

   Measure in inches or feet?

   ____

2. the length of a soccer field

   Tool ____________

   Measure in inches or yards?

   ____

3. the amount of water in a tub

   Tool ____________

   Measure in cups or gallons?

   ____

4. the weight of a letter

   Tool ____________

   Measure in ounces or pounds?

   ____

Go to the next side.

Name ______________________ Skill ______________________

Practice on Your Own

Skill 56

Choose the **tool** and **unit** you would use to measure each.

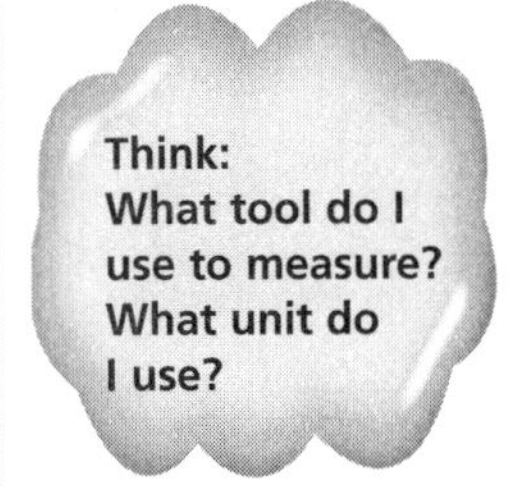

the length of a soccer field	the weight of a truck	the amount of water in a bathtub

Use a **measuring tape.**	Use a **scale**.	Use a **gallon container.**
Measure in **yards.**	Measure in **tons.**	Measure in **gallons.**

Tools:

inch ruler, yardstick, scale, odometer, gallon container, measuring cup, pint container

Units: inches, feet, yards, miles, ounces, pounds, tons, cups, pints, gallons.

Complete. Choose the tool and unit that you would use to measure each.

1. the length of a pencil
   Tool ______________
   Units ______________

2. the width of a window
   Tool ______________
   Units ______________

3. the weight of a whale
   Tool ______________
   Units ______________

4. the amount of water needed to make a glass of lemonade
   Tool ______________
   Units ______________

5. the weight of a coin
   Tool ______________
   Units ______________

6. the height of a fence
   Tool ______________
   Units ______________

7. the weight of a watermelon
   Tool ______________
   Units ______________

8. the amount of juice in a juice box
   Tool ______________
   Units ______________

9. the distance between two state capitals
   Tool ______________
   Units ______________

Check

Complete. Choose the tool and units that you would use to measure each.

10. the amount of milk in a large bucket
    Tool ______________
    Units ______________

11. the width of a computer
    Tool ______________
    Units ______________

12. the weight of a bicycle
    Tool ______________
    Units ______________

Using Skill 57

OBJECTIVE Choose the correct unit to measure, and choose the correct tool to measure

MATERIALS classroom objects, meterstick

Direct students' attention to the chart at the top of the page. Discuss each of the metric measures. Review the centimeter and decimeter units and how to read them on a meterstick. Stress that the symbols for only three of the units listed are written with a capital letter: milliliter (mL), liter (L), and degrees Celsius (°C).

Have the students choose an appropriate measure (linear, mass, capacity) for measuring various classroom objects. Next, ask the students to choose a metric unit for measuring each object. For example, a flower pot may be measured with a linear measure, a mass measure, or a capacity measure. Useful units for the flower pot would be centimeters or decimeters (height or diameter of the pot), grams or kilograms (mass of the empty or full pot), and milliliters or liters (capacity of small or large pot).

TRY THESE Exercises 1–4 provide practice for choosing either the unit or the tool to measure.

- **Exercises 1–2** Choose the unit.
- **Exercises 3–4** Choose the tool.

PRACTICE ON YOUR OWN Direct students' attention to the examples at the top of the page. Discuss each example. You may wish to have volunteers write the symbols for the units in each example.

CHECK Determine if students can identify an appropriate metric tool and unit to measure each item. Success is indicated by 3 out of 3 correct responses.

Students who successfully complete the **Practice on Your Own** and **Check** are ready to move to the next skill.

COMMON ERRORS

- Students may choose an inappropriate unit.
- Students may not understand the metric prefixes.

Students who made more than 1 error in the **Practice on Your Own**, or who were not successful in the **Check** section, may benefit from the **Alternative Teaching Strategy** on the next page.

Optional

Alternative Teaching Strategy

What's in a Prefix?

15 Minutes

OBJECTIVE Introduce the linear measures and their prefixes, choose an appropriate linear unit to measure an object

MATERIALS metric length chart, meterstick, centimeter ruler, objects with an easily defined linear dimension

Distribute copies of the metric length chart and discuss it with the students. Point out that the meter is the basic unit of length. The word *meter* is part of the name of every other metric unit of length. The prefix tells you the size of the unit relative to the meter. *Kilo-* means *thousand,* so a kilometer is 1000 meters. *Deci-* means *tenth,* so a decimeter is 1/10 of a meter. *Centi-* means *hundredth,* so a centimeter is 1/100 of a meter.

Ask: **How do you think *centi* relates to our word *cent?*** (A cent is 1/100 of a dollar.)

What happens to the length of a unit if you move to the right on the chart? (The length of the unit decreases.)

What happens to the length of the unit if you move to the left on the chart? (The length of the unit increases.)

If you measure the width of your desk in centimeters, and then measure it again in decimeters, would you have more centimeters, more decimeters, or the same number of centimeters as decimeters? (more centimeters) **Why?** (It takes more of a smaller unit to show the same length.)

What would be a more appropriate unit to measure the distance from here to your house, centimeters or kilometers? (kilometers) **Why?** (The number of centimeters would be very large and it would be hard to accurately measure that distance in centimeters.)

Distribute the metric length charts, the objects, and the measuring tools. Have students work in pairs in this activity. One partner chooses an object, then decides on a linear unit to measure it. The other student records the choice on a chart. The partners then measure and discuss whether another unit might have been more appropriate. Partners take turns choosing an object and a unit, and then measuring to confirm their choices.

If needed, repeat this exercise for capacity (basic unit: liter) and mass (basic unit: gram).

Name ______________________ Skill ______________________

Metric Units and Tools

Grade 5 Skill 57

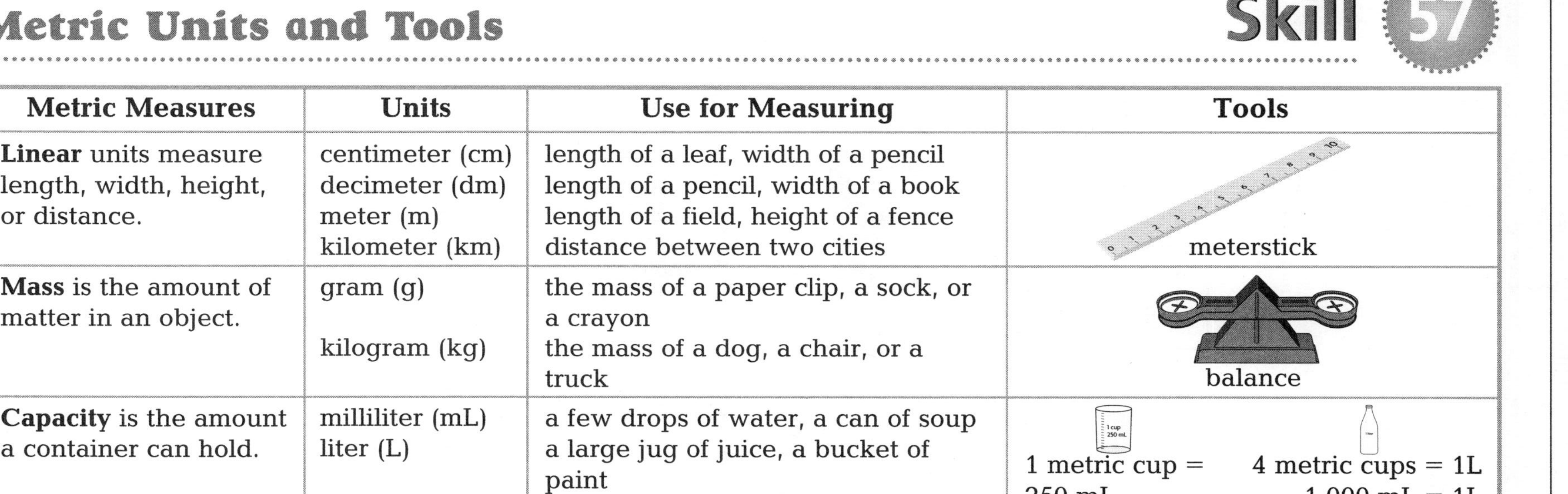

Metric Measures	Units	Use for Measuring	Tools
Linear units measure length, width, height, or distance.	centimeter (cm) decimeter (dm) meter (m) kilometer (km)	length of a leaf, width of a pencil length of a pencil, width of a book length of a field, height of a fence distance between two cities	meterstick
Mass is the amount of matter in an object.	gram (g) kilogram (kg)	the mass of a paper clip, a sock, or a crayon the mass of a dog, a chair, or a truck	balance
Capacity is the amount a container can hold.	milliliter (mL) liter (L)	a few drops of water, a can of soup a large jug of juice, a bucket of paint	1 metric cup = 250 mL 4 metric cups = 1L 1,000 mL = 1L
Temperature is a measure of warmth or coolness.	degrees Celsius (°C)	how warm or cold the day is A warm day is 26°C. A cool day is 10°C.	Celsius thermometer

Try These

Complete. Choose the unit or tool you would use to measure each.

1. the width of a button

   Unit: meters or centimeters?

   ____

2. a glass of water

   Unit: milliliters or liters?

   ____

3. the temperature in an oven

   Tool: meterstick or thermometer?

   ____

4. the mass of a muffin

   Tool: balance or metric cup?

   ____

Go to the next side.

Name ______________________ Skill ______________________

Practice on Your Own

Skill 57

Choose the **metric tool** and **unit** you would use to measure each.

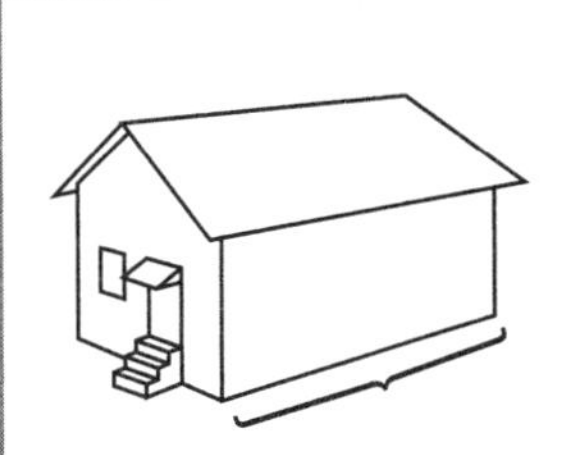

length of a house	bowl of soup	mass of six bricks	temperature on an iceberg
tool: meterstick	tool: metric cup	tool: balance	tool: thermometer
unit: meters	unit: milliliters	unit: kilograms	unit: degrees Celsius

Tools:

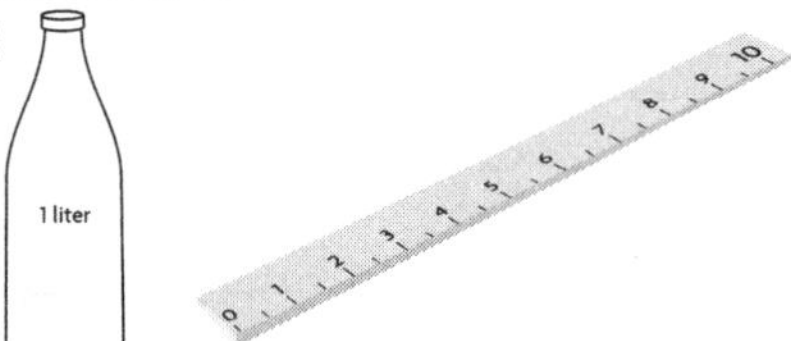

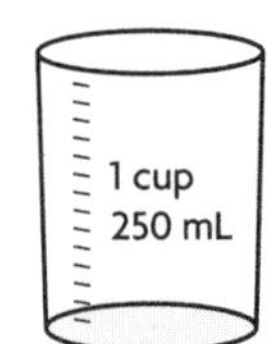

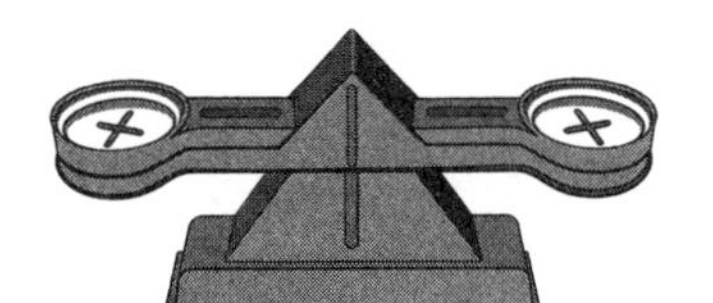

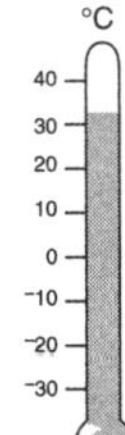

Units: centimeters, meters, kilometers, grams, kilograms, milliliters, liters, degrees Celsius

Complete. Choose the metric tool or unit that you would use to measure each.

1. distance across your room
   meterstick or balance
   tool: ______________

2. a glass of juice
   thermometer or metric cup
   tool: ______________

3. temperature of bath water
   balance or thermometer
   tool: ______________

4. a few drops of honey
   liters or milliliters
   unit: ______________

5. the mass of a grape
   grams or kilograms
   unit: ______________

6. the length of your hand
   centimeters or kilometers
   unit: ______________

Check

Choose the metric tool you would use to measure each, then choose the unit.

7. the amount of milk in a big container
   ____
   ____

8. the mass of a computer
   ____
   ____

9. how tall you are
   ____
   ____

Read a Thermometer (Negative Numbers)

Using Skill 58

OBJECTIVE Read a temperature above and below zero on a Fahrenheit thermometer

MATERIALS number line

Begin the lesson by reminding students how to read a horizontal number line. Recall with them that numbers are positive to the right of zero and are negative to the left of zero. Next, rotate the number line 90°. Highlight the movement of positive numbers upward from zero, and negative numbers downward from zero.

Draw students' attention to the thermometers in Step 1. Be sure students understand how to assign a value for each interval. Ask: **The thermometer on the left shows how many intervals from 0° to 10°?** (2) **How many degrees does each interval represent?** (5°) **How do you know?** (10° divided by 2 equals 5°.)

Continue with similar questions for the thermometer on the right. Be sure students recognize the difference in the number of intervals and how many degrees each interval represents.

Direct students' attention to Step 2. Guide students to recognize that one thermometer shows a temperature above zero, and the other below zero.

Continue with Step 3. Be sure students can read the scale and assign a temperature. Start with the thermometer on the left. Ask: **How many intervals above 0° are shaded?** (3) **How many degrees does each interval represent?** (5°) **What is the temperature represented?** (15° F)

Continue with similar questions for the thermometer on the right.

Guide students to recognize that all temperatures below zero are named with negative numbers and written with a negative sign (−).

TRY THESE In Exercises 1–3 students read temperature on a thermometer.

- **Exercise 1** Name a temperature below zero.
- **Exercises 2–3** Name a temperature above zero.

PRACTICE ON YOUR OWN Review the example at the top of the page. Have students name the number of degrees represented by each interval, the direction from 0°, and the temperature reading for the thermometer.

CHECK Determine if students can name the correct temperature in degrees Fahrenheit.

Success is indicated by 2 out of 2 correct responses.

Students who successfully complete **Practice on Your Own** and **Check** are ready to move to the next skill.

COMMON ERRORS

- Students may not name a temperature below 0° with a negative number.
- Students may determine intervals incorrectly.
- Students may not include the unit of measure (degree) and symbol for Fahrenheit (F) in the name of a temperature.

Students who made more than 3 errors in the **Practice on Your Own**, or who were not successful in the **Check** section, may benefit from the **Alternative Teaching Strategy** on the next page.

Optional

Alternative Teaching Strategy

15 Minutes

Model Distance from Zero on a Number Line

OBJECTIVE Use a number line model to determine distance from zero

MATERIALS flip chart

The goal of this activity is to have students apply positional thinking to an understanding of reading the intervals on a thermometer (or any type of vertical number line).

Draw and label a vertical number line like the one shown.

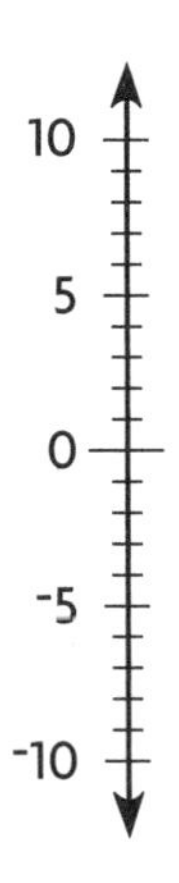

Tell students that each mark on the number line represents one number. Guide students to understand that the marks for the numbers 5 and $^-5$ represent the same distance from zero, but in *opposite* directions.

Explain that the negative sign for $^-5$ is used to identify the location of the mark in relation to zero.

Remind students that sometimes the marks above zero are labeled with a plus sign. Sometimes they are not labeled. When a number is not labeled with any sign, it is assumed that the number is positive.

Ask students to name other marks above and below zero.

Name ______________________ Skill ______________________

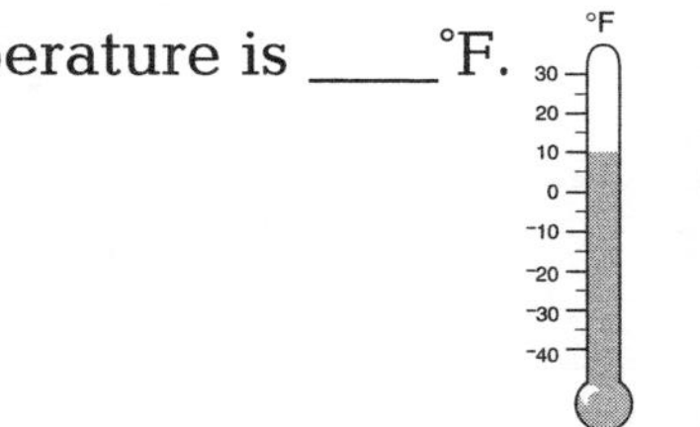

Grade 5 Skill 58

Read a Thermometer (Negative Numbers)

Fahrenheit is the measure of temperature in the customary system.
Name the temperature in degrees Fahrenheit.

Step 1
Find the number of degrees from one mark to the next. This is an interval.

Thermometer 1 — Each interval is 5°F.

Thermometer 2 — Each interval is 1°F.

Step 2
Is the temperature above zero or below zero?

Thermometer 1 — Temperature is *above* zero.

Thermometer 2 — Temperature is *below* zero.

Step 3
Read the scale to name the temperature.

Temperatures below zero are expressed using negative numbers.

Thermometer 1 — Count 3 intervals *above* zero. The temperature is 15°F.

Thermometer 2 — Count 4 intervals *below* zero. The temperature is $^{-}4$°F.

So, the temperatures are 15°F and $^{-}4$°F.

Try These

Complete.

1. Each interval is ____ °F.
   The temperature is ____ zero.
   The temperature is ____°F.

2. Each interval is ____ °F.
   The temperature is ____ zero.
   The temperature is ____°F.

3. Each interval is ____ °F.
   The temperature is ____ zero.
   The temperature is ____°F.

Go to the next side.

Name ____________________ Skill ____________________

Practice on Your Own

Skill 58

Think:
Find the scale. How many degrees is it from one mark to the next?

Is the temperature *above* or *below* zero?

Name the temperature in degrees Fahrenheit.

The interval is 5°F.

The temperature is below zero.

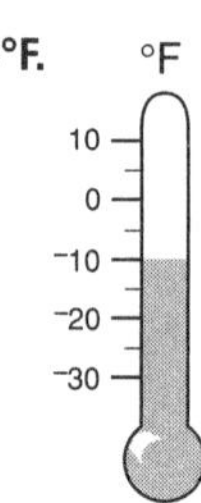

The temperature is ⁻10°F.

Complete.

1. Each interval is 2°F.
   The temperature is *above* zero.
   The temperature is ____°F.

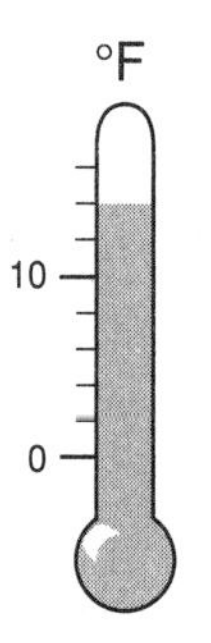

2. Each interval is ____ °F.
   The temperature is *below* zero.
   The temperature is ____°F.

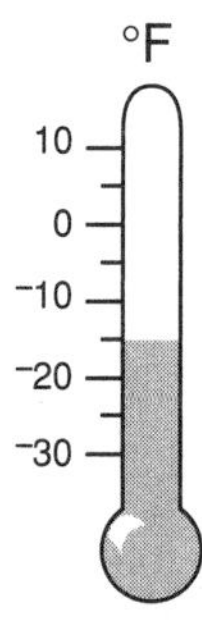

3. Each interval is ____ °F.
   The temperature is ________ zero.
   The temperature is ____°F.

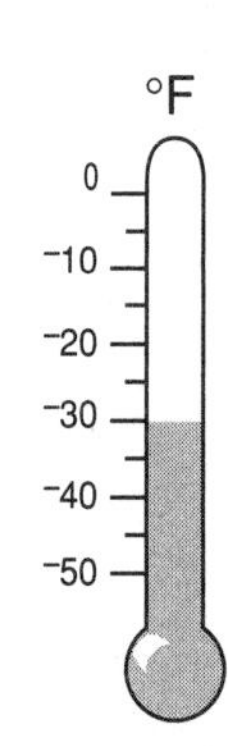

4. Each interval is ____ °F.

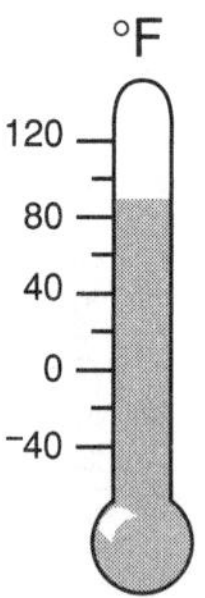

Check

Complete.

5. Thermometer

°F
20
10
0
⁻10
⁻20

The temperature is ____ °F.

6. Thermometer

°F
40
20
0
⁻20
⁻40

The temperature is ____ °F.

Name ______________________ Skill ______________________

Answer Card

Measurement

Grade 5

Skill 56

TRY THESE

1. Inch ruler, inches
2. Measuring tape, yards
3. Gallon container, gallons
4. Scale, ounces

PRACTICE

1. Inch ruler, inches
2. Yardstick, inches
3. Scale, tons
4. Measuring cup, cup
5. Scale, ounces
6. Yardstick, yards or feet
7. Scale, pounds
8. Measuring cup, cups or ounces
9. Odometer, miles

CHECK

10. Gallon container, gallons
11. Inch ruler or yardstick, inches
12. Scale, pounds

Skill 57

TRY THESE

1. centimeters (cm)
2. milliliters (mL)
3. thermometer
4. balance

PRACTICE

1. meterstick
2. metric cup
3. thermometer
4. milliliters
5. grams
6. centimeters

CHECK

7. metric cup or 1-liter container; liters
8. balance; kilograms
9. meterstick; meters

Name ______________________ Skill ______________________

SKILL 58

TRY THESE

1. 10; below, $^{-}10$
2. 2; above; 8
3. 5; above; 10

PRACTICE

1. 14
2. 5; $^{-}15$
3. 5; below; $^{-}30$
4. 20

CHECK

5. 15
6. $^{-}20$

Answer Card

Measurement

Grade 5

Measurement and Geometry

Geometry

Skill 59 Grade 5

Classify Plane Figures

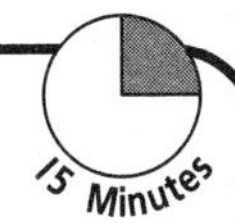

Using Skill 59

OBJECTIVE Classify a plane figure as either open or closed

MATERIALS diagrams of open and closed plane figures

Begin by demonstrating how to decide whether a plane figure is open or closed. Tell students to place their fingers at a point, then see if they can trace around the figure without lifting their fingers, ending at the starting point. If they can do that, then the figure is a closed figure.

Ask students to find examples of plane figures that do not begin and end at the same point. Tell them that these figures are called open plane figures.

Call students' attention to the *Closed Figures* section at the top of the page. Have them choose a point on each figure and trace around it with their fingers.

For the *Open Figures* section, have students use the same procedure.

TRY THESE Exercises 1–2 provide the students with a question to help them decide whether a figure is open or closed.

- **Exercise 1** Closed figure.
- **Exercise 2** Open figure.

PRACTICE ON YOUR OWN Review the definitions of closed and open plane figures as students look at the figures at the top of the page.

Ask: **How can you check to see whether a figure is open or closed?** (If you can trace around it and return to the same point, it is closed.)

CHECK Determine if students can draw an example of an open figure and a closed figure. Success is indicated by 2 out of 2 correct responses.

Students who successfully complete the **Practice on Your Own** and **Check** are ready to move to the next skill.

COMMON ERRORS

- Students may not classify figures that are not polygons as closed.

Students who made more than 1 error in the **Practice on Your Own**, or who were not successful in the **Check** section, may benefit from the **Alternative Teaching Strategy** on the next page.

Optional

Alternative Teaching Strategy

15 Minutes

Use Models to Classify Plane Figures

OBJECTIVE Use models to classify plane figures

MATERIALS open and closed figures made from chenille stems, chenille stems

Distribute the chenille-stem figures. Start with the figures shown.

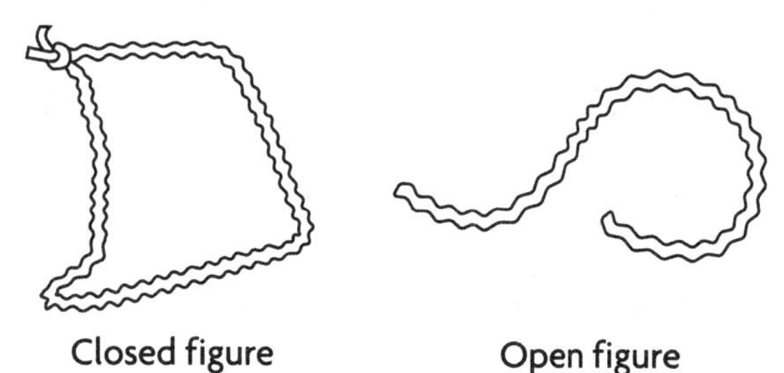

Ask: **How can you test to see whether these figures are open or closed?** (Trace to see whether you can return to the starting point without lifting your finger.)

Have students work in pairs. One partner traces the figures with his or her finger while the other describes the figures as open or closed. Partners take turns tracing and describing each figure.

Distribute the chenille stems. Ask each pair of students to make a simple closed figure. Next, ask students to change the shape of the figure and decide whether it is still closed.

Finally, ask each partner to make a figure and write whether it is open or closed. Have them exchange figures and decide whether they agree about the classification of the figures.

Name ______________________ Skill ______________________

Grade 5 Skill 59

Classify Plane Figures

A **plane figure** is a flat figure that lies in one plane. It can be an **open** or **closed** figure.

Closed Figures

A closed figure begins and ends at the same point.

Open Figures

An open figure does not begin and end at the same point.

Try These

Circle *yes* or *no*. Write *open* or *closed* to describe the figure.

1. Does the figure begin and end at the same point?

   yes no

   It is a(n) ____________ figure.

2. Does the figure begin and end at the same point?

   yes no

   It is a(n) ____________ figure.

Go to the next side.

Name ______________________ Skill ______________________

Practice on Your Own

Skill 59

Think:
A closed figure begins and ends at the same point.

Plane Figures

Closed

Open

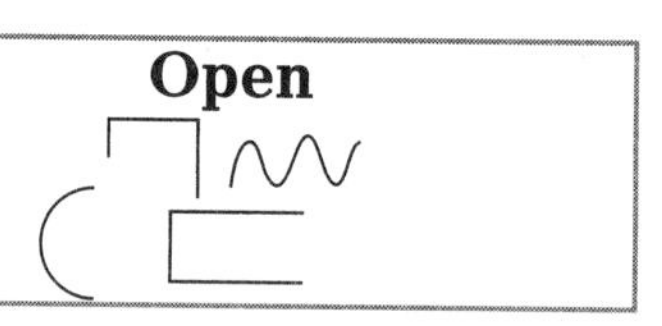

Circle *yes* or *no*. Write *open* or *closed* to describe the figure.

Does the figure begin and end at the same point?

yes no

It is a(n) ____________ figure.

2

Does the figure begin and end at the same point?

yes no

It is a(n) ____________ figure.

Write *open* or *closed* to describe the figure.

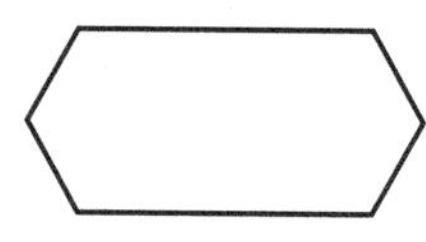

____

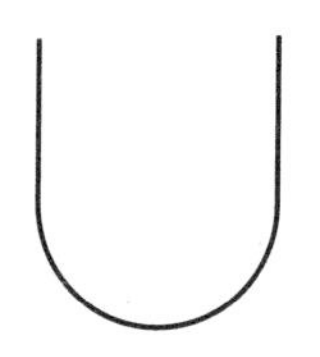

____

5

____

6

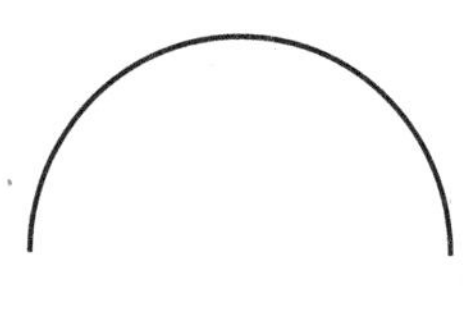

____

7

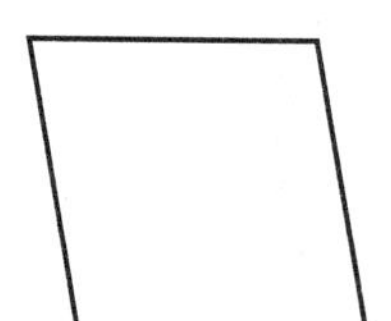

____

8

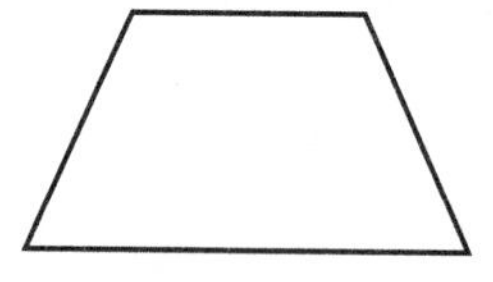

____

Check

Complete.

9 Draw an **open** figure.

10 Draw a **closed** figure.

Skill 60 Grade 5

Name Polygons

Using Skill 60

OBJECTIVE Identify and name triangles, quadrilaterals, pentagons, hexagons, and octagons

Begin the lesson by having students read the definition of a polygon. You may wish to have students suggest some facts they know about a polygon.

- It is a closed figure.
- It has straight sides.
- Its sides meet to form angles.
- It has the same number of sides and angles.

Emphasize that a polygon is named by its number of sides and angles. Compare the regular and irregular pentagons. Point out that even though the figures look different, they are both pentagons—each figure has five sides and five angles.

Direct students' attention to the triangles. Ask: **How many sides form a triangle?** (3)

How many angles do you see in each triangle? (3)

How are the two triangles alike? How are they different? (Alike: They both have 3 sides and 3 angles. Different: One triangle has congruent sides, and the other triangle does not.)

Is a triangle a polygon? How do you know? (Yes, it is a closed figure with 3 straight sides that form 3 angles.)

Continue to ask similar questions as you discuss the other polygons.

TRY THESE In Exercises 1–3, students count sides and angles to name polygons.

- **Exercise 1** 4 sides and angles.
- **Exercise 2** 6 sides and angles.
- **Exercise 3** 5 sides and angles.

PRACTICE ON YOUR OWN Review the examples at the top of the page. Ask students to use the figures to review the definition of a polygon.

CHECK Determine if students can name a polygon based on the number of its sides and angles. Success is indicated by 3 out of 3 correct responses.

Students who successfully complete the **Practice on Your Own** and **Check** are ready to move to the next skill.

COMMON ERRORS

- Students may not count the number of sides and/or angles correctly.
- Students may lose track counting angles and sides on an irregular figure.
- Students may find the names of the polygons difficult to remember.

Students who made more than 3 errors in the **Practice on Your Own**, or who were not successful in the **Check** section, may benefit from the **Alternative Teaching Strategy** on the next page.

Optional

Alternative Teaching Strategy

Use Flash Cards to Name Polygons

15 Minutes

OBJECTIVE Use flash cards to recognize and name triangles, quadrilaterals, pentagons, hexagons, and octagons

MATERIALS large index cards, markers

Prepare four cards for each polygon. Make one card to show each of the following: a diagram of the polygon, the number of sides, the number of angles, and the name of the polygon.

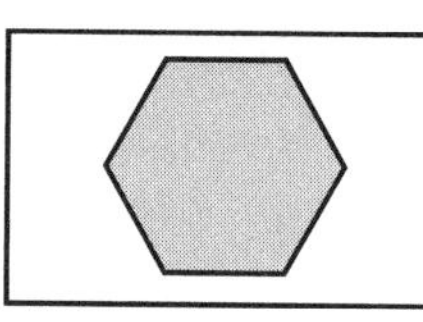

6 sides

6 angles

hexagon

As you display each diagram of a polygon, discuss the number of sides and angles. Suggest memory clues to help students remember the name of each polygon. For example, a triangle has three sides, a tricycle has three wheels; a pentagon has fives sides, the Pentagon building has five sides; an octagon has eight sides, an octopus has eight arms, and so forth.

Place the diagrams of the polygons face up on a table or desk. Give out the remaining cards and have students take turns matching each card to the appropriate polygon.

Then collect the cards that describe the number of angles and display them face up. Distribute all the remaining cards. Have students take turns matching their cards to the angle cards.

Repeat the activity having students also match the cards describing the number of sides. Finally, have students match the name cards.

When students are confident that they can recognize these polygons by diagram, name, angles, and sides, have each student draw one card from the pile and describe to the group the polygon that the card matches.

Name ______________________ Skill ______________________

Grade 5 Skill 60

Name Polygons

A **polygon** is a closed plane figure with straight sides. The sides are line segments that meet to form angles. A polygon has the same number of sides and angles. A polygon is named by the number of its sides and angles.

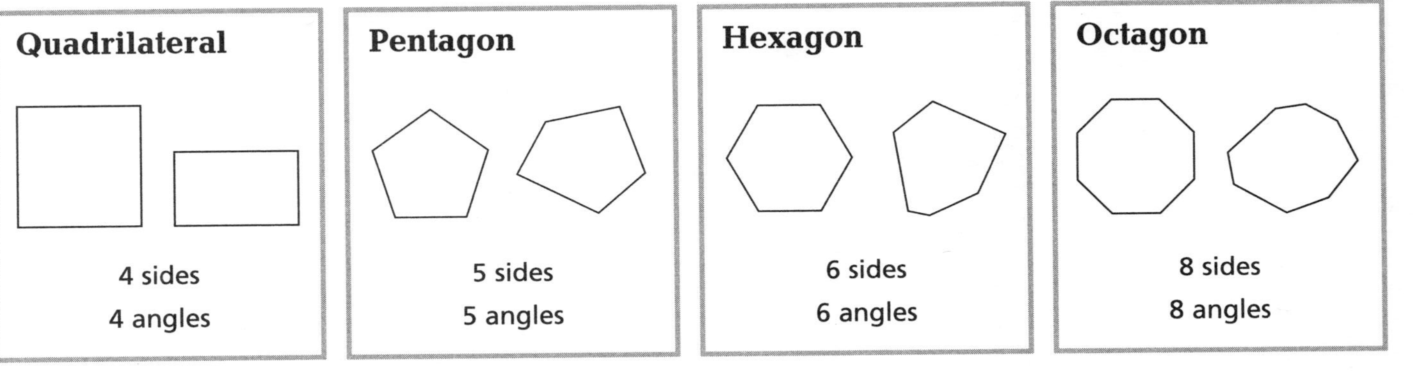

Try These

Count the sides and angles. Name the polygon.

1. _____ sides
   _____ angles
   ____

2. _____ sides
   _____ angles
   ____

3. _____ sides
   _____ angles
   ____

Go to the next side.

Name ______________________ Skill ______________________

Practice on Your Own

Skill 60

Think:
A polygon has the same number of sides and angles.

Polygons

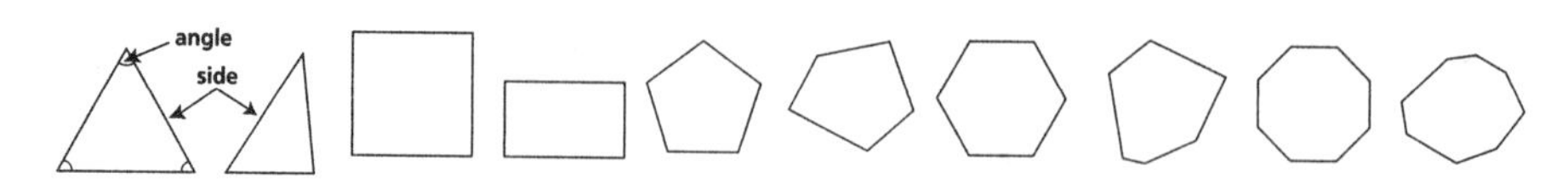

	Triangle	Quadrilateral	Pentagon	Hexagon	Octagon
Sides	3	4	5	6	8
Angles	3	4	5	6	8

Tell how many sides and angles. Name the polygon.

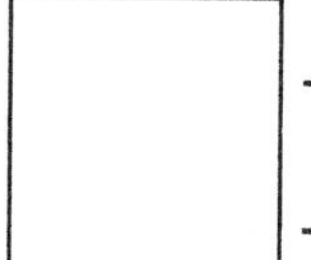
____ sides
____ angles
____

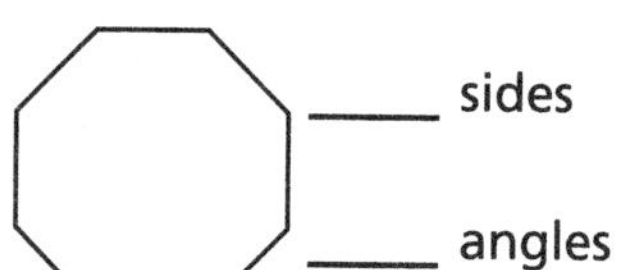
____ sides
____ angles
____

____ sides
____ angles
____

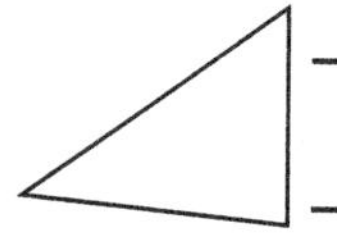
____ sides
____ angles
____

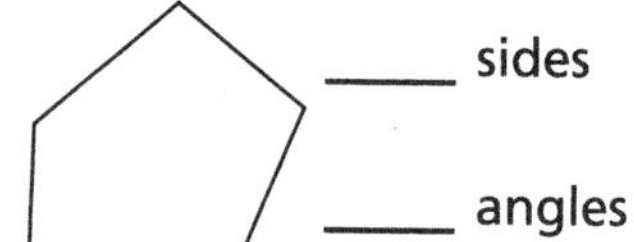
____ sides
____ angles
____

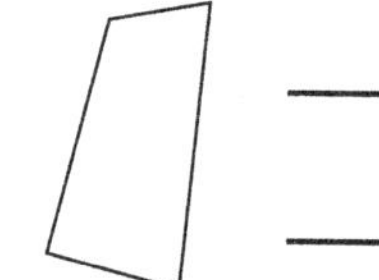
____ sides
____ angles
____

Name the polygon.

____

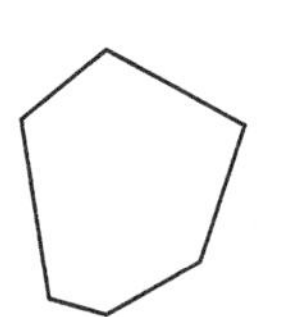
____

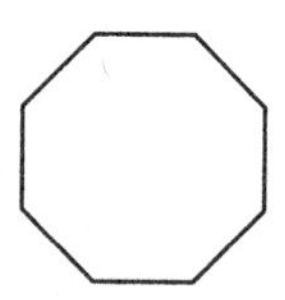
____

Check

Name the polygon.

10
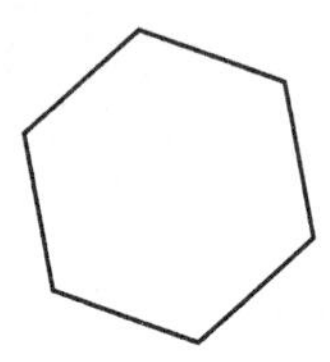
____

11
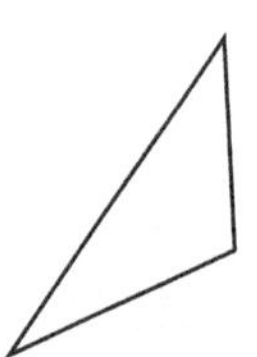
____

12
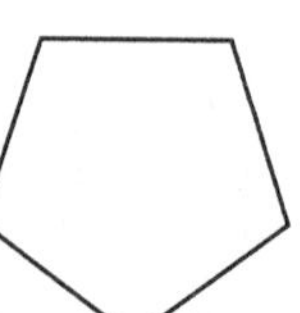
____

Using Skill 61

OBJECTIVE To recognize and name right angles, acute angles, and obtuse angles

On Skill 61, draw attention to the three types of angles. Be sure students are familiar with the attributes of right, acute, and obtuse angles. It may help some students to use the right angle as a reference to gauge the other angle types. For example, a right angle can be associated with the corner of a book shelf or the letter "L." So, when they need a reference, students can think of those examples to help name angle types.

Draw attention to the example for right angles. Ask: **Do the angles shown form square corners?** (yes) **What is the measure of a an angle that forms a square corner?** (90°) **Does the measure of a right angle change when it is placed in different positions?** (no) **Why?** (The size of the angle does not change.)

Draw attention to the example for acute angles. Help students see that the opening between the two rays is less than that of a right angle. Ask: **What are the possible measures for acute angles?** (greater than 0° but less than 90°) Point out to students that an acute angle will always fit inside a right angle.

Draw attention to the example for obtuse angles. Help students recognize that the opening between the rays is greater than that of a right angle. Ask: **Can an obtuse angle fit inside a right angle?** (no) **Why?** (Its opening is greater than 90°.)

TRY THESE Exercises 1–3 model the three angles: acute, right, and obtuse.

- **Exercise 1** Name an acute angle.
- **Exercise 2** Name a right angle.
- **Exercise 3** Name an obtuse angle.

PRACTICE ON YOUR OWN Review the example at the top of the page. Help students refer to a right angle when evaluating other angle types.

CHECK Determine if students can classify each of the angle types correctly.

Success is indicated by 3 out of 3 correct responses.

Students who successfully complete the **Practice on Your Own** and **Check** are ready to move to the next skill.

COMMON ERRORS

- Students may not recognize an angle because of a rotation of the position of the angle.
- Students may confuse the attributes of *acute* and *obtuse* angles.
- Students may be using an inappropriate object as a concrete reference for a right angle, and, as a result, may make errors in naming other angle types.

Students who made more than 3 errors in the **Practice on Your Own,** or who were not successful in the **Check** section, may benefit from the **Alternative Teaching Strategy** on the next page.

Optional

Alternative Teaching Strategy

Clock Face Angles

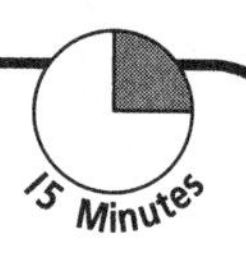

OBJECTIVE Use a clock face to classify angles

MATERIALS clock face, index cards

Provide students with a clock face model and an index card. Remind students that the corner of the card represents a right angle.

Position the clock hands for 3 o'clock. Ask a student to place the index card over the clock hands so that the edges of the card coincide with the hour and minute hands as shown.

Guide students to understand that a right angle is formed by the clock hands whenever the edges of the index card coincide this way.

Have a student re-set the clock face to 7:30. Slide the index card so that one edge of the card coincides with a hand and the other hand lies under the card.

Inform students that an acute angle is formed when one hand coincides with an edge of the card and the other hand lies under the index card.

Direct students to think of an example to illustrate an obtuse angle. Remind students to use the index card.

Help students to position the clock hands so that the edge of the index card coincides with one clock hand while the other clock hand lies outside the card. An example of an obtuse angle is shown below.

Repeat examples where students classify the angle formed by different times of the day. One student may set the position of the clock hands and another may use the index card to help name the angle.

Name ______________________ Skill ______________________

Grade 5 Skill 61

Classify Angles

An angle is formed by two rays that have the same endpoint.

Right Angles

A **right angle** forms a square corner.The measure of a right angle is 90 degrees, or 90°.

A right angle can be placed in any position.

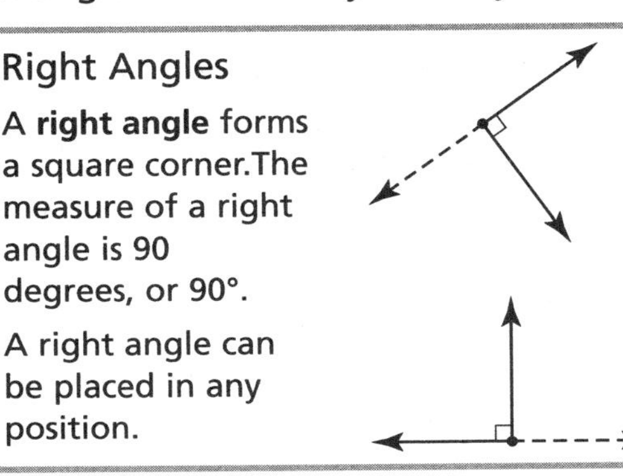

Acute Angles

An **acute angle** is *less than* a right angle. Its angle measure is less than 90°.

An acute angle can also be placed in any position.

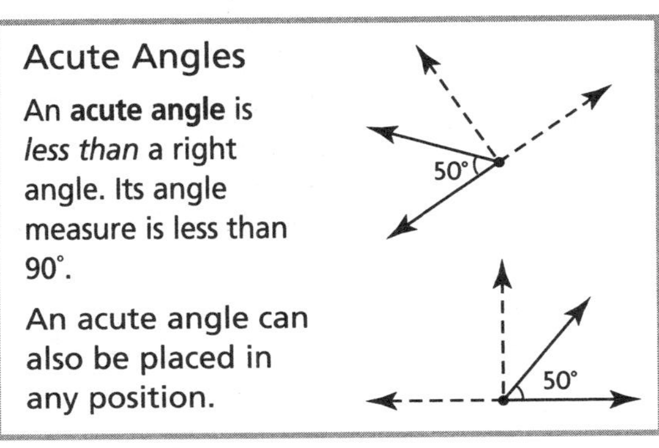

Obtuse Angles

An **obtuse angle** is *greater than* a right angle. Its angle measure is greater than 90° but less than 180°.

A straight, or 180° angle forms a straight line.

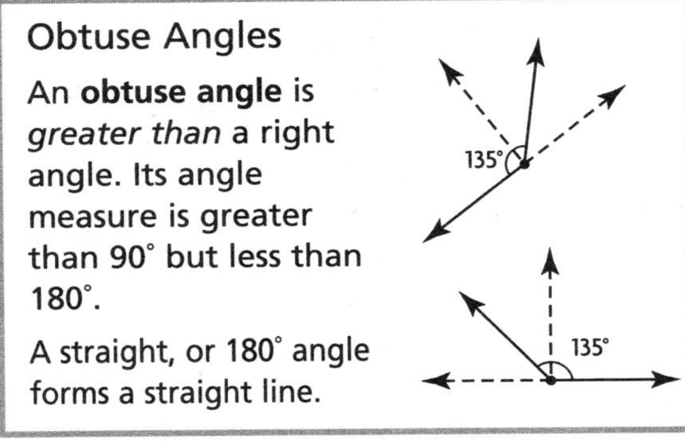

Remember: A point marks an exact location in space. A ray is part of a line that has one endpoint. A ray extends without end in one direction.

Try These

Classify the angle. Write *acute*, *right*, or *obtuse*.

1

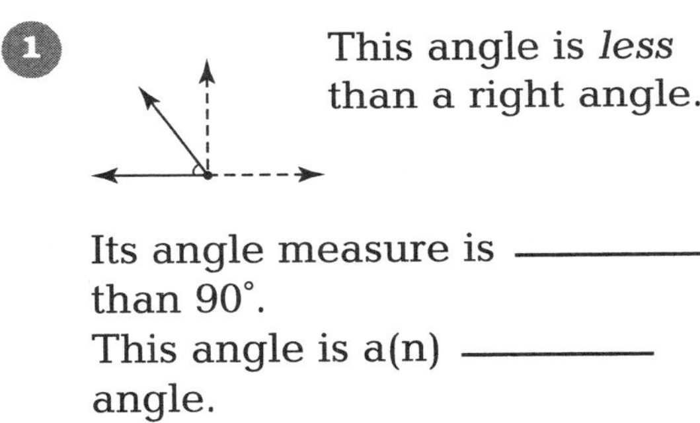

This angle is *less* than a right angle.

Its angle measure is ________ than 90°.

This angle is a(n) ________ angle.

2

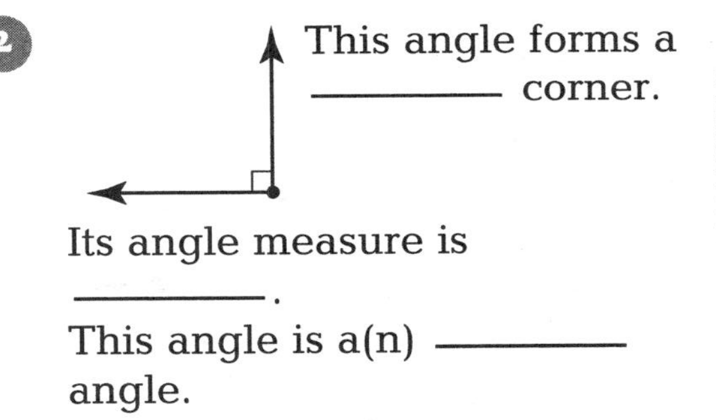

This angle forms a ________ corner.

Its angle measure is ________.

This angle is a(n) ________ angle.

3

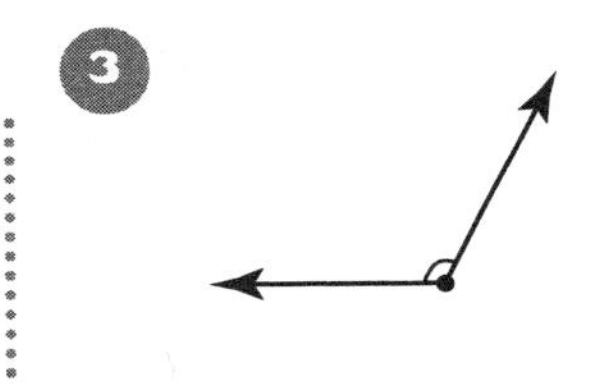

This angle is a(n) ________ angle.

Go to the next side.

Name ______________________ Skill ______________________

Practice on Your Own

Skill 61

Think:
An **acute angle** or a **right angle** can fit inside an **obtuse angle**.

Classify the angle.

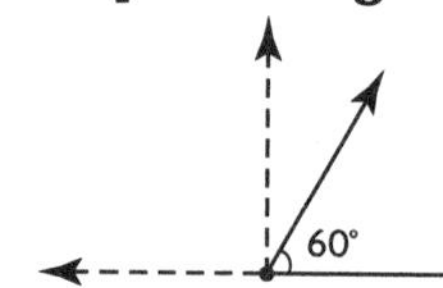

This angle is less than a right angle. The angle measure is less than 90°. So, it is an acute angle.

Think:
How does this angle compare with the corner of a sheet of paper?

Classify the angle. Complete.

1

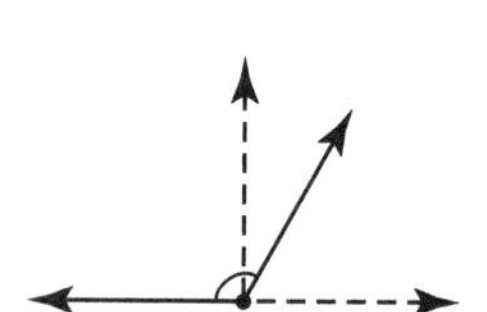

Think:
How does this angle compare to a right angle?

This angle is a(n) ________ angle.

2

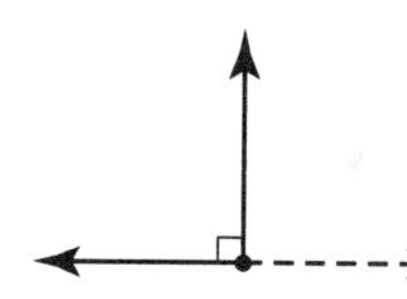

Think:
How does this angle compare to a right angle?

This angle is a(n) ________ angle.

3

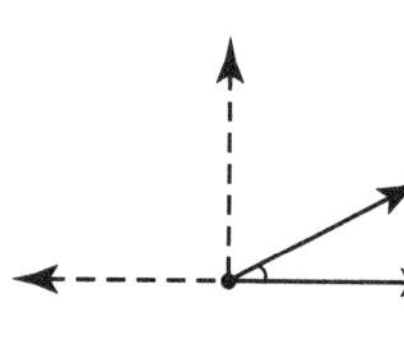

Think:
How does this angle compare to a right angle?

This angle is a(n) ________ angle.

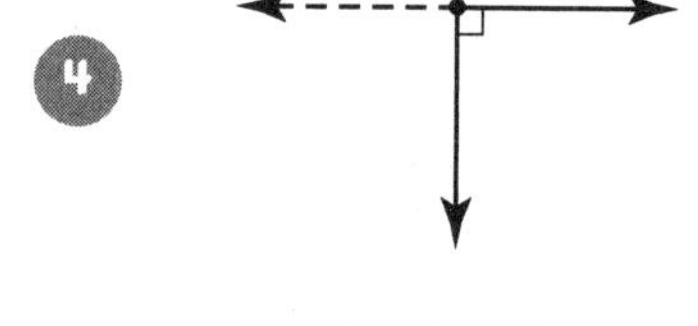

____

____

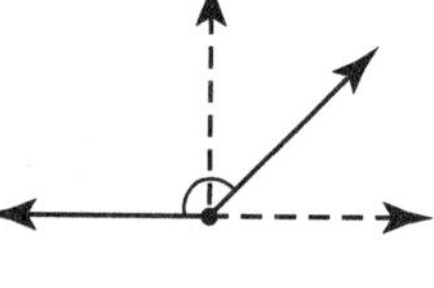

____

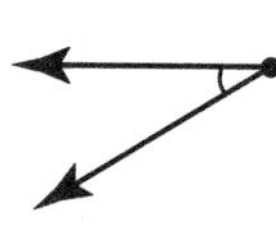

____

____

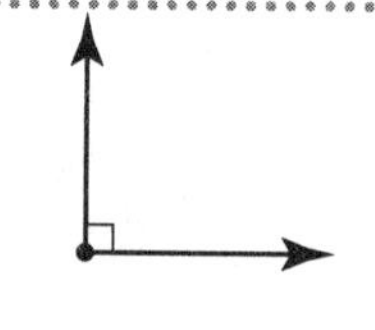

____

Check

Classify the angle. Complete.

10

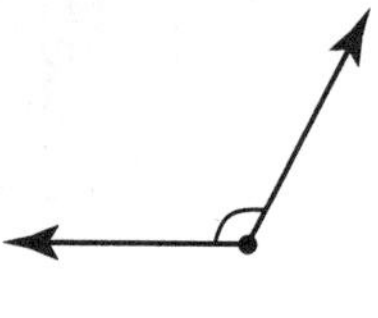

____

11

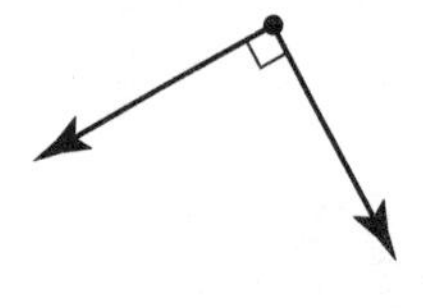

____

12

____

Faces of Solid Figures

Using Skill 62

OBJECTIVE Identify and name the faces on a solid figure

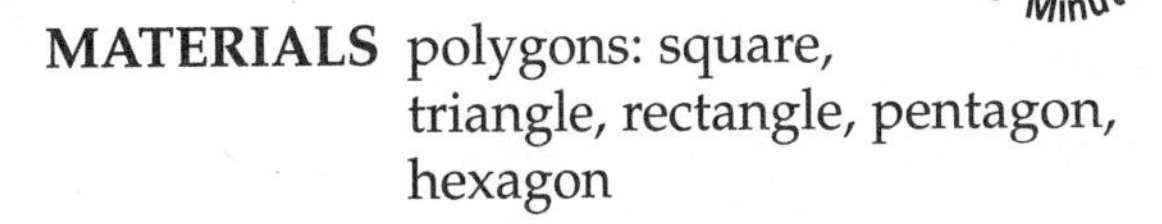

MATERIALS polygons: square, triangle, rectangle, pentagon, hexagon

Draw students' attention to the first example. Show the model of the square. Discuss the characteristics of a square: congruent sides, congruent right angles. Ask: **What does congruent mean?** (same size, same shape) **What does it mean when the sides of a figure are congruent?** (The sides have the same length.) **What does it mean when the angles are congruent?** (The angles have the same measure.) **What is a right angle?** (An angle that measures 90°)

Draw attention to the second example. Point out the triangular faces on the triangular prism, the triangular pyramid and the square pyramid. Help students recognize the difference between a prism and a pyramid. Ask: **How are a prism and a pyramid different?** (A prism has 2 parallel bases; a pyramid has only one base.)

Draw attention to the third example. Review with students the definition of a polygon. Ask: **What is a polygon?** (a closed plane figure made up of line segments) You may wish to review the names of the polygons with 3, 4, 5, and 6 sides. Point out the shaded face on the pentagonal pyramid. Ask: **How many sides does this face have?** (5) **What polygon names this face?** (pentagon)

Now point to the shaded face on the hexagonal pyramid. Ask: **How many sides does this face have?** (6) **What polygon names this face?** (hexagon)

Next, point to the shaded face on the rectangular prism. Ask: **How many sides does this face have?** (4) **Are all the sides congruent?** (no) **What polygon names this face?** (rectangle)

TRY THESE In Exercises 1–3, students count sides and angles to name the shaded face of a solid figure.

- **Exercise 1** Square face of a cube.
- **Exercise 2** Triangular face of a triangular pyramid.
- **Exercise 3** Pentagonal face of a pentagonal pyramid.

PRACTICE ON YOUR OWN Review the example at the top of the page. Have students count the number of sides and angles on each shaded face. Then have them name the shaded face.

CHECK Determine if students can correctly name shaded faces on solid figures.

Success is indicated by 3 out of 3 correct responses.

Students who successfully complete the **Practice on Your Own** and **Check** are ready to move to the next skill.

COMMON ERRORS

- Students may confuse prism and pyramid.
- Students may not be able to identify the name of the shape of a face.

Students who made more than 2 errors in the **Practice on Your Own,** or who were not successful in the **Check** section, may benefit from the **Alternative Teaching Strategy** on the next page.

Optional

Alternative Teaching Strategy

Trace Faces of Solid Figures

15 Minutes

OBJECTIVE Trace faces of geometric solid figures to name their polygonal faces

MATERIALS geometric solid figures; unlined paper

Provide students with geometric solids of prisms and pyramids. Remind students that the names of polygons are used to name the faces or flat surfaces of solid figures.

Have a student place a rectangular prism on a sheet of paper. Direct the student to outline one of the faces of the solid figure with a pencil.

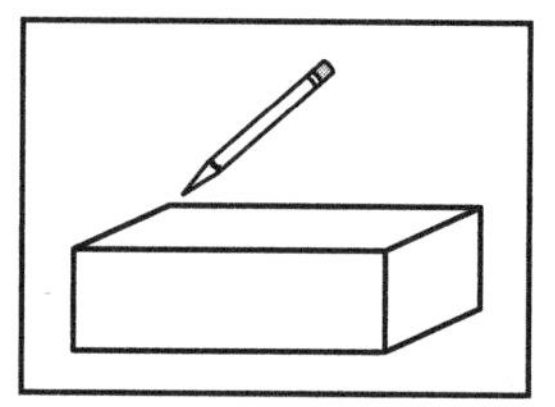

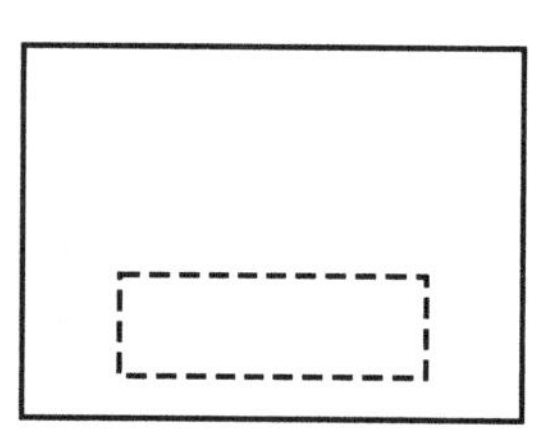

Have the student remove the solid figure and show the result, then measure sides and test angles with an index card.

Guide students to name the outlined face.

Have another student outline a different face of the rectangular prism.

Again, have the student remove the solid figure. Ask students to identify the number of sides and angles shown. Ask students to name the polygon outlined.

Next have a student outline a face on a pyramid. Help students realize that for any pyramid, only triangular faces meet at a common point above its base.

Continue with other solid figures. Review the number and measure of angles and sides on the faces of the solid figures. Have students name the polygonal faces.

Name ______________________ Skill ______________________

Grade 5
Skill 62

Faces of Solid Figures

Polygons are two-dimensional figures. They have length and width.
Solid figures are three-dimensional figures. They have length, width, and height.
You can identify the **faces** on some solid figures as polygons.

Some faces on solid figures are squares.

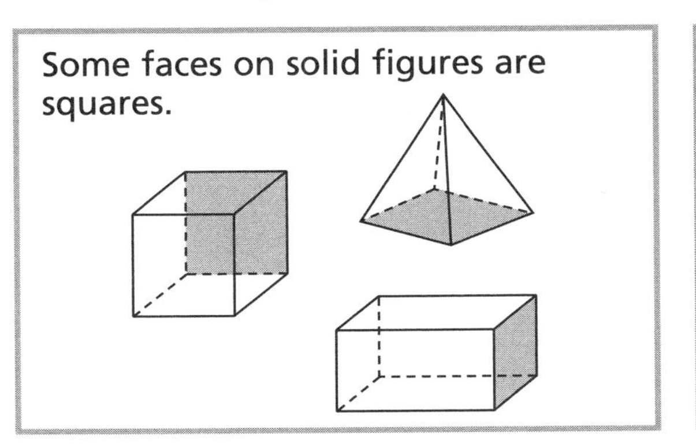

Some faces on solid figures are triangles.

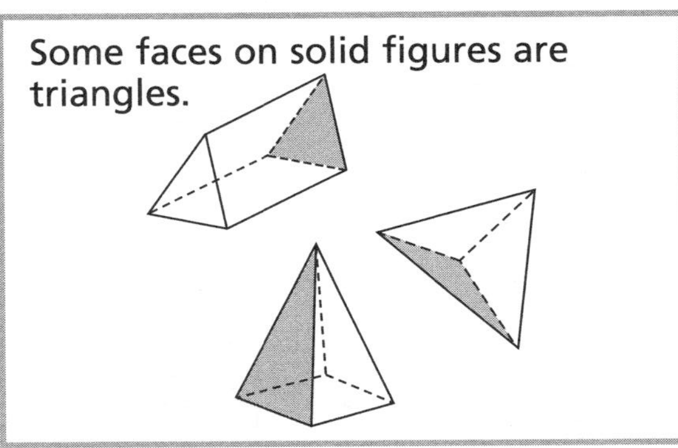

Some faces on solid figures are shaped like other polygons.

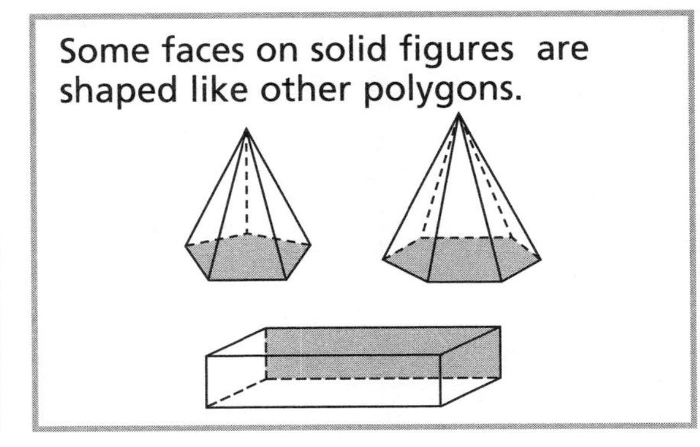

Try These

Write the number of faces. Then name the polygon that is the shaded face of each figure.

1.

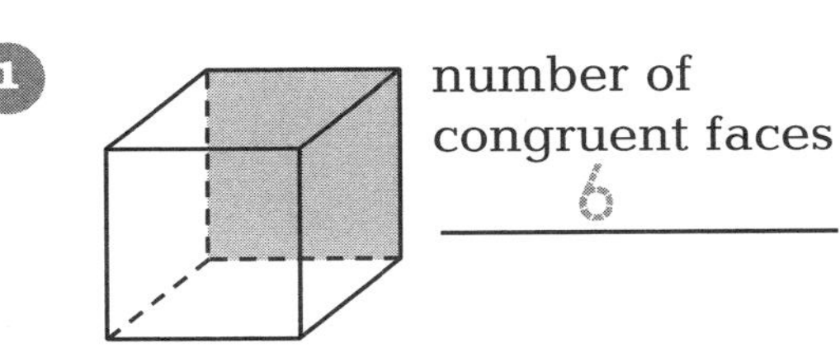

number of congruent faces 6

The shaded polygon is a ______________.

2.

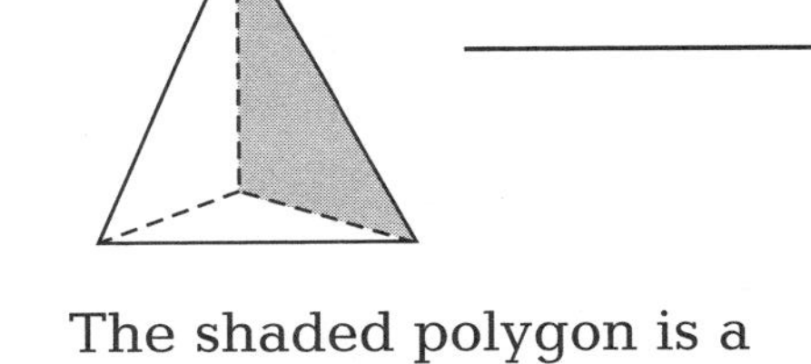

number of faces ______________

The shaded polygon is a ______________.

3.

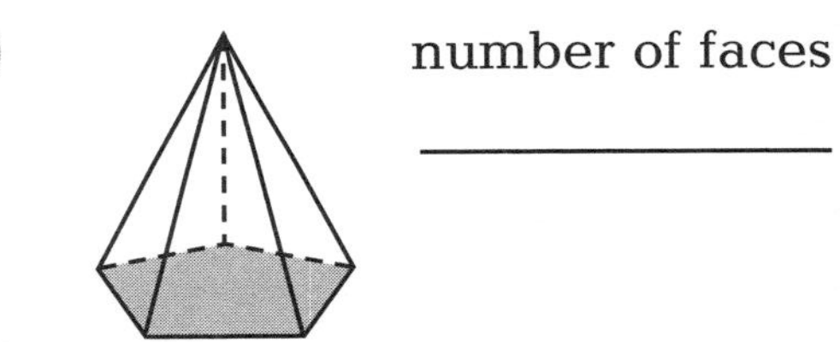

number of faces ______________

The shaded polygon is a ______________.

Go to the next side.

Name ______________________ Skill ______________________

Practice on Your Own

Skill 62

Name the polygon that is shaded.
Remember: A polygon is named by the number of its sides and angles.

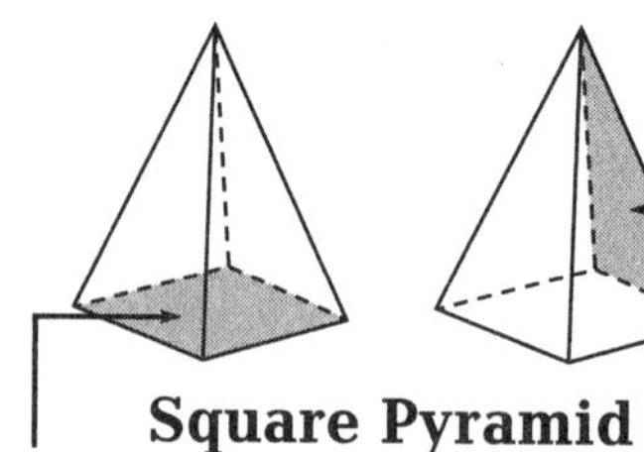

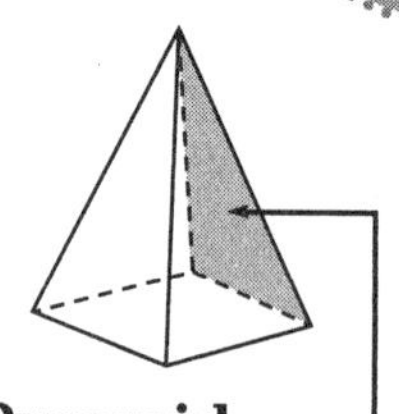

Square Pyramid

The shaded polygon is a square.

The shaded polygon is a triangle.

Write the number of faces. Name the polygon that is the shaded face of each figure.

1. number of faces ________

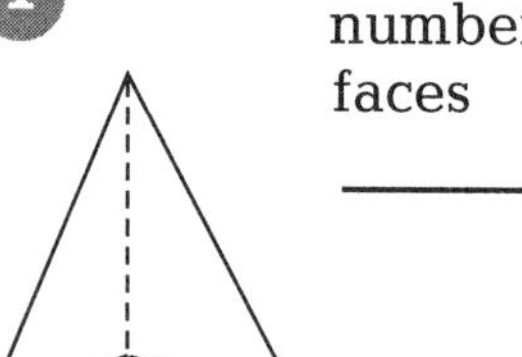

Shaded polygon: ______________________

2. number of faces ________

Shaded polygon: ______________________

3. number of faces ________

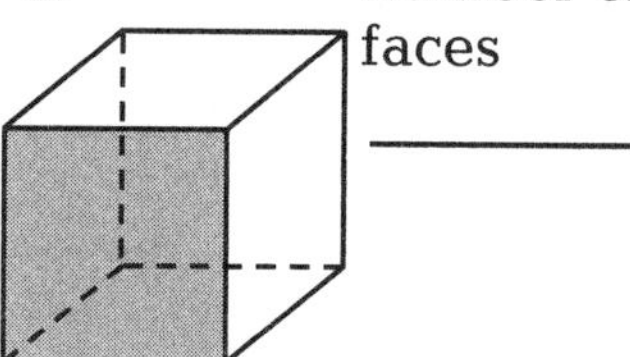

Shaded polygon: ______________________

4. faces ________

Shaded polygon: ______________________

5. faces ________

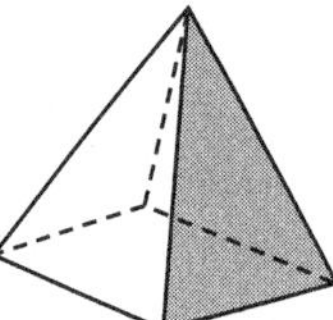

Shaded polygon: ______________________

6. faces ________

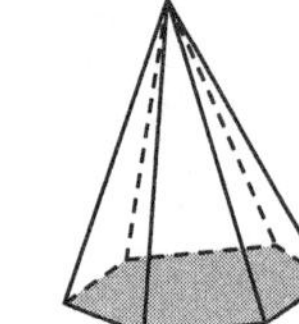

Shaded polygon: ______________________

Check

Name the polygon that is the shaded face of each figure.

7. 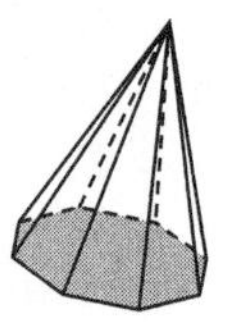

Polygon: ________

8. 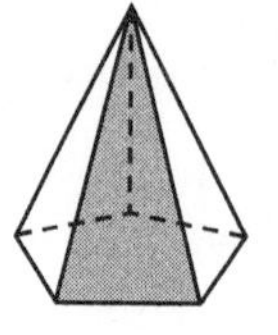

Polygon: ________

9. 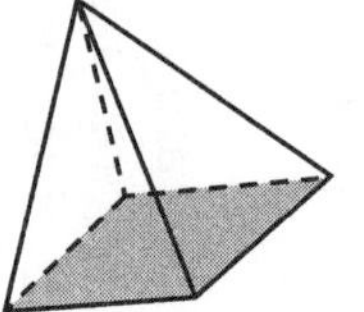

Polygon: ________

Using Skill 63

OBJECTIVE Find perimeter by counting units around a figure; find area by counting square units inside a figure

MATERIALS graph paper

Draw students' attention to the rectangular figures. Be sure students understand the difference between perimeter and area. Ask: **What is the measure around the edge of a figure called?** (perimeter) **What is the measure inside a figure called?** (area)

Draw attention to finding the perimeter of the rectangle. Ask: **How do you find the perimeter of a figure?** (Add the lengths of all the sides.) Point out to students that they can add all the sides of the rectangle to find the perimeter. Show them that they can write this another way, as the sum of two times the length and two times the width.

Next, draw attention to finding the area of the rectangle. Emphasize to students that area is the measure of the space inside the figure. On graph paper, draw a rectangle. Ask: **How can you find the area of the rectangle?** (Possible answers: count the squares; count the squares in one row, then multiply by the number of rows.)

Using the example shown, work through the steps to find the area of the rectangle. Emphasize to students that area is expressed in square units.

Help students recognize that perimeter, P, is the distance around a figure expressed in linear units and, that area, A, is the number of squares inside a figure, expressed in square units.

TRY THESE In Exercises 1–3 students find perimeter and area.

- **Exercise 1** Find the perimeter of a rectangle.
- **Exercise 2** Find the area of a rectangle.
- **Exercise 3** Find the perimeter and area of an irregular figure.

PRACTICE ON YOUR OWN Review the example at the top of the page. Ask a student to demonstrate the difference between perimeter and area.

CHECK Determine if students can find perimeter and area and label them with the correct units.

Success is indicated by 2 out of 4 correct responses.

Students who successfully complete the **Practice on Your Own** and **Check** are ready to move to the next skill.

COMMON ERRORS

- Students may use linear units instead of square units for area.
- Students may confuse the definitions of *area* and *perimeter.*
- Students may not know multiplication facts, or may make errors in addition.

Students who made more than 2 errors in the **Practice on Your Own,** or who were not successful in the **Check** section, may benefit from the **Alternative Teaching Strategy** on the next page.

Optional

Alternative Teaching Strategy

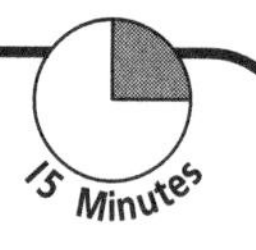

Model Perimeter and Area

OBJECTIVE Find the perimeter and area of figures

MATERIALS ruler or measuring tape, yarn, 36 square tiles, paper

The goal is for students to recognize that perimeter, P, is the distance around a figure expressed in linear units and area, A, is the number of units inside a figure expressed in square units.

Prior to the lesson, draw a diagram of rectangle with an area of 36 tiles. Put it aside for later use.

Provide students with enough yarn to outline a rectangular shape found in the classroom.

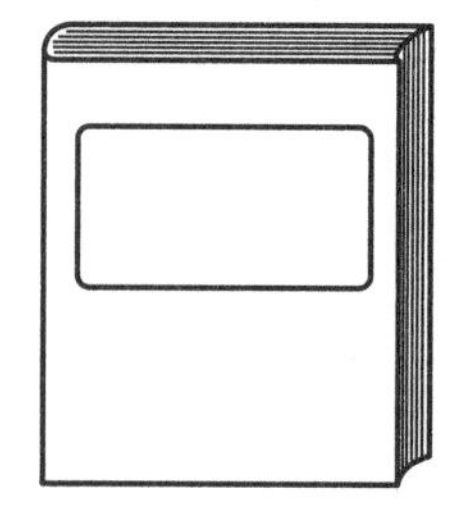

Ask students to use the yarn to outline the object. Direct them to tie a knot to indicate the end of the measure.

Then have students use the ruler or measuring tape to measure the length of the yarn. Record the length on paper. Make sure students label the length with the correct units.

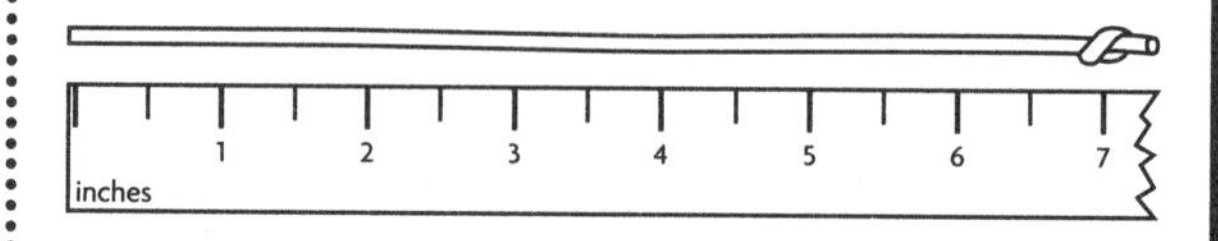

Next, direct students to use the ruler or measuring tape to find the length of each side of their object. Have them record each measure.

Now have students add the measures of the sides. Have them compare this sum with the measure they recorded for the length of yarn. Help students recognize that both measures are the same.

Use this opportunity to point out to students that the perimeter of a figure is the distance around the figure.

Next, show students the diagram of the rectangle. Direct them to find the number of square tiles that will fit inside.

Explain to students that the number of square tiles that covers the interior of the rectangle is called the area of the rectangle. Point out that this measurement is always expressed in square units.

Name ______________________ Skill ______________________

Grade 5 Skill 63

Perimeter and Area

Perimeter is the distance around a figure.
Area is the number of square units needed to cover a surface.

Find the perimeter of the rectangle.

Count the units *around* the figure.

Perimeter = side + side + side + side
or
= (2 × length) + (2 × width)

Perimeter = 4 + 5 + 4 + 5
or
= (2 × 4) + (2 × 5)

Perimeter = 18 units

Find the area of the rectangle.

Count the square units *inside* the figure.

Area = number of square units counted
or
= length × width

Area = 3 units × 7 units

Area = 21 square units

Find the perimeter and area of the rectangle.

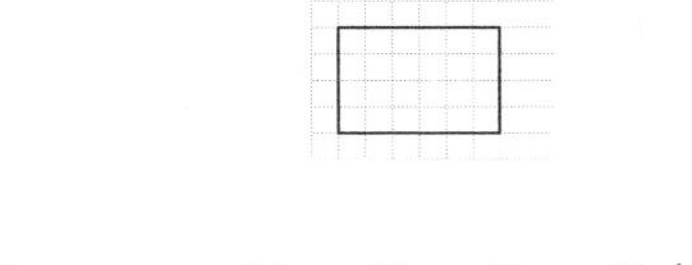

Perimeter = (2 × 6) + (2 × 4)
= 12 + 8
= 20 units

Area = 4 × 6
= 24 square units

Try These

1. Find the perimeter of the rectangle.

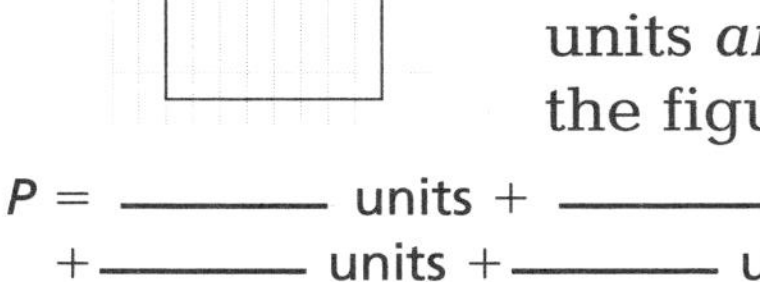

Count the units *around* the figure.

P = ______ units + ______ units
+ ______ units + ______ units

P = (2 × ______ units) +
(2 × ______ units)

The perimeter equals ______ units.

2. Find the area of the rectangle.

Count the square units *inside* the figure.

A = ______ units × ______ units

A = ______ × 3

The area equals ______ square units.

3. Find the perimeter and area of the figure.

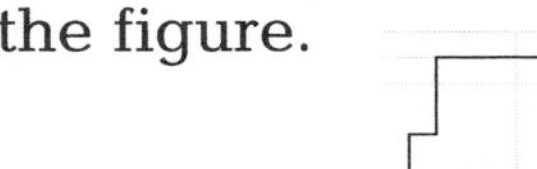

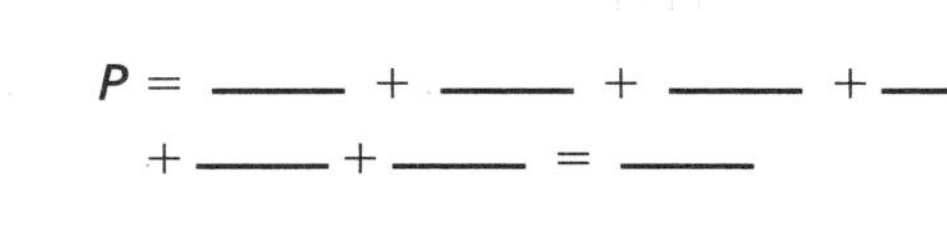

P = ____ + ____ + ____ + ____
+ ____ + ____ = ____

A = ______________

Go to the next side.

Name ______________________ Skill ______________________

Practice on Your Own

Skill 63

Think:
Find the **perimeter** by counting the units around a figure.

Find the **area** by counting the square units inside a figure or by multiplying length times width if the figure is a rectangle.

Find the perimeter and area of the rectangle.

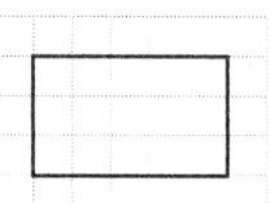

$P = 5 + 3 + 5 + 3$, or 16 units
The perimeter equals 16 units.
$A = 3 \times 5$, or 15 square units
The area equals 15 square units.

Find the perimeter and area of each figure.

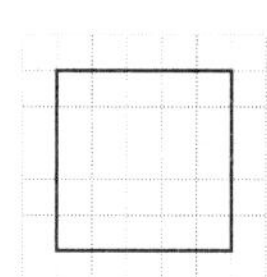

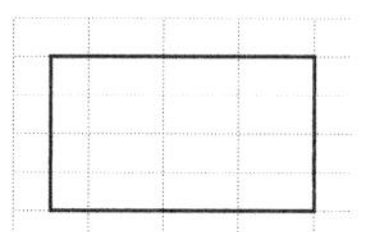

1. $P =$ ______ + ______ + ______ + ______
   $P =$ ______ units

2. $A =$ ______ $\times$ ______
   $A =$ ______ square units

3. $P = (2 \times$ ______ $) + (2 \times$ ______ $)$
   $P =$ ______ units

4. $A =$ ______ $\times$ ______
   $A =$ ______ square units

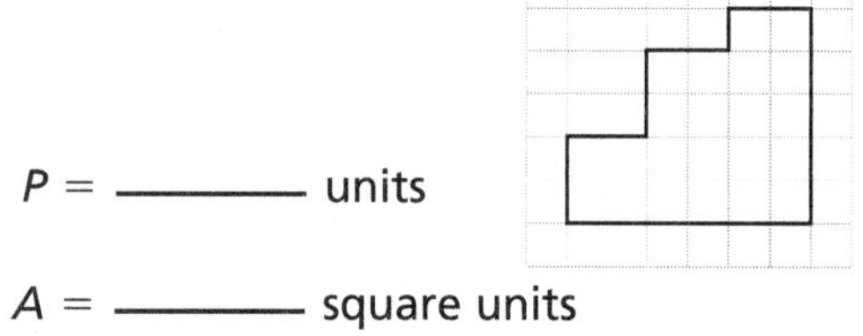

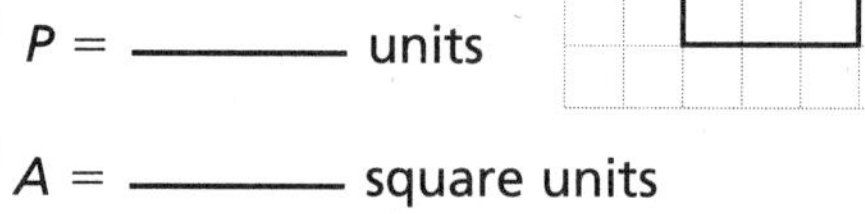

5. $P =$ ______ units

6. $A =$ ______ square units

7. $P =$ ______ units

8. $A =$ ______ square units

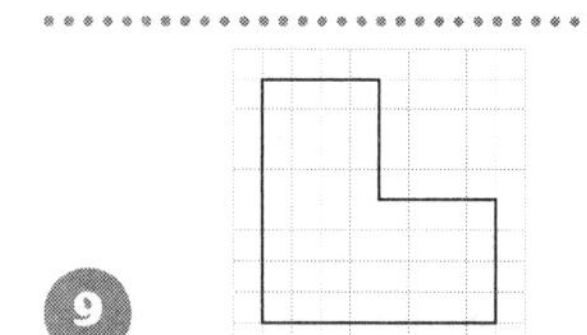

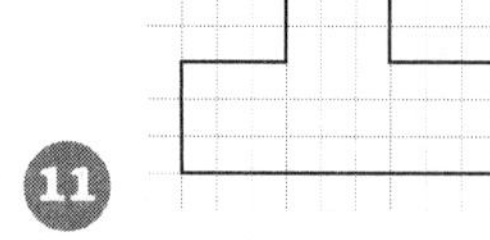

9. $P =$ ______

10. $A =$ ______

11. $P =$ ______

12. $A =$ ______

Check

Complete.

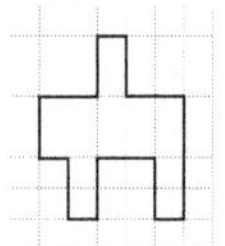

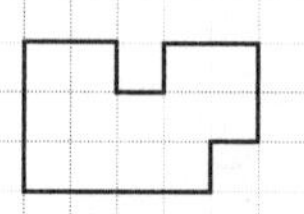

13. $P =$ ______

14. $A =$ ______

15. $P =$ ______

16. $A =$ ______

Skill 64 Grade 5 — Faces, Edges, Vertices

15 Minutes

Using Skill 64

OBJECTIVE Count the number of faces, edges, and vertices on solid figures

MATERIALS geometric solids (optional)

Begin the lesson by reviewing the difference between a plane figure and a solid figure. Ask: **How many dimensions does a sheet of paper have?** (2) **What are the dimensions?** (length, width) **How many dimensions does a box have?** (3) **What are the dimensions?** (length, width, and height)

Manipulating solid models will help students visualize three-dimensional figures. If geometric solids are available, show students a cube, a triangular prism, and a square pyramid. Encourage students to feel the faces, edges, and vertices of each solid figure.

Draw attention to the flat surfaces of the solid figures. Ask: **What do you call a flat surface on a solid figure?** (face)

Focus on the cube. Ask: **What is unique about the faces on this solid figure?** (All 6 faces are squares.) **How is the *square pyramid* different from the *cube?*** (There is only one square face on the square pyramid, and it has 4 triangular faces that meet at a point.)

Be sure students recognize the difference between a prism and a pyramid. Ask: **How is a *prism* different from a *pyramid?*** (A prism has 2 parallel bases; a pyramid has 1 base and all the other faces meet at a point.)

Draw attention to the examples of edges. If available, use a geometric figure to show students the line where two faces meet. Ask: **What do you call the line where 2 faces meet?** (an edge)

Draw attention to the examples of vertices. Ask: **What is the difference between an edge and a vertex?** (An edge is a line; a vertex is a point where three or more faces meet.)

TRY THESE In Exercises 1–3 students find the number of faces, edges, and vertices on solid figures.

- **Exercise 1** Faces on a rectangular prism.
- **Exercise 2** Edges on a triangular pyramid.
- **Exercise 3** Vertices of a cube.

PRACTICE ON YOUR OWN Review the example at the top of the page. Have students identify and count the faces, edges, and vertices of a rectangular pyramid. Suggest to students that they mark starting points on the figure to help them keep track while counting.

CHECK Determine if students can count the number of faces, edges, and vertices on various solid figures.

Success is indicated by 3 out of 3 correct responses.

Students who successfully complete the **Practice on Your Own** and **Check** are ready to move to the next skill.

COMMON ERRORS

- Students may miscount because they cannot visualize three-dimensional figures that are rendered in two dimensions.
- Students may lose count because they forget where they started counting.

Students who made more than 3 errors in the **Practice on Your Own,** or who were not successful in the **Check** section, may benefit from the **Alternative Teaching Strategy** on the next page.

Optional

Alternative Teaching Strategy

Making Marshmallow Solids

15 Minutes

OBJECTIVE Construct and use models of solid figures to identify and count faces, vertices, and edges

MATERIALS miniature marshmallows (or balls of clay), toothpicks, plastic wrap, masking tape, paper, pencils

Have students work with partners. Provide each pair with at least 20 toothpicks, 20 miniature marshmallows, masking tape, and a large piece of plastic wrap.

Guide students through the construction of a model for a cube. Use toothpicks for edges and marshmallows for corners (vertices).

Each partner can create a base using 4 toothpicks and 4 marshmallows; then they can work together using 4 more toothpicks to connect the two faces.

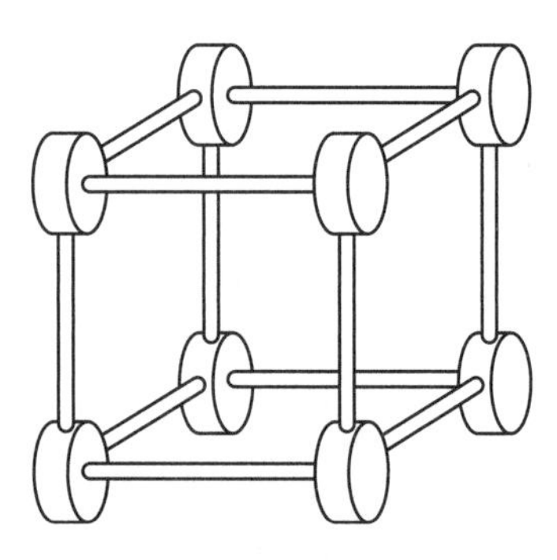

To simulate the faces, have students wrap the cube in plastic. Using plastic will allow them to see through the cube.

Have one student mark 1 on a piece of masking tape; the other student places the piece of tape on one of the faces. Have students work together labeling the faces as they count all the faces on the cube.

Direct students to create the recording sheet shown below.

Faces	Edges	Vertices

First have students record the number of faces on the cube under the column *Faces.*

Next, direct students to remove the toothpicks from the marshmallows. Have students count the number of toothpicks. Explain that the toothpicks model the sides of the cube. Ask: **What do you call the sides of a solid figure?** (edges) Have students record the number of toothpicks under the column *Edges.*

Finally, explain that the marshmallows model the corners of the cube. Ask: **What do you call the point on a solid figure where three or more faces meet?** (vertex)

Have them count the marshmallows and record the number under the column *Vertices.* If necessary, explain that *vertices* is the plural of *vertex.*

Encourage students to model and list the number of faces, vertices, and edges of other prisms. Expand their experience to include pyramids.

Name ____________________ Skill ____________________

Grade 5 **Skill 64**

Faces, Edges, Vertices

Count the number of faces, edges, and vertices on solid figures.

A **face** is a flat surface on a solid figure.

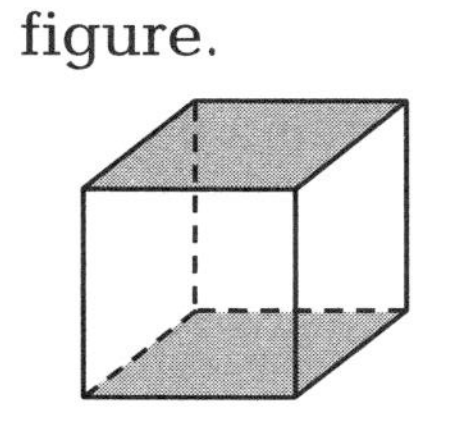

1 top square face
4 side square faces
1 bottom square face
total: 6 faces
Cube

2 triangular faces
3 rectangular faces
total: 5 faces
Triangular Prism

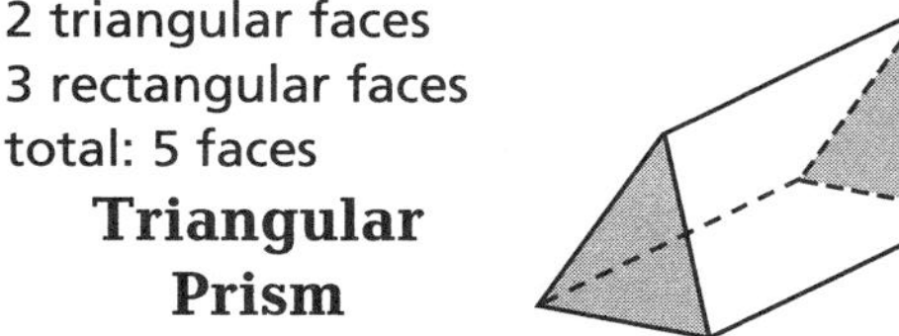

1 square face
4 triangular faces
total: 5 faces
Square Pyramid

An **edge** on a solid figure is where 2 faces meet.

4 top edges
4 side edges
4 bottom edges
total: 12 edges
Cube

3 back edges
3 side edges
3 front edges
total: 9 edges
Triangular Prism

4 side edges
4 bottom edges
total: 8 edges
Square Pyramid

A **vertex** is a corner where 3 or more faces meet.

4 top vertices
4 bottom vertices
total: 8 vertices
Cube

3 front vertices
3 back vertices
total: 6 vertices
Triangular Prism

1 top vertex
4 bottom vertices
total: 5 vertices
Square Pyramid

Try These

Write the number of faces, edges, or vertices.

1. How many faces?
   top rectangular face _____
   side rectangular faces _____
   bottom rectangular face _____
   Rectangular Prism **Total faces:** _____

2. How many edges?
   bottom edges _____
   side edges _____
   Total edges: _____
   Triangular Pyramid

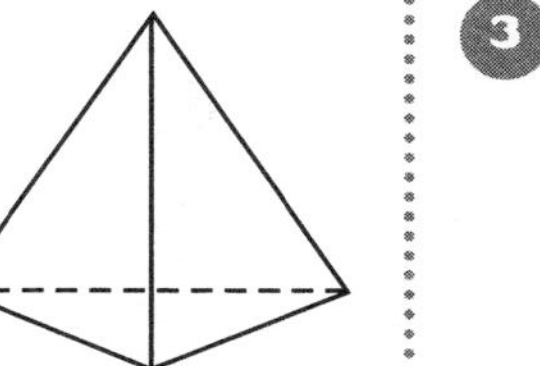
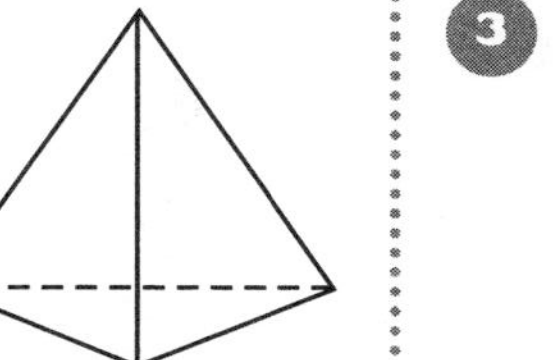
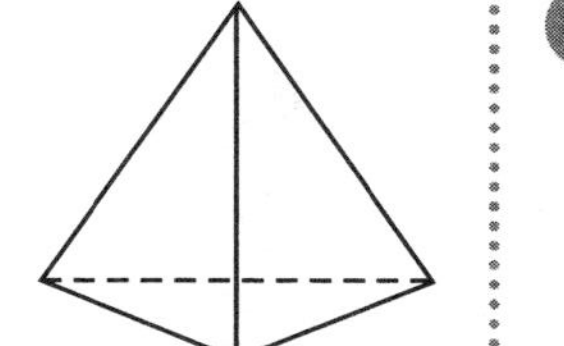
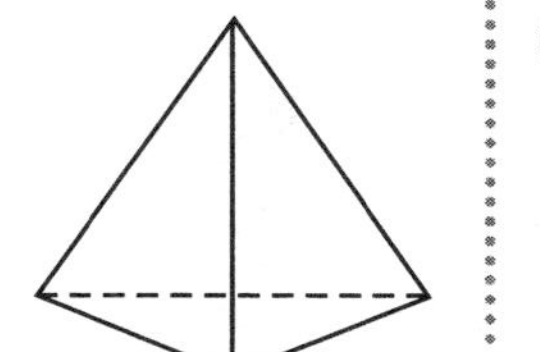
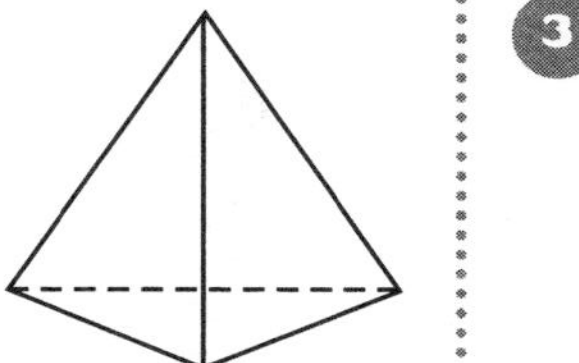

3. How many vertices?
   top vertices _____
   bottom vertices _____
   Total vertices: _____
   Cube

Go to the next side.

Name ______________________ Skill ______________

Practice on Your Own

Skill 64

Think:
Count flat surfaces to find **faces**.
Count corners to find **vertices**.
2 faces meet to form an **edge**.

Find the total number of faces, edges and vertices.

5 faces

8 edges

5 vertices

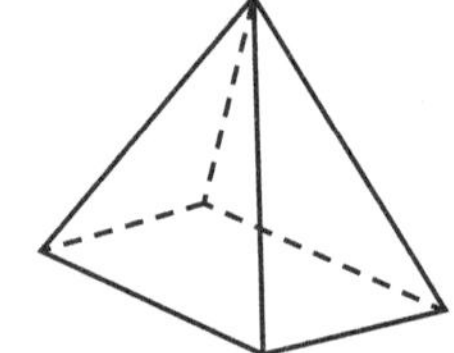

Rectangular Pyramid

Write the number of faces, edges, or vertices.

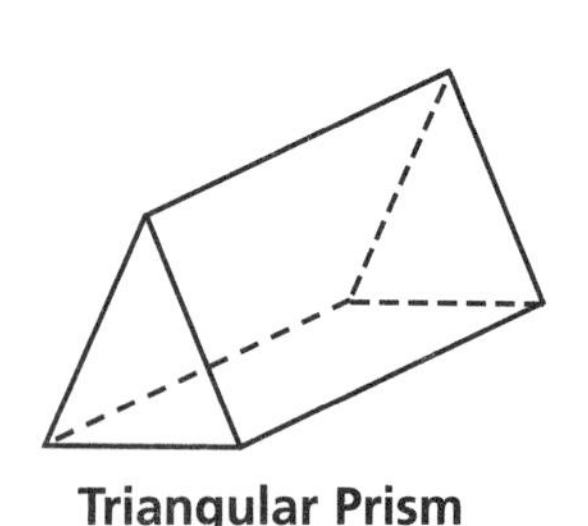

Triangular Prism

1 Faces
Back face ___
Side faces ___
Front face ___
Total faces: ___

2 Edges
Back edges ___
Side edges ___
Front edges ___
Total edges: ___

3 Vertices
Back vertices ___
Front vertices ___
Total vertices: ___

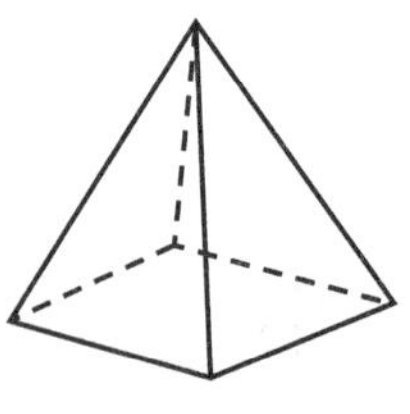

Rectangular Pyramid

4 Faces
Side ___
Bottom ___
Total faces: ___

5 Edges
Side ___
Bottom ___
Total edges: ___

6 Vertices
Top ___
Bottom ___
Total vertices: ___

Write the number of faces, edges, or vertices.

7

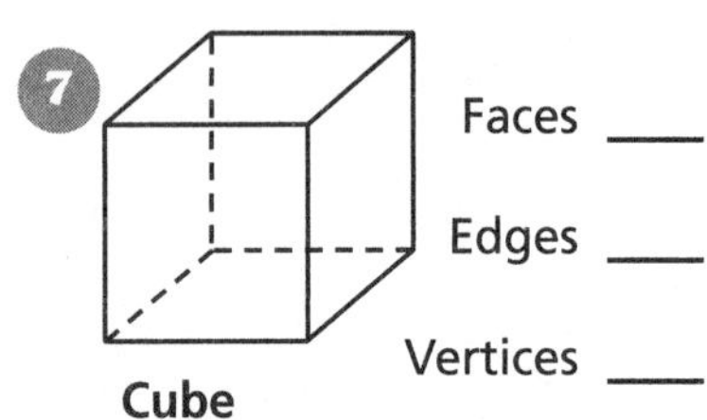

Cube

Faces ___
Edges ___
Vertices ___

8 **Rectangular Prism**

Faces ___
Edges ___
Vertices ___

9 **Square Pyramid**

Faces ___
Edges ___
Vertices ___

Check

Write the number of faces, edges, or vertices.

10

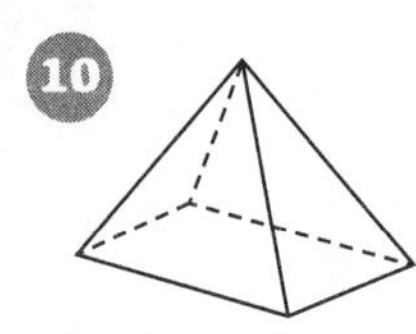

Rectangular Pyramid

Faces ___
Edges ___
Vertices ___

11

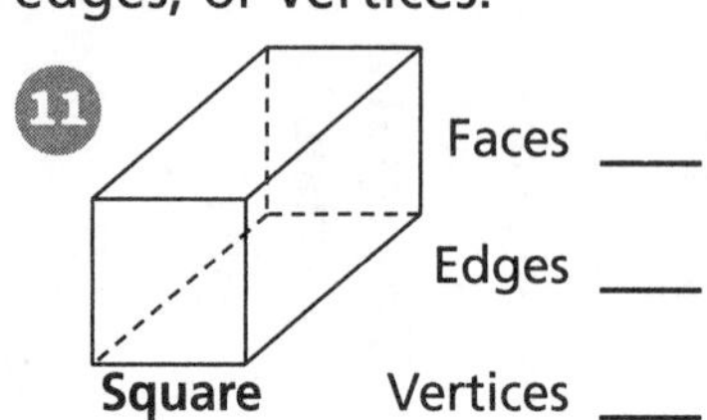

Square Prism

Faces ___
Edges ___
Vertices ___

12 **Triangular Pyramid**

Faces ___
Edges ___
Vertices ___

Name ______ Skill ______

Answer Card

Geometry

Grade 5

SKILL 59

TRY THESE

1. yes, closed
2. no, open

PRACTICE

1. yes, closed
2. no, open
3. closed
4. open
5. open
6. open
7. closed
8. closed

CHECK

9. Figures will vary; check that students have drawn an open figure.
10. Figures will vary; check that students have drawn a closed figure.

SKILL 60

TRY THESE

1. 4, 4, quadrilateral
2. 6, 6, hexagon
3. 5, 5, pentagon

PRACTICE

1. 4, 4, quadrilateral
2. 8, 8, octagon
3. 5, 5, pentagon
4. 3, 3, triangle
5. 5, 5, pentagon
6. 4, 4, quadrilateral
7. quadrilateral
8. hexagon
9. octagon

CHECK

10. hexagon
11. triangle
12. pentagon

SKILL 61

TRY THESE

1. less; acute
2. square; 90°; right
3. obtuse

PRACTICE

1. obtuse
2. right
3. acute
4. right
5. acute
6. obtuse
7. acute
8. obtuse
9. right

CHECK

10. obtuse
11. right
12. acute

SKILL 62

TRY THESE

1. 6; square
2. 4; triangle
3. 6; pentagon

PRACTICE

1. 4; triangle
2. 6; rectangle
3. 6; square
4. 6; pentagon
5. 5; triangle
6. 7; hexagon

CHECK

7. octagon
8. triangle
9. Possible answers: square, quadrilateral

Name ______________ Skill ______________

Answer Card

Geometry

Grade 5

SKILL 63

TRY THESE

1. 5, 8, 5, 8; 5, 8; 26
2. 9, 3; 9; 27
3. 5, 6, 6, 3, 1, 3; 24 units; 33 square units

PRACTICE

1. 5, 5, 5, 5; 20
2. 5, 5; 25
3. 7, 4; 22
4. 7, 4; 28
5. 22
6. 22
7. 16
8. 10
9. 32 units
10. 48 square units
11. 30 units
12. 36 square units

CHECK

13. 26 units
14. 16 square units
15. 18 units
16. 13 square units

SKILL 64

TRY THESE

1. 1, 4; 1; 6
2. 3; 3; 6
3. 4; 4; 8

PRACTICE

1. 1, 3, 1, 5
2. 3, 3, 3, 9
3. 3, 3, 6
4. 4; 1; 5
5. 4; 4; 8
6. 1; 4; 5
7. 6; 12; 8
8. 6; 12; 8
9. 5; 8; 5

CHECK

10. 5; 8; 5
11. 6; 12; 8
12. 4; 6; 4

Statistics, Data Analysis, and Probability

Using Skill 65

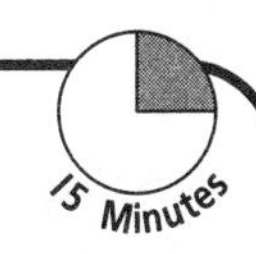

OBJECTIVE Read and answer questions from a frequency table

Draw attention to the tables. Be sure students recognize that the Tally Table and the Frequency Table display the same set of data.

Ask: **How is the frequency table different from the tally table?** (The frequency table uses numbers instead of tally marks.) **How are these two tables similar?** (They both record the same data.) **What is the purpose of using a tally or frequency table?** (to see how often a particular activity occurs)

Guide students to understand that many questions can be answered by examining the data from a tally or frequency table. Often, that information can be used to create graphs or to make predictions.

In business, frequency tables play an important role in inventory control and the pricing of consumer products. For example, in supermarkets, frequency tables might help determine the quantity of produce to be re-ordered. In department stores, data from frequency tables might be used to price items for special discounts during sales.

TRY THESE In Exercise 1 students use data in a tally table to complete a frequency table.

- **Exercise 1** Complete frequency table; answer questions about data.

PRACTICE ON YOUR OWN Review the example at the top of the page. Have students explain how to use the data in the tally table to complete the frequency table. Check to be sure that students understand *less than* and *more than* as they apply to the numbers in the table.

CHECK Determine if students can read the questions carefully and get the correct data from the frequency table.

Success is indicated by 3 out of 3 correct responses.

Students who successfully complete the **Practice on Your Own** and **Check** are ready to move to the next skill.

COMMON ERRORS

- Students might be unfamiliar with counting tally marks in groups of 5.
- Students might fail to choose the proper operation because they do not understand certain words or phrases such as *more, less, sum, greatest, least, more than, less than, fewest number.*
- Students might need to review basic computation facts.

Students who made more than 3 errors in the **Practice on Your Own,** or who were not successful in the **Check** section, may benefit from the **Alternative Teaching Strategy** on the next page.

Optional

Alternative Teaching Strategy

15 Minutes

Frequency Tables and Related Questions

OBJECTIVE Read data from a frequency table

MATERIALS flip chart, index cards

On a flip chart prepare the frequency table below.

Students with at least 8 hours sleep	Students with more than 8 hours sleep	Total number of students surveyed
75	45	100

On index cards prepare the following related questions.

1. According to the frequency table, how many students had at least 8 hours of sleep?
2. How many students had more than 8 hours of sleep?
3. What is the total number of students surveyed?

Explain to students that the frequency table represents data regarding the number of hours of sleep students had on Saturday night.

Ask a volunteer to read one of the questions. Allow time for group discussion.

For question 1, guide students to understand that the phrase *at least* means 8 hours of sleep *or more.*

For question 2, point out that the phrase *more than* does not include students having exactly 8 hours of sleep.

Remind students of the importance of reading questions carefully and understanding key words or phrases.

Continue with similar examples that require students to think carefully about what is being asked. Also, provide opportunities for students to interpret the data.

Answers to index card questions:

1. 75 students had at least 8 hours of sleep
2. 45 students had more than 8 hours of sleep
3. 100 students

Name ____________________ Skill ____________________

Grade 5 Skill 65

Frequency Tables

Use a **frequency table** to organize data from a tally table.

Both tables show the frequency of phone calls received during five days. The first column tells the day of the week. The second column shows how many phone calls were received each day. The tally table shows one tally for each call received.

Tally Table
Phone Calls Received
5:00–7:00 p.m.

Day	Phone Calls
Monday	卌 |
Tuesday	||
Wednesday	||||
Thursday	|||
Friday	卌 ||||

| one phone call

卌 five phone calls

On which days were 4 or more phone calls received?

Think: Find the days that have 4 or more tally marks.
Monday 6, Wednesday 4, Friday 9

Answer: Monday, Wednesday, Friday

How often something happens, or how often a number occurs, is called the **frequency**. The **frequency table** lists the **total number** of phone calls.

Frequency Table
Phone Calls Received
5:00–7:00 p.m

Day	Phone Calls
Monday	6
Tuesday	2
Wednesday	4
Thursday	3
Friday	9

A frequency table shows the total number instead of tally marks.

How many phone calls were received before Thursday?

Think: Find the sum for Monday, Tuesday, and Wednesday.

$6 + 2 + 4 = 12$

Answer: 12 phone calls

Try These

Complete the frequency table and answer the question.

1. How many students have either a spring or summer birthday?
   Find the sum for Spring and ____________. 9 + ____ = ____
   Answer: ______________________________

Season	Number of Birthdays
Fall	卌 ||||
Winter	卌
Spring	卌 ||||
Summer	卌 卌

Season	Number of Birthdays
Fall	
Winter	
Spring	
Summer	

Go to the next side.

Name ______________________ Skill ______________________

Practice on Your Own — Skill 65

Use the frequency table to answer the questions.

How many students spend more than 2 hours studying?

Think: 3 hours: 7 students 7 + 5 = 12
4 hours: 5 students

Answer: 12 students

Tally Table

Hours Daily	Number of Students
1	|||
2	~~||||~~ |
3	~~||||~~ ||
4	~~||||~~

Frequency Table

Hours Daily	Number of Students
1	3
2	6
3	7
4	5

Use the frequency table to answer the questions.

Quarter	Points Scored
1	22
2	14
3	18
4	21

1. How many points were scored in the last 2 quarters of the game?
   Think: Find the sum of 18 and ____. Answer: ____ points
2. How many points were scored before the fourth quarter?
   Think: Find the sum of ____, ____, and ____. Answer: ____ points
3. How many points were scored in the game?
   Think: Find the total: ______________ Answer: ____ points

Type	Frequency
plain	10
corn	24
apple	15

4. How many of the muffins sold were not plain? ____________
5. For which type of muffin were the least sold? ____________
6. How many muffins were sold altogether? ____________

Score	Frequency
75	22
85	17
95	15
100	8

7. How many students scored either 95 or 100 points? ____________
8. How many students scored less than 95 points? ____________
9. How many students took the test? ____________

Check

Use the frequency table to answer the questions.

Model	Frequency
10-Speed	110
Mountain	125
3-Speed	98

10. How many 10-Speed and Mountain bikes were sold? ________
11. For which model were the greatest number sold? ________
12. How many bikes were sold altogether? ____________

Skill 66 Grade 5

Read Pictographs

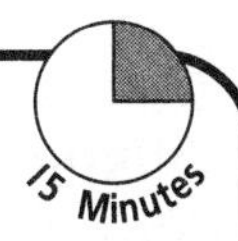

Using Skill 66

OBJECTIVE Read and compare data using pictographs

Direct students' attention to the key in Step 2.

Ask: **What does each symbol represent?** (10 votes)

Have volunteers multiply by 10 to find the number of votes for sledding (30) and ice fishing (30).

Point out the half symbol.

Ask: **What does each half symbol represent?** (5 votes)

Point out that finding the number of votes for skiing involves two steps and two operations. First, multiply the number of whole symbols by 10; then add 5 to find the total number of votes.

Have students multiply by 10 and add 5 to find the number of votes for skiing (45).

Ask: **Why is it easy to compare votes at a glance when reading a pictograph?** (You can easily find which group has more symbols.)

TRY THESE Exercises 1–5 require students to read a pictograph and answer questions about it. Lead students to realize that, when finding how many more peppers than cheese in Exercise 4, they only need to find the value of the one extra symbol next to peppers.

- **Exercises 1 and 2** Read and count data.
- **Exercises 3 and 4** Compare data.
- **Exercise 5** Add data.

PRACTICE ON YOUR OWN Focus on the examples at the top of the page. Be sure students understand half symbols.

Ask: **What is the value of half of a container?** (2 votes)

CHECK Make sure that students can calculate the correct value of symbols. Success is indicated by 3 out of 4 correct responses.

Students who successfully complete **Practice On Your Own** and **Check** are ready to move to the next skill.

COMMON ERRORS

- Students may have trouble finding the value of the half symbols and then adding that value to the total value of the whole symbols.

Students who made more than 3 errors in the **Practice On Your Own,** or who were not successful in the **Check** section, may benefit from the **Alternative Teaching Strategy** on the next page.

Optional

Alternative Teaching Strategy
Read Pictographs

20 Minutes

OBJECTIVE Read pictographs

MATERIALS a simple pictograph of favorite foods, in which each symbol represents 1 vote, as shown below.

FAVORITE FOODS	
Spaghetti	🍴🍴🍴🍴🍴🍴
Chicken Fajitas	🍴🍴🍴🍴🍴🍴🍴
Pancakes	🍴🍴
Veggie Pizza	🍴🍴🍴🍴

Have students examine the graph.

Tell them that each symbol represents 1 vote. Add a key under the graph.

🍴 = 1 vote

Ask: **What is the title of the graph?** (Favorite Foods) **How many students voted for spaghetti?** (6) **How do you find the number of votes?** (Count the symbols.) **Which food was the most popular?** (chicken fajitas) **How many more students voted for veggie pizza than for pancakes?** (2) **How many students voted in all?** (19)

Call on volunteers to ask a few more similar questions about the graph.

Now tell students to look at the graph again. Tell them that this time each symbol represents 4 votes. Change the key to reflect this change.

Ask: **How do you find the number of votes?** (Count the symbols and then multiply the number of symbols by 4.)

Ask questions about the information in the graph; have students say the multiplication aloud.

Finally, add some half symbols to the graph. Tell students that each symbol still represents 4 votes.

Ask: **How do you find the number of votes for a half symbol?**

(Think: If 🍴 = 4 votes,

then (half symbol) = 2 votes.)

Have partners repeat the activity, assigning other values for the whole symbol.

Name ______________________ Skill ______________________

Grade 5 Skill 66

Read Pictographs

A **pictograph** uses pictures to show and compare information.

Which sport has the most votes?

Step 1
Find the title.

This is the title.

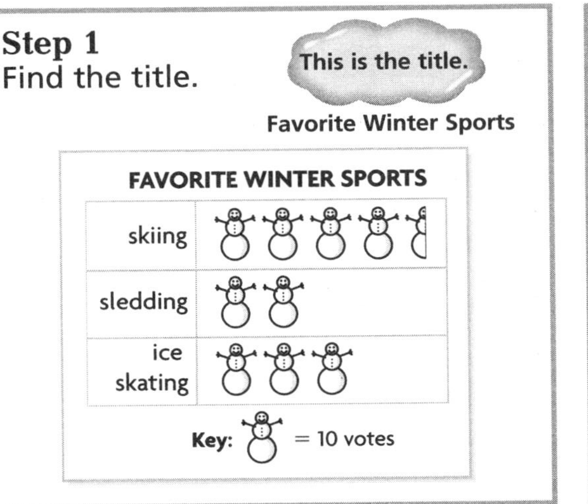

Step 2
Read the key.
The key shows what each picture in the graph represents.

Each picture represents 10 votes.

Key: = 10 votes

So, multiply by 10 to find the number of votes for each sport.

Step 3
Find the sport with the most votes.
Skiing has the most pictures.
Skiing:

Multiply by 10 to find the number of votes.

Multiply by 10:
$4.5 \times 10 = 45$

So, skiing has the most votes. It has 45.

Think: If = 10 votes, then is half as many votes.

Try These

Read the pictograph. Answer the questions.

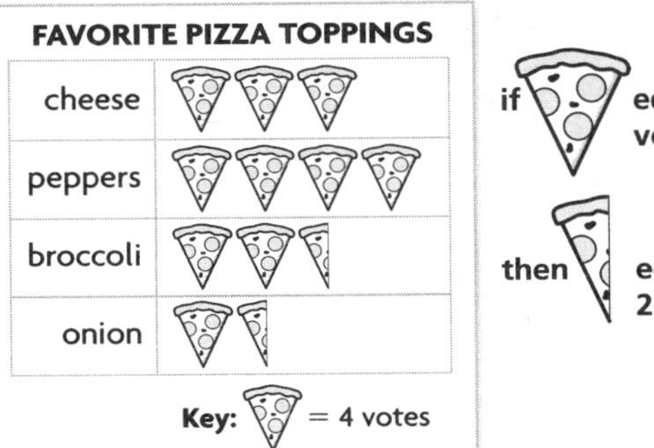

1. What is the title of the pictograph? ______________________
2. How many students voted for the broccoli topping? __________
3. Which topping was the most popular? ______________________
4. How many more students voted for peppers than cheese? __________
5. How many students voted in all? __________

Go to the next side.

Name ____________________ Skill ____________________

Practice on Your Own

The graph is about **favorite juice drinks.**

Each picture represents **4 votes.**
A half picture represents **2 votes.**
The flavor with the fewest votes is **grape.**

FAVORITE JUICE DRINK	
apple	
grape	
cranberry	

Key: = 4 votes

Read the pictograph. Answer the questions.

1. What is the title of the pictograph?
   ____

2. How many books does each picture represent ? __________

3. How many books does a half picture represent? __________

4. Who read the most books? __________
   How many? __________

5. How many students voted for each type of game?
   board games __________ video games __________
   puzzles __________ tag games __________

6. If 35 students liked the board games best, how many pictures would you need? __________

Read the pictograph. Answer the questions.

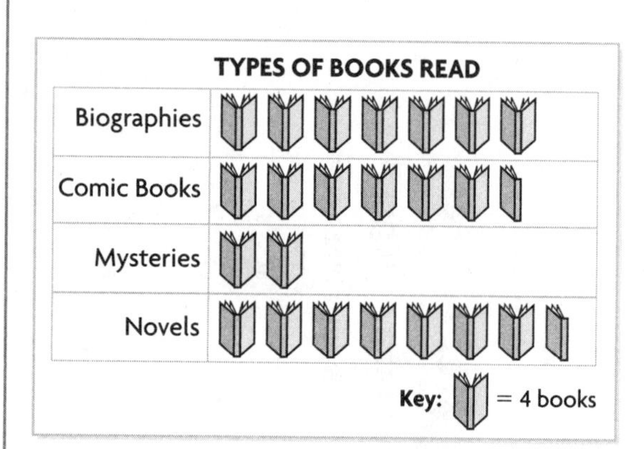

7. How many biographies were read? __________

8. How many more novels than mysteries were read? __________

9. How many books were read in all?
   ____

10. If 4 more mysteries were read, how many pictures would there be for mysteries in all? __________

Skill 67 Grade 5 — Read Bar Graphs

Using Skill 67

OBJECTIVE Read horizontal and vertical bar graphs

Direct students' attention to the horizontal bar graph. As you read through the skill, have students point to each part of the bar graph: the title, the two labels, the scale. Say: **The *scale* shows you the number of students. On this bar graph, the scale has an *interval* of one.**

Make sure students understand that an *interval* is the space *between*. Explain why each space represents 1 student: **The *interval* is the space between the numbers labeled on the scale. What is the pattern of numbers on the scale?** (Possible response: counting numbers: 0, 1, 2, 3, 4, …)

Have a student read the title of the graph. *(Books Students Read)* Then ask students to use the graph to answer questions such as: **How many students read Science Fiction?** (7) **How did you find the number of students who read Science Fiction?** (Possible response: I looked at the bar for Science Fiction, saw where it ended on the graph, and traced down that line see where it was on the scale.) **Which type of book is read the most? How do you know?** (Novels; it's the longest bar on the graph.) **How many students are counted in all? Explain.** (33; I added together the number of students who read each type of book.)

TRY THESE Students use the data in the bar graph to answer questions.

- **Exercises 1–3** Analyze and interpret data in a bar graph.

PRACTICE ON YOUR OWN Review the example at the top of the page. Make sure students understand the meaning of *horizontal* and *vertical.* Ask students to explain how the horizontal and vertical bar graphs are different and how they are alike.

CHECK Determine if students can read a graph and compare its data. Success is indicated by 3 out of 3 correct responses.

Students who successfully complete the **Practice On Your Own** and **Check** are ready to move to the next skill.

COMMON ERRORS

- Some students may read the wrong number on the scale, or may not know how to interpret the scale when a bar falls between two numbers.
- Some students may find it difficult to read a graph when the orientation is vertical; others may find it difficult when it is horizontal.

Students who made more than 2 errors in the **Practice On Your Own,** or who were not successful in the **Check** section, may benefit from the **Alternative Teaching Strategy** on the next page.

Optional

Alternative Teaching Strategy
Read Bar Graphs

OBJECTIVE Practice reading a vertical bar graph

MATERIALS copies of graph (shown below) prepared on grid paper

Distribute the grid paper. First make sure students can read the bar graph correctly. Point out the parts of the graph: *title, side label, bottom label, scale.*

Say: **The length of each bar shows you how many of each pet were sold.**

Demonstrate how the length of each bar is related to a number on the scale.

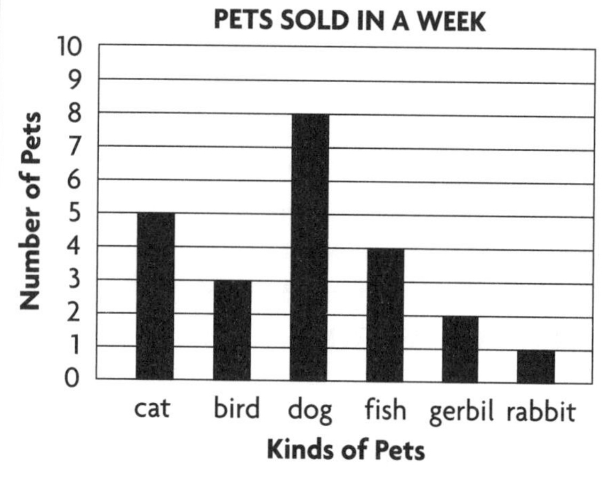

Then ask: **How many different kinds of pets are shown on the graph?** (6) **How many dogs were sold?** (8) **How many fish were sold?** (4)

If necessary model for students how to trace along the line at each bar's end to find the number on the scale.

Then help students use the graph to compare data. Ask: **How many more birds were sold than rabbits?** (2) Guide students to read the bar graph for *bird* first, to find out how many were sold, and then to read the bar for *rabbit* to find the difference.

Continue: **What does a short bar on this graph tell you?** (Few of that type of pet were sold.) **What does a tall bar tell you?** (Many of that type of pet were sold.) **Which pet did the store sell the least number of?** (rabbit) **How do you know?** (The bar labeled rabbit is the shortest bar.) **How many pets in all were sold during one week?** (23) **How do you know?** (Possible response: I added the number of pets shown for each bar.)

When students show an understanding of reading the vertical bar graph, have them write a question about the data. Let students trade their questions with each other and discuss answers.

Name ____________ Skill ____________

Grade 5 Skill 67

Read Bar Graphs

Bar graphs use bars to show data.
Horizontal bar graphs have bars that go across.

The title tells what the graph is about.
There are two labels: Types of Books and Number of Students.
The scale is labeled from 0 to 10 in intervals of 1. Each interval represents 1 student.
Read the graph.

- How many students read poetry?
  The bar for poetry stops at 4. So, 4 students read poetry.
- How many students were counted altogether? Add to find the number of students.

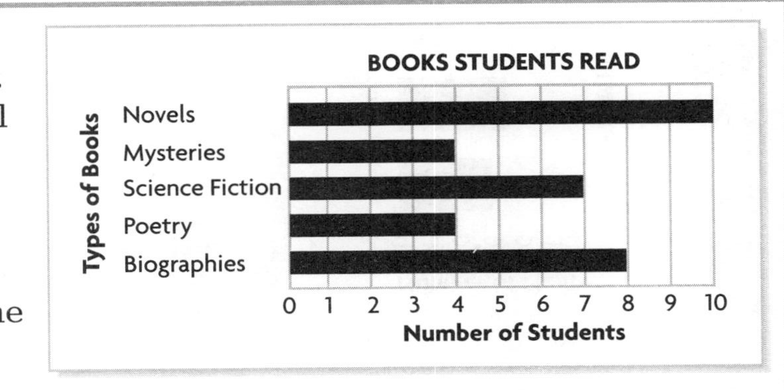

10 (Novels) + 4 (Mysteries) + 7 (Science Fiction) + 4 (Poetry) + 8 (Biographies) = 33 (students counted)

So, 33 students were counted.

Try These

Use the graph above.

1. How many students read mysteries? ________

2. Which type of book did the greatest number of students read? ________

3. How many more students read biographies than science fiction? ________

Go to the next side.

Name ______________________ Skill ______________________

Practice on Your Own

Skill 67

This is a **vertical** bar graph.
It has bars that go up.

- The graph shows vegetable plants in a garden.
- The scale shows that each interval stands for 10 vegetables.

The bar for pumpkins stops halfway between 10 and 20.
The number halfway between 10 and 20 is 15.
So, 15 pumpkin plants are in the garden.
If each pumpkin plant occupies 3 square feet, the pumpkins in all occupy 45 square feet.

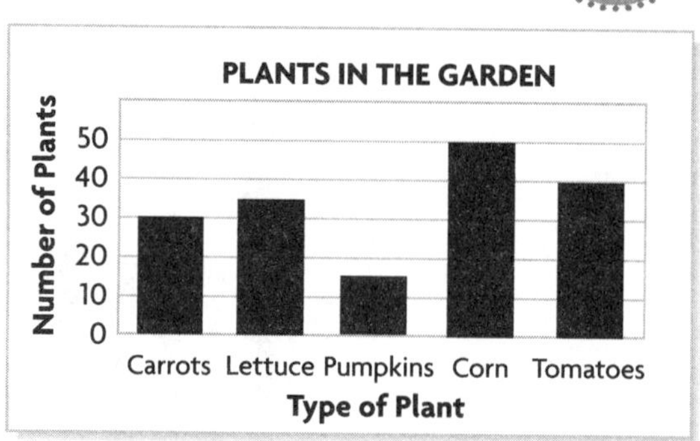

Use the graph above to answer the questions.

1. How many types of plants are there? ________

2. How many carrot plants are in the garden? ________

3. Which vegetable has the most plants? ________

4. How many tomato and lettuce plants are there? ________

5. Which vegetable has the fewest plants? ________

6. How many more tomato plants are there than pumpkin? ________

7. How many more corn plants are there than lettuce? ________

8. If a corn plant occupies 0.5 square feet of soil, how many square feet do the corn plants occupy in all? ________

9. How many more carrot plants are there than pumpkin? ________

Check

Use the graph above to answer the questions.

10. How many more lettuce plants are there than carrot plants? ________

11. How many pumpkin and carrot plants are there? ________

12. How many plants are there in the garden? ________

Skill 68 Grade 5

Read Line Graphs

Using Skill 68

OBJECTIVE Read line graphs

Read the information at the top of the page about line graphs. Discuss the fact that graphs show data in a way that helps you see relationships. Line graphs show data that changes over time.

Have students examine the graph. Focus on the title and the labels on the horizontal and vertical axes.

Ask: **What does this graph show?** (changes in ticket sales during one week) **By just looking at the graph, how can you tell when ticket sales increase?** (line goes up) **How can you tell when sales decrease?** (line goes down) **What does the scale 0–600 represent?** (number of tickets) **What is the interval?** (50 tickets)

Follow the steps for gathering data from the graph. Help students interpolate the number of tickets for Wednesday as shown in Step 2. Then practice locating a few more points.

TRY THESE Exercises 1–4 provide practice in reading line graphs.

- **Exercise 1** Find the number of tickets sold on a particular day.
- **Exercise 2** Find the day on which the fewest tickets were sold.
- **Exercise 3** Find the day on which the most tickets were sold.
- **Exercise 4** Find how many more tickets were sold on one day than on another.

PRACTICE ON YOUR OWN Focus on the graph at the top of the page.

Ask: **What does the graph show?** (change in number of library members from 1990 to 2000)

Discuss the vertical scale of 0–20,000 with an interval of 2,000. Have students first find the number of library members in 1999 (about 17,000); then help them interpolate the number for 1998 (about 15,000).

CHECK Determine that students are carefully reading from both axes and accurately comparing data points. Success is determined by 3 out of 3 correct responses.

Students who successfully complete the **Practice On Your Own** and **Check** are ready to move to the next skill.

COMMON ERRORS

- Students may not understand how to interpolate data points that fall within intervals.
- Students may not be able to visually track up or across the graph.

Students who made more than 2 errors in the **Practice On Your Own,** or who were not successful in the **Check** section, may benefit from the **Alternative Teaching Strategy** on the next page.

Optional

Alternative Teaching Strategy

Read Line Graphs

20 Minutes

OBJECTIVE Read a line graph

MATERIALS newsprint or dry-erase board, yardstick, yarn and tape

Prepare and display a line graph showing number of school lunches ordered for 1 week.

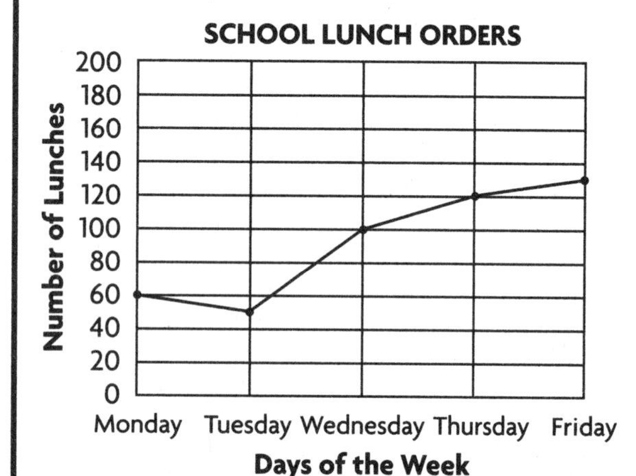

Ask: **What is the title of the graph?** (School Lunch Orders) **What does the graph show?** (daily number of school lunches ordered) **What period of time does the graph cover?** (1 week, Monday–Friday) **What is the vertical scale?** (0–200) **What is the interval?** (20)

Practice locating a few points on the graph, using the yardstick to help read up and across. To locate a point within an interval, tape a piece of yarn from the point that shows the number of lunches ordered across to the vertical scale.

Have students gather information from the graph.

1. **How many lunches were ordered on Wednesday?** (100) **On Thursday?** (120) **On Friday?** (130)
2. **On which day were the fewest lunches ordered?** (Tuesday)
3. **On which day were the most lunches ordered?** (Friday)
4. **How many more lunches were ordered on Thursday than on Monday?** (60)
5. **How many fewer lunches were ordered on Tuesday than on Wednesday?** (50)
6. **Which day shows a decrease in lunches ordered?** (Tuesday)

If your school has a cafeteria, have students make their own school lunch graph covering one week. Then repeat this exercise using their data.

Name ______ Skill ______

Grade 5 **Skill 68**

A line graph shows how data changes over time.

Read Line Graphs

This graph shows the normal daily ticket sales at a small movie theater. Find how many movie tickets were sold on Wednesday.

On the graph, the line connecting the points shows the changes in ticket sales during one week.
The *scale* of the graph is 0–600 with an *interval* of 50.

Step 1 Look across the bottom of the graph. Find the line labeled for **Wednesday**. Follow that line up to the point on the line graph.
Step 2 Move to the scale on the left of the graph to locate the number of tickets sold.

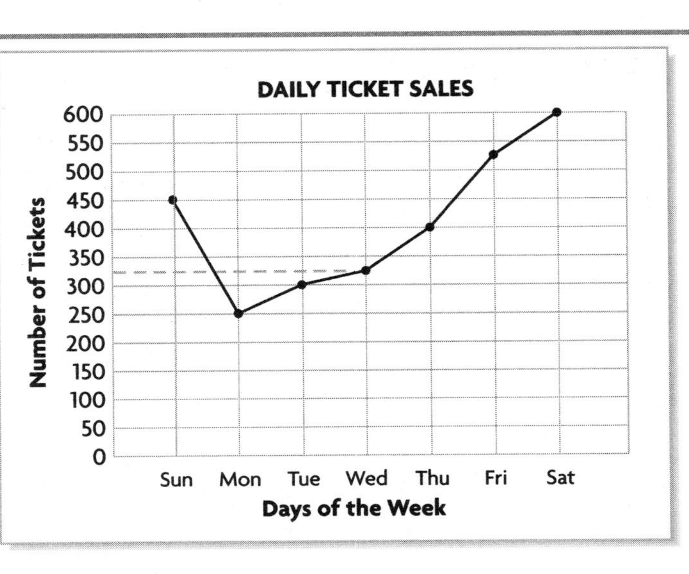

The point on the line for Wednesday appears to be halfway between 300 and 350.

So, the daily ticket sales for Wednesday is 325.

Try These

Use the graph above.

1. How many tickets were sold on Sunday? ______
2. On which day were the fewest number of tickets sold? ______
3. On which day were the most tickets sold? ______
4. How many more tickets were sold on Saturday than on Friday? ______

Go to the next side.

Name ______________________ Skill ______________________

Practice on Your Own

Skill 68

Think:

To find out how many new library members there were in 1999, subtract the number of members in 1998 from the number in 1999.

17,000 − 15,000 = 2,000
↓ 1999 ↓ 1998

There were 2,000 new library members in 1999.

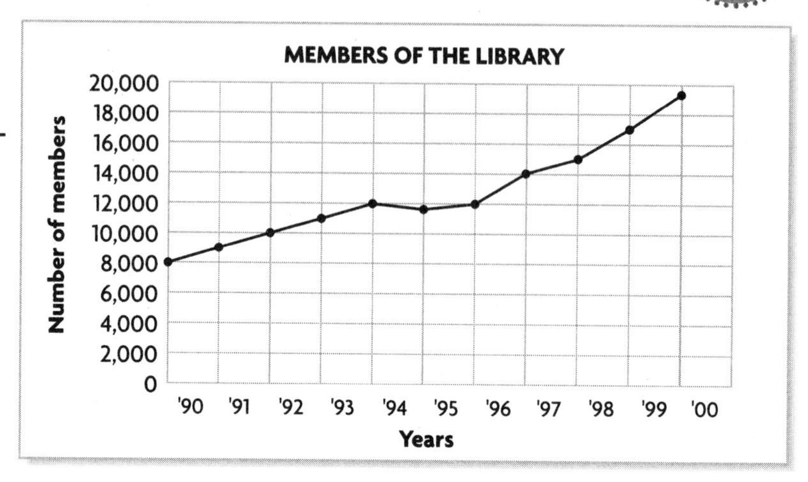

The scale of the graph is 0–20,000 with an *interval* of 2,000.

Use the graph above to answer the questions.

1. How many years are shown on this graph? ______

2. How many library members were there in 1992? ______

3. How many library members were there in 1997? ______

4. Which year had the greatest number of library members? How many? ______ ______

5. Which year had the fewest number of library members? How many? ______ ______

6. How many more library members were there in 1998 than 1997? ______

7. Which year shows a drop, or a decrease, in library members? ______

8. For how many years was there a decrease in membership? ______

9. During which three–year period did the number of members increase the most? ______

Check

Use the graph above to answer the questions.

10. Was there a greater increase in members from 1992–1993 or from 1998–1999? ______

11. In which two years were there the same number of members? ______

12. How many new members were there from the year 1990 to the year 2000? ______

Skill 69 Grade 5

Certain or Impossible Events

Using Skill 69

OBJECTIVE Use probability to name events that are certain or impossible

Begin this lesson by naming some events that are *certain* to happen or be true. For example, Tuesday follows Monday. Or, if you have five nickels, you have 25¢. Also name some events that are *impossible.* For example, you find a triangle with four vertices, or you will be 100 years old on your next birthday.

Draw attention to the *Certain Events* section. Help students recognize that all the examples cited are events that always happen. There are no exceptions.

Ask: **Is there any example where the sum of 6 and 2 is not 8?** (no) **Is the sum of 6 and 2 always 8?** (yes)

Continue with similar questions for the other examples in this section. The goal is for students to see that certain events will always occur.

Draw attention to the *Impossible Events* section. Contrary to the previous section, help students recognize that all the examples cited are situations that cannot happen. There are no exceptions.

Ask: **What is the product of 4 and 5?** (20) **When is 4 × 5 any number other than 20?** (never)

Continue with similar questions for the other examples in this section. The goal is for students to see that impossible events will never occur.

TRY THESE Exercises 1–2 model certain or impossible events.

- **Exercise 1** Certain event.
- **Exercise 2** Impossible event.

PRACTICE ON YOUR OWN Review the example at the top of the page. Have students find the correct sum and explain why it is impossible for the sum of the numbers on the two cubes to be greater than 20.

CHECK Determine if students can analyze the situation and decide whether an event or situation is certain or impossible.

Success is indicated by 2 out of 3 correct responses.

Students who successfully complete the **Practice on Your Own** and **Check** are ready to move to the next skill.

COMMON ERRORS

- Students might not have enough real-life experiences and so be unable to decide whether the event is certain or impossible.

Students who made more than 4 errors in the **Practice on Your Own,** or who were not successful in the **Check** section, may benefit from the **Alternative Teaching Strategy** on the next page.

Optional

Alternative Teaching Strategy
Certain or Impossible Events

OBJECTIVE Identify situations as impossible or certain

MATERIALS flip chart; self-stick note pad; shoebox (optional)

Prepare the table below on a flip chart.

Certain	Impossible

On the self-stick notes, write several examples of certain and impossible events. Be sure to select situations that are within students' real-life experiences. For example:

The sun will rise tomorrow morning.

You will be 2 years older tomorrow.

Roll a 9 on a cube numbered 1–6.

You are a mammal.

Earth rotates around an axis.

There will be 2 Mondays next week.

The month of May has 31 days.

$1 + 1 = 6$

It will rain cats and dogs in the desert.

Put the finished notes in a shoebox (optional).

Ask a volunteer to pick an event from the shoebox and read the event aloud. Encourage students to discuss whether the event is certain or impossible. They should cite examples to support their ideas. Once students reach a consensus, have the volunteer place the self-stick note in the correct location on the flip chart.

Remind students that an impossible event can *never* happen and a certain event will *always* happen.

Continue the activity with other examples. Encourage students to verbalize their thinking to support their decisions.

Repeat the exercise with events that students write on self-stick notes.

Name ______________________ Skill ______________________

Grade 5 Skill 69

Certain or Impossible Events

Decide if the event is *certain* or *impossible*.

Certain Events

An event is **certain** if it will *always* happen.

- The sum of 6 + 2 equals 8.
  This event is certain.
  The sum of 6 and 2 *always* equals 8.
- Flag Day is June 14.
  This event is certain.
  Flag Day is *always* June 14.
- The spinner at the right lands on an even number.
  This event is certain.
  The spinner will *always* land on an even number.

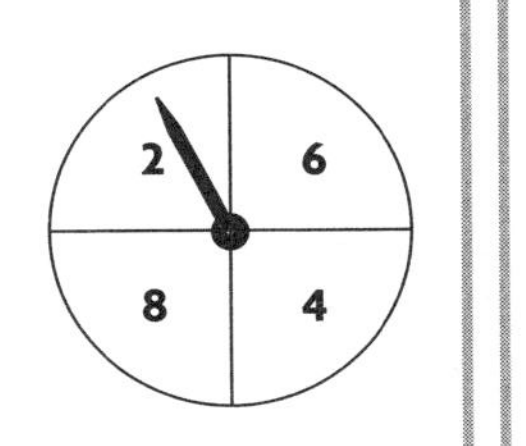

Impossible Events

An event is impossible if it will *never* happen.

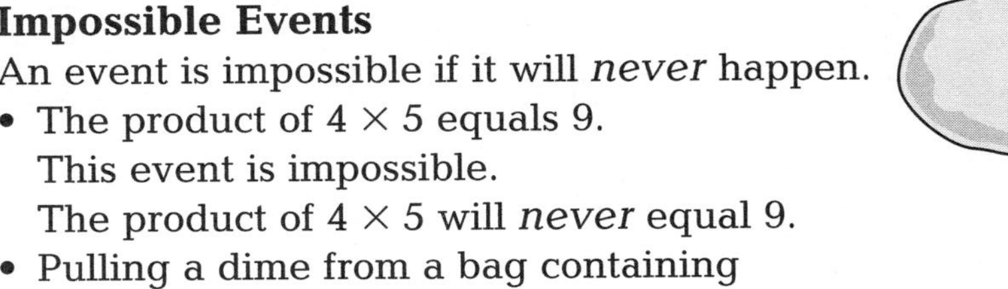

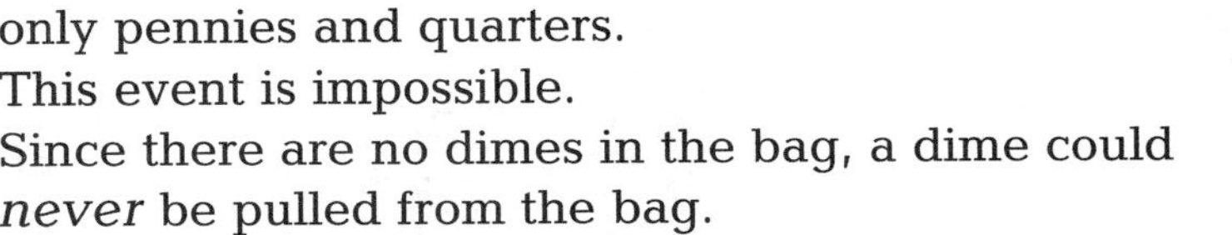

- The product of 4 × 5 equals 9.
  This event is impossible.
  The product of 4 × 5 will *never* equal 9.
- Pulling a dime from a bag containing only pennies and quarters.
  This event is impossible.
  Since there are no dimes in the bag, a dime could *never* be pulled from the bag.
- The spinner at the right lands on the number 5.
  This event is impossible.
  The spinner *never* lands on the number 5.

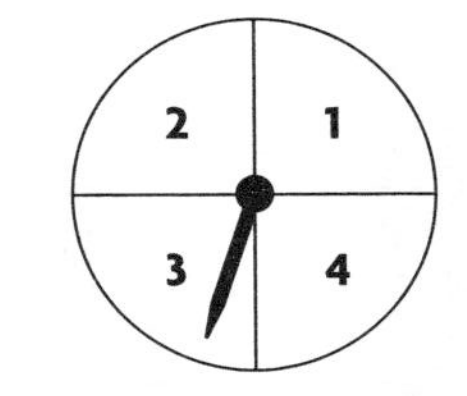

Try These

Tell whether the event is *certain* or *impossible*.

1. The spinner at the right lands on a number less than 9.
   Think: Will this event always happen?
   Will this event never happen?
   This event is ______________.

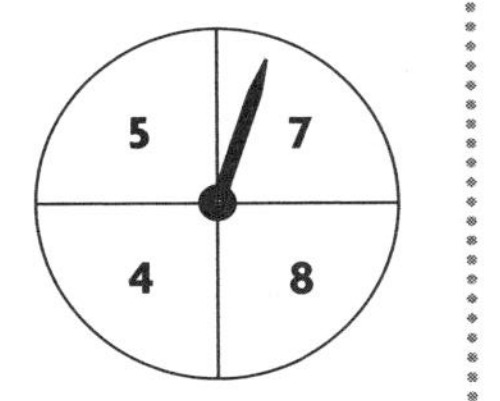

2. Pull a blue marble from a bag containing 2 red marbles and 3 green marbles.
   Think: Will this event always happen?
   Will this event never happen?
   This event is ______________.

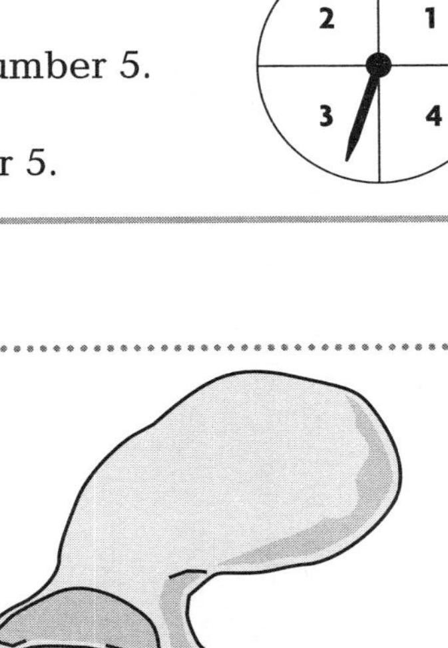

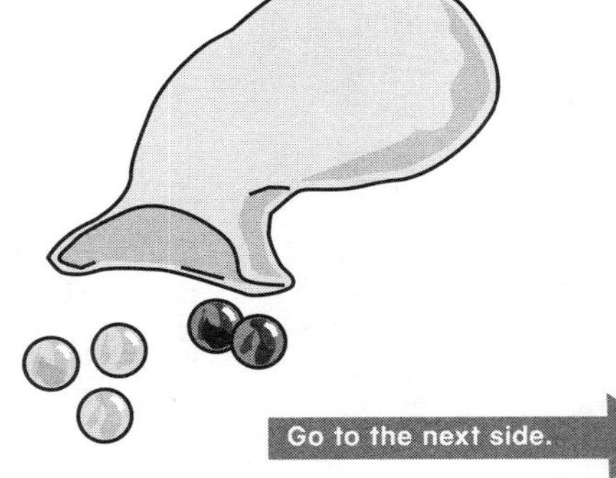

Go to the next side.

Name ______________________ Skill ______________________

Practice on Your Own

Skill 69

Think:
A **certain** event will *always* happen.
An **impossible** event will *never* happen.

The sum of the numbers rolled on two cubes, numbered 1, 2, 3, 4, 5, and 6, is greater than 20.

- The greatest sum is 6 + 6, or 12.
- This event is impossible. It will never happen.

Tell whether an event is certain or impossible.

1. Fly to the moon for vacation tomorrow.
Think: Will this event *always* happen? never happen?
This event is __________.

2. The quotient of 12 ÷ 3 equals 4.
Think: Will this event *always* happen? *never* happen?
This event is __________.

3. You pull a nickel from a bag containing 5 nickels.
Think: Will this event *always* happen? *never* happen?
This event is __________.

Use the spinner for 4–6.

4. Spin an odd number.
Think: Will this event *always* happen? *never* happen?
This event is __________.

5. Spin a number divisible by 2.
Think: Will this event *always* happen? *never* happen?
This event is __________.

6. Spin a number greater than 10.
Think: Will this event *always* happen? *never* happen?
This event is __________.

7. Pull a vowel from a bag of tiles labeled with *A, E, I, O, U*.
This event is __________.

8. Spring follows winter.
This event is __________.

9. Pull a green marble from a bag of blue marbles.
This event is __________.

Check

Tell whether an event is certain or impossible.

10. You will be one year younger tomorrow.
This event is __________.

11. Pull a vowel from a bag with tiles labeled N, D, T, and W.
This event is __________.

12. 4:30 follows 1 hour after 3:30.
This event is __________.

Using Skill 70

OBJECTIVE Use probability to decide whether events are likely or unlikely

Begin this lesson by naming some events that are likely to happen and some that are unlikely to happen. Encourage students to discuss why it is that a particular event is either likely or unlikely. Distinguish between likely and certain, and between unlikely and impossible.

Draw attention to the *Likely Events* section. Ask: **Why is it that picking a green tile is more likely than picking a blue tile?** (There are more green tiles than blue tiles.)

Direct students to present other examples of likely events. Allow time to discuss each event. Encourage students to justify their interpretations. Help them to see that if there are more ways to pick green tiles than blue tiles, then picking green is more likely. The greater the difference between ways events can happen, the more likely one event is to occur.

Ask: **How could you make it more likely that a blue tile would be picked?** (Add blue tiles and/or take out green tiles until there are more blue tiles than green tiles.) **With 5 green tiles and 2 blue tiles, is it impossible that you will pick a blue tile?** (no) **Is it certain that you will pick a green tile?** (no)

Draw attention to the *Unlikely Events* section.

Ask: **What are the chances of rolling a 4?** (1 out of 6) Allow time for students to discuss why 1 chance out of 6 makes an event unlikely. Point out to students that an event is unlikely to happen when the number of ways it can occur (1) is less than the number of ways it will not occur (5).

TRY THESE Exercises 1–2 model a likely or unlikely event.

- **Exercise 1** Likely event.
- **Exercise 2** Unlikely event.

PRACTICE ON YOUR OWN Review the statements at the top of the page.

CHECK Determine if students can list whether an event is likely or unlikely.

Success is indicated by 2 out of 4 correct responses.

Students who successfully complete the **Practice on Your Own** and **Check** are ready to move to the next skill.

COMMON ERRORS

- Students might fail to list all possible outcomes.
- Students might need to review strategies for comparing and ordering numbers.
- Students might not understand what particular outcomes are.

Students who made more than 3 errors in the **Practice on Your Own,** or who were not successful in the **Check** section, may benefit from the **Alternative Teaching Strategy** on the next page.

Optional

Alternative Teaching Strategy
Drawing with Colored Counters

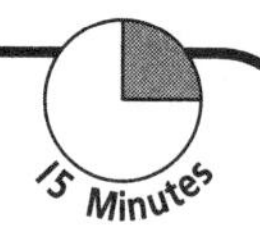

OBJECTIVE Model likely or unlikely events

MATERIALS colored counters, flip chart, shoebox

Out of view of the students, put 2 blue, 6 yellow, and 12 red counters in a shoebox.

Ask a volunteer to remove the counters from the box and sort the counters by color.

Ask another volunteer to count the number of counters of each color and to list the numbers from least to greatest. Record the results on the flip chart.

Yellow	卌 I	6
Blue	II	2
Red	卌 卌 II	12
	2, 6, 12	

Return all the counters to the box.

Ask a third volunteer to come forward. Direct the student to reach into the box, with eyes closed, and draw a counter.

Before revealing the color of the counter, ask: **Which color would you be most likely to draw from the container? Why do you think so?**

Guide students to the realization that the color with the greatest number of counters is most likely to be drawn. There are 12 red counters out of a total of 20 counters. A red counter is most likely to be drawn.

Encourage students to explain which color counter they would be least likely to draw. Remind students to refer to the data on the flip chart. Emphasize that unlikely does *not* mean impossible and that likely does *not* mean certain.

Provide students with similar experiences in which they can count the number of outcomes to determine the likelihood of an event.

Name ______________________ Skill ______________________

Grade 5 Skill 70

Likely, Unlikely

Use probability to name events that are likely or unlikely to happen.

Likely Events
An event is likely to happen if the number of times it could happen is *greater than* the number of times it will not.

- **Pull a green tile from a bag containing 5 green tiles and 2 blue tiles.**
  There are 5 green tiles. So there are 5 chances out of 7 to pull a green tile. This event is *likely*.

Unlikely Events
An event is unlikely to happen if the number of times it could happen is *less than* the number of times it will not.

- **Rolling a 4 on a number cube labeled 1, 1, 1, 4, 1, and 1.**
  There is only one 4 on the cube. So, there is only 1 outcome out of 6 possibilities.
  The event is *unlikely*.

Try These

Tell whether the event is likely or unlikely.

1. Pull a green marble from a bag containing 2 red marbles, 6 green marbles, and 1 blue marble.
   How many outcomes are green? ________
   How many outcomes are not green? ________
   This event is ______________.

2. The spinner at the right lands on a number less than 3.
   How many outcomes are greater than 3? ________
   How many outcomes are less than 3? ________
   This event is ______________.

0 4 8 9 10 18 20 12

Go to the next side.

Name ______________________ Skill ______________________

Practice on Your Own

Skill 70

Think:
An event is **likely** if the number of times it could happen is greater than the number of times it will not.
An event is **unlikely** if the number of times it could happen is less than the number of times it will not.

Complete. Then tell whether the event is *likely* or *unlikely*.

1. Roll a number less than 5 on a cube numbered 1, 2, 3, 4, 6, and 7.
Number of outcomes less than 5 ________
Number of outcomes greater than 5 ________
This event is ______________.

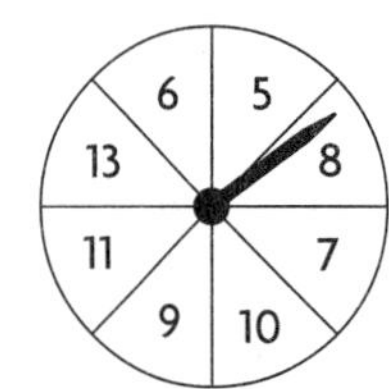

2. Pull a blue marble from a bag containing 6 blue, 4 red, and 5 green marbles.
Number of blue outcomes ________
Number of other outcomes ________
This event is ______________.

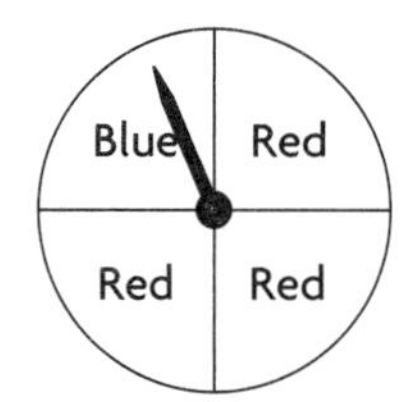

3. The spinner lands on an even number
________ outcomes are even.
________ outcomes are not even.
This event is ______________.

4. The spinner lands on red.
________ outcomes are red.
________ outcomes are not red.
This event is ______________.

5. Pull a dime from a bag containing 6 pennies, 8 dimes, and 4 nickels.
This event is ______________.

6. Spin an odd number on a spinner numbered: 1, 3, 4, 5, 8, 9, 11, 13.
This event is ______________.

Check

Complete. Then tell whether the event is *likely* or *unlikely*.

7. Draw a blue tile from a bag containing 3 blue tiles, 6 green tiles, and 1 red tile.
This event is ___________.

8. Pull number 7 from a hat with tickets numbered 6, 7, 7, 6, 7, 8, 9, and 8.
This event is ___________.

9. Roll 2 cubes each numbered 1–6 and getting a sum greater than 10.
This event is ___________.

10. Pull a red marble from a bag containing 7 red and 4 black marbles.
This event is ___________.

Name ______________ Skill ______________

Answer Card

Statistics, Data Analysis and Probability

Grade 5

SKILL 65

TRY THESE

1. summer, 10, 19; 19 students; Fall 9, Winter 5, Spring 9, Summer 10

PRACTICE

1. 21, 39
2. 22, 14, 18; 54
3. 22 + 14 + 18 + 21; 75
4. 39 muffins
5. plain
6. 49 muffins
7. 23 students
8. 39 students
9. 62 students

CHECK

10. 235
11. Mountain bike
12. 333 bikes

SKILL 66

TRY THESE

1. Favorite Pizza Toppings
2. 10 students
3. Peppers
4. 4 students
5. 44 students

PRACTICE

1. Books Read This Month
2. 2 books
3. 1 book
4. Suki, 12 books
5. 40, 35, 25, 10
6. Three and one–half pictures

CHECK

7. 28
8. 22
9. 92 books
10. 3

SKILL 67

TRY THESE

1. 4
2. novels
3. 1

PRACTICE

1. 5
2. 30
3. corn
4. 75
5. pumpkins
6. 25
7. 15
8. 25
9. 15

CHECK

10. 5
11. 45
12. 170

SKILL 68

TRY THESE

1. 450 tickets
2. Monday
3. Saturday
4. 75

PRACTICE

1. 11 years
2. 10,000 members
3. 14,000 members
4. 2000; about 19,500 members
5. 1990; 8,000 members
6. 1,000 more members
7. 1995
8. 1 year
9. 1998, 1999, 2000

CHECK

10. from 1998-1999
11. 1994 and 1996
12. about 11,500

Name ______________________ Skill ______________________

SKILL 69

TRY THESE

1. Certain
2. Impossible

PRACTICE

1. Impossible
2. Certain
3. Certain
4. Impossible
5. Certain
6. Impossible
7. Certain
8. Certain
9. Impossible

CHECK

10. Impossible
11. Impossible
12. Certain

SKILL 70

TRY THESE

1. 6, 3, likely
2. 7; 1; unlikely

PRACTICE

1. 4,2, likely
2. 6; 9; unlikely
3. 3; 5; unlikely
4. 3, 1, likely
5. unlikely
6. likely

CHECK

7. unlikely
8. unlikely
9. unlikely
10. likely

Answer Card

Statistics, Data Analysis and Probability

Grade 5